Fundamentals of Photovoltaic Modules and Their Applications

RSC Energy Series

Series Editor:
Julian Hunt FRS, *University College London, London, UK*

Titles in the Series:
1: Hydrogen Energy: Challenges and Prospects
2: Fundamentals of Photovoltaic Modules and Their Applications

How to obtain future titles on publication:
A standing order plan is available for this series. A standing order will bring delivery of each new volume immediately on publication.

For further information please contact:
Book Sales Department, Royal Society of Chemistry, Thomas Graham House, Science Park, Milton Road, Cambridge, CB4 0WF, UK
Telephone: +44 (0)1223 420066, Fax: +44 (0)1223 420247,
Email: books@rsc.org
Visit our website at http://www.rsc.org/Shop/Books/

Fundamentals of Photovoltaic Modules and Their Applications

G. N. Tiwari and Swapnil Dubey
Centre for Energy Studies, Indian Institute of Technology (IIT) Delhi, New Delhi, India

RSC Publishing

RSC Energy Series No. 2

ISBN: 978-1-84973-020-4
ISSN: 1757-6741

A catalogue record for this book is available from the British Library

Published by The Royal Society of Chemistry,
Thomas Graham House, Science Park, Milton Road,
Cambridge CB4 0WF, UK

Registered Charity Number 207890

For further information see our web site at www.rsc.org

Dedication

Our respected teacher and guruji
Padmashri Professor M. S. Sodha F.N.A.
on his 78th birthday (8 February, 2010)

Preface

The word 'energy' has been continuously in the news since 1973 due to the shortages of oil in many parts of the world and the price of this commodity has increased steeply. It is now clear that the fossil-fuel era of non-renewable resources is gradually coming to an end. The renewable sources of energy derived from the Sun are one of the promising options. Solar energy can be used both directly and indirectly. It can be used directly in a variety of thermal applications like heating air or water, drying, distillation and space heating *etc*. A second way in which solar energy can be used directly is through the photovoltaic effect, in which it is converted to electrical energy. Indirectly, the Sun causes winds to blow, plants to grow, rain to fall and temperature differences to occur from the surface to the bottom of the oceans. Useful energy can be obtained for commercial and non-commercial purposes through all these renewable sources.

In this book, we are primarily concerned with the collection and storage of solar energy for thermal and electrical applications. The purpose of writing this book is to provide a suitable text for teaching the subject to engineering and science students, as well as a reference book for scientists and professionals. The material is based on the author's research experience and his experience of teaching the subject for a number of years to postgraduate and undergraduate engineering students. We assume that the reader of this book has a basic background in physics, mathematics, thermodynamics, heat transfer, electrical and electronics. This book is quantitative and applications-oriented, with an emphasis on resource estimation, system sizing and economic evaluation.

The objective of the book is to provide a platform to disseminate the knowledge regarding fundamentals of photovoltaic thermal systems, namely:

- fundamentals of solar energy and basic heat transfer;
- characteristics of solar cells and their materials;
- use of photovoltaic modules and arrays in solar systems;

RSC Energy Series No. 2
Fundamentals of Photovoltaic Modules and Their Applications
By G. N. Tiwari and Swapnil Dubey

Published by the Royal Society of Chemistry, www.rsc.org

- importance of batteries;
- thermal modelling of solar systems;
- energy and exergy analysis;
- CO_2 mitigation and carbon credit;
- economic analysis of PV/T systems, *etc.*

to undergraduate and post-graduate students, learners, scientists, professionals, practitioners and designers. To understand the above objectives a large number of figures, solved examples and tables have been provided. At the end of each chapter, problems/exercises have also been given, along with hints to solve them.

We have drawn the material for inclusion in the book from a number of references, which are cited at the appropriate places. These include: *Solar Energy, Fundamentals, Design, Modelling and Applications* by G. N. Tiwari; *Fundamentals of Solar Dryers* by G. N. Tiwari and P. Barnwal; *Solar Engineering of Thermal Processes* by J. A. Duffie and W. A. Beckman; research papers by Prof. H. A. Zondag, Prof. S. D. Hendrie, Prof. P. Raghuraman, Prof. T. T. Chow, Prof. J. Prakash, Prof. Y. Tripanagnostopoulos, Prof. D. Infield, Prof. K. Nagano, Prof. L. W. Florschuetz, Prof. E. C. Kern Jr. and Prof. M. C. Russell, Prof. D. L. Evans, Prof. S. A. Kalogirou, Prof. B. J. Huang, Prof. J. Ji, Prof. H. P. Garg, Prof. A. D. Jones and Prof. C. P. Underwood, Prof. A. A. Hegazy, Prof. K. Sopian, Prof. J. K. Tonui, Prof. J. Mumba, Prof. B. K. Bala, Prof. I. Dincer, *etc.* We are highly appreciative of the courtesy of authors Prof. T. T. Chow, China; Prof. Ivan Katic, Denmark; Prof. Niccolò Aste, Italy; Prof. Gilles Notton, France; Prof. G. Fraisse, France; Prof. Abraham Kribus, Israel; Prof. Y. B. Assoa, France; Prof. B. Robles-Ocampo, Mexico; Prof. H. Yang, Hong Kong; Prof. Emmanuel Kymakis, Greece, for providing the photographs of different PV/T systems. This list is incomplete and we apologize to anyone we have omitted.

The present book has been divided into 10 chapters to study the basic knowledge of photovoltaic thermal (PV/T) systems from thermal and electrical points of view. Chapter 1 deals with availability of solar radiation emitted from the Sun and its propagation through the atmosphere, as well as concepts of greenhouse gases. It also includes importance and basics of solar radiation such as atmosphere and Sun–Earth angles, cloudiness/haziness factor and total solar radiation *etc.* Chapter 2 deals with the history/review of work done on photovoltaic (PV) integrated systems by various researchers. It includes air and water systems, building integrated photovoltaic systems (BIPV) systems, temperature-dependent electrical performance and market potential *etc.* The basics of semiconductors and their characteristics, characteristics of solar cells in dark and daylight situations and fundamentals of characteristic curves of semiconductors have been given in Chapter 3. The fundamentals of PV modules, various combinations of solar cells and PV modules and array analyses have been discussed in Chapter 4. The various types and working principles of batteries with life and economics of batteries have been highlighted in Chapter 5. Chapter 6 provides the various case studies on BIPV and PV/T systems related with field exposures. The thermal modelling and results of various configurations of PV/T systems, including air collectors, water heaters, distillation systems and dryers, have been discussed in Chapter 7. The energy and exergy analysis on the basis of embodied energy of materials used for fabrication of different components of PV/T systems has been highlighted in Chapter 8. Chapter 9 deals with the net CO_2 mitigation, carbon credit and climate change. The techno-economics of the solar systems has been discussed in Chapter 10.

SI units have been used throughout. Appendices have been given at end of the book.

This book aims to provide a great insight into the subject, particularly to learning students/professionals doing self-study. In spite of our best efforts, some errors might have crept into the text. We fully welcome valuable suggestions and comments from all readers for further improvement of the book in the next edition.

It is our immense pleasure to express our heartfelt gratitude to Director (IIT Delhi), Head (CES, IIT Delhi) and Prof. S. K. Dube, former director, IIT Kharagpur, for their kind encouragement.

We acknowledge with thanks the financial support by the Curriculum Development Cell, IIT Delhi, for preparation of the book.

We are also thankful to Dr P. C. Pant, Scientist, Solar Energy Center, MNRE, New Delhi, for providing the material on batteries and to Dr V. K. Kaul, Central Electronics Limited, Sahibabad (UP), for providing the details on SPV water pumping systems. We owe a special note of thanks to Dr Arvind Tiwari, Dr P. Barnwal, Dr Shiv Kumar, Dr V. K. Dwivedi, Mrs Sujata Nayak, Mr S. C. Solanki, Mr M. K. Gaur, Mr Basant Agarwal, Mr Jamil Ahmad, Mr Rajeev Mishra, Mr Gaurav Singh, Mr. Abhishek Ranjan, Sh. Lakhmi Chand and all the members of our group for their valuable support during preparation of the manuscript.

Full credit is due to our publishers, RSC Publishing, Cambridge, UK, for producing a nice print of the book.

Last, but not least, we express out deep gratitude to our respected parents, Late Smt. Bhagirathi Tiwari, Late Sh. Bashisht Tiwari, Smt. Vandana Dubey and Sh. Shailendra Kumar Dubey for their blessings, which helped us to reach our target.

G. N. Tiwari
Swapnil Dubey

Contents

RSC Energy Series No. 2
Fundamentals of Photovoltaic Modules and Their Applications
By G. N. Tiwari and Swapnil Dubey

Published by the Royal Society of Chemistry, www.rsc.org

About the Authors

Prof. G. N. Tiwari
Professor
Centre for Energy Studies
Indian Institute of Technology Delhi

Professor Gopal Nath Tiwari was born on 1 July, 1951, at Adarsh Nagar, Sagerpali, Ballia (UP), India. He received postgraduate and doctoral degrees in 1972 and 1976, respectively, from Banaras Hindu University (BHU). Since 1977, he has been actively involved in the teaching programme at the Centre for Energy Studies, IIT Delhi. His research interests in the field of Solar Thermal Applications are solar distillation, water/air heating systems, greenhouse technology for agriculture as well as for aquaculture, Earth to air heat exchangers, passive building design and hybrid photovoltaic thermal (HPVT) systems, climate change, energy security, *etc*. He has guided about 60 PhD students and published over 400 research papers in journals of repute. He has authored 18 books associated with respected publishers, namely Pergamon Press UK, CRC Press USA and Narosa Publishing House. He was a co-recipient of the 'Hariom Ashram Prerit S.S. Bhatnagar' Award in 1982. Professor Tiwari has been recognized at both national and international levels. His contribution to the successful implementation of a hot water system in the IIT campus has been highly appreciated. He went to the University of Papua, New Guinea, in 1987–1989 as Energy and Environment Expert. He was also a recipient of the European Fellow in 1997 and went to the University of Ulster (UK) in 1993. He has also been nominated for the IDEA award in the past. He is responsible for development of the 'Solar Energy Park' at IIT Delhi and the Energy Laboratory at the University of Papua, New Guinea, Port Moresby. Professor Tiwari has visited many countries, namely Italy, Canada, USA, UK, Australia, Greece, Thailand, Singapore,

PNG and Taiwan *etc.* for invited talks, chairing international conferences, providing expertise in renewable energy, presenting research papers, *etc.* He has successfully co-coordinated various research projects on solar distillation, water heating systems, greenhouse technology, hybrid photovoltaic thermal (HPVT) systems, *etc.* funded by the government of India in the recent past.

Professor Tiwari has been offered the post of Associate Editor for *Solar Energy Journal (SEJ)* in the area of solar distillation. He has also been the Editor of the *International Journal of Agricultural Engineering* since 2006.

Professor Tiwari organized *SOLARIS 2007*, the third international conference on 'Solar Radiation and Day Lighting', held at IIT Delhi, New Delhi, India, from February 7–9, 2007.

Recently, Professor G. N. Tiwari was conferred as 'Vigyan Ratna' by the government of UP, India, on 26 March, 2008, and Valued Associated Editor by the *Journal of Solar Energy*.

Dr. Swapnil Dubey
Research Scholar
Centre for Energy Studies
Indian Institute of Technology (IIT) Delhi

Dr. Swapnil Dubey was born on 20 July, 1981, at Indore (MP). He received his Bachelor of Engineering degree in Mechanical Engineering from the Institute of Engineering and Technology, Devi Ahilya Vishwavidyalaya, Indore, in 2003. He received his postgraduate degree (MTech) in Energy Studies from the Centre for Energy Studies, Indian Institute of Technology (IIT) Delhi, in 2006. Based on his MTech project, he has presented two papers in international conferences.

Presently, he has obtained his PhD degree under the supervision of Professor G. N. Tiwari. During his PhD, he also worked as an organizing member of the third International Conference on 'Solar Radiation and Day Lighting', *SOLARIS 2007*, held at IIT Delhi during February 7–9, 2007. He has also participated in the UK–India–Sri Lanka Young Scientists Networking Conference on 'Towards sustainable energy technologies and low-carbon buildings for climate change mitigation' organized by the British Council during February 6–8, 2007, New Delhi. He visited City University of Hong Kong, Hong Kong, during December 2008.

Dr. Dubey has published 12 research papers in international journals, *viz.* *Solar Energy*, *Applied Energy*, *Energy Research*, *Energy and Buildings* and *Renewable Energy* and four research papers in international conferences.

His areas of research interest are solar thermal, photovoltaics, thermodynamics, heat and mass transfer, exergy, CO_2 mitigation, climate change and carbon trading.

CHAPTER 1

Solar Radiation

1.1 Introduction

Sunlight, in the broad sense, is the total spectrum of the electromagnetic radiation given off by the Sun. On Earth, sunlight is filtered through the atmosphere, and the solar radiation is obvious as daylight when the Sun is above the horizon. This is usually during the day hours. Near the poles in summer, sunlight also occurs during the night hours and in the winter at the poles sunlight may not occur at any time. When the direct radiation is not blocked by clouds, it is experienced as sunshine, a combination of bright light and heat. Radiant heat directly produced by the radiation of the Sun is different from the increase in atmospheric temperature due to the radiative heating of the atmosphere by the Sun's radiation. Sunlight may be recorded using a sunshine recorder, pyranometer or pyrheliometer. The World Meteorological Organization (WMO) defines sunshine as direct irradiance from the Sun measured on the ground of at least $120\,W\,m^{-2}$. Direct sunlight gives about 93 lux of illumination per watt of electromagnetic power, including infrared, visible and ultraviolet. Bright sunlight provides illumination of approximately 100 000 lux per square metre at the Earth's surface. Sunlight is a key factor in the process of photosynthesis.

1.1.1 The Sun

The Sun is the star at the centre of the solar system. The Earth and other matter (including other planets, asteroids, meteoroids, comets and dust) orbit the Sun, which by itself accounts for about 99.8% of the solar system's mass. Energy from the Sun, in the form of sunlight, supports almost all life on Earth *via* photosynthesis, and drives the Earth's climate and weather.

The Sun has an effective black-body temperature T_S of 5777 K and it is the largest member of the solar system. The Sun is a sphere of intensely hot, gaseous matter with a diameter of 1.39×10^9 m and is, on average, 1.5×10^{11} m away from the Earth. The Sun is, effectively, a continuous fusion reactor. It is estimated that

RSC Energy Series No. 2
Fundamentals of Photovoltaic Modules and Their Applications
By G. N. Tiwari and Swapnil Dubey

Published by the Royal Society of Chemistry, www.rsc.org

90% of the Sun's energy is generated in the region 0 to $0.23R$ (R being the radius of the Sun $= 6.95 \times 10^8$ m); the average density (ρ) and the temperature (T) in this region are $10^5\,\mathrm{kg\,m^{-3}}$ and about $8\text{-}40 \times 10^6$ K respectively. At a distance of about $0.7R$ from the centre, the temperature drops to about 1.3×10^5 K and the density to 70 kg $\mathrm{m^{-3}}$. Hence for $r > 0.7R$ convection begins to be important and the region $0.7R < r < R$ is known as the convective zone. The outer layer of this zone is called the photosphere. The maximum spectral intensity occurs at about 0.48 μm wavelength (λ) in the green portion of the visible spectrum. About 8.73% of the total energy is contained in the ultraviolet region ($\lambda < 0.40$ μm); another 38.15% in the visible region (0.40 μm $< \lambda <$ 0.70 μm) and the remaining 53.12% in the infrared region ($\lambda > 0.70$ μm).

1.1.2 The Earth

Earth is the third planet from the Sun. Earth is the largest of the terrestrial planets in the solar system in diameter, mass and density. The Earth, almost round in shape with a diameter of about 13 000 km, came into existence some 4.6×10^9 years ago. The Earth's inner core is a solid made of iron and nickel. The eruption of volcanoes generally occurs at the plate boundary of the Earth. During eruption of volcanoes, various greenhouse gases, namely carbon dioxide (CO_2), methane (CH_4), nitrous oxide (NO_x), ozone (O_3) and water vapour (H_2O) *etc.*, existing inside the ground, are also discharged through the plate boundary. These discharged gases, at the boundary of the plate, move upwards towards the Sun due to its low density. These gases form a layer between the Sun and Earth (Figure 1.1). This layer is generally referred to as the Earth's atmosphere. The Earth revolves around the Sun once in about a year. Nearly two-thirds of the Earth is covered by water and the remaining one-third is land. Half of the Earth is lit by sunlight at a time. It reflects one-third of the sunlight that falls on it. This is known as Earth's albedo. The Earth is spinning at a constant rate about its axis, inclined at an angle of 23.5°. As a result, the lengths of days and nights are constantly changing. The heat flux at Earth's surface due to heat conduction from the centre is 0.04–0.06 $\mathrm{W\,m^{-2}}$ with a temperature gradient of 30–40 $\mathrm{°C\,km^{-1}}$.

1.1.3 Earth's Atmosphere

The temperature of the Earth's atmosphere varies with altitude among five different atmospheric layers:

Exosphere: from 500–1000 km up to 10 000 km, free-moving particles that may migrate into and out of the magnetosphere or the solar wind.

Ionosphere: the part of the atmosphere that is ionized by solar radiation. It plays an important part in atmospheric electricity and forms the inner edge of the magnetosphere. It has practical importance because, among other functions, it influences radio propagation to distant places on the Earth. It is located in the thermosphere and is responsible for auroras.

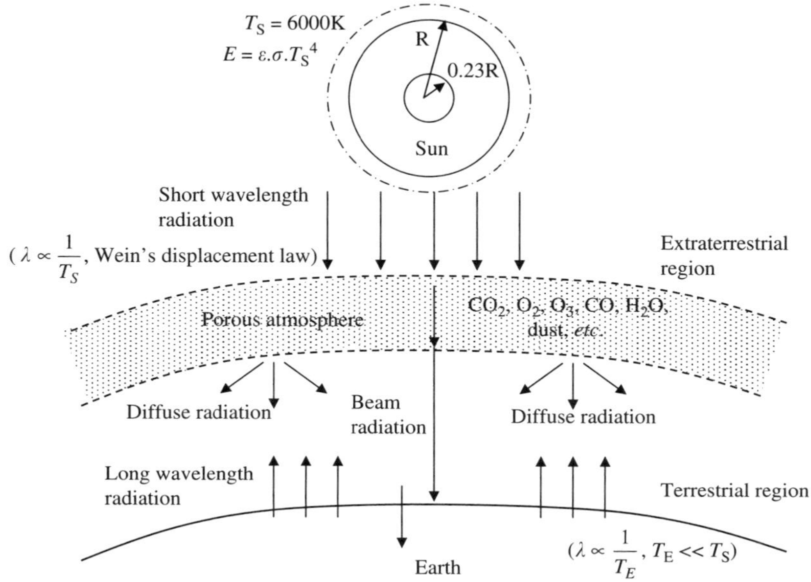

Figure 1.1 Positions of the Sun, the atmosphere and the Earth.

Thermosphere: from 80–85 km to 640 + km, the temperature increasing with height.

Mesosphere: extends from about 50 km to the range of 80–85 km, the temperature decreasing with height. This is also where most meteors burn up when entering the atmosphere.

Stratosphere: extends from the troposphere's 7- to 17-km range to about 50 km. Temperature increases with height. The stratosphere contains the ozone layer, the part of the Earth's atmosphere which contains relatively high concentrations of ozone. 'Relatively high' means a few parts per million (ppm) – much higher than the concentrations in the lower atmosphere but still small compared to the main components of the atmosphere. It is mainly located in the lower portion of the stratosphere from approximately 15 to 35 km above Earth's surface, though the thickness varies seasonally and geographically.

Troposphere: the lowest layer of the atmosphere; it begins at the surface and extends to between 7 km at the poles and 17 km at the equator, with some variation due to weather factors. The troposphere has a great deal of vertical mixing because of solar heating at the surface. This heating warms air masses, which makes them less dense so they rise. When an air mass rises, the pressure upon it decreases so it expands, doing work against the opposing pressure of the surrounding air. To do work is to expend energy, so the temperature of the air mass decreases. As the temperature decreases, water vapour in the air mass may condense or solidify, releasing latent heat that further uplifts the air mass. This process determines the maximum rate of decline of temperature with height, called the adiabatic lapse rate. The troposphere contains roughly 80% of the total mass of the atmosphere. Fifty percent of the total mass of the atmosphere is located in the lower 5.6 km of the troposphere.

The average temperature of the atmosphere at the surface of Earth is 15 °C. The average atmospheric pressure, at sea level, is about 101.3 kilopascals with a scale height of about 8.5 km; total atmospheric mass is 5.1480×10^{18} kg. Atmospheric pressure is a direct result of the total weight of the air above the point at which the pressure is measured. This means that air pressure varies with location and time, because the amount (and weight) of air above the Earth varies with location and time.

The density of air at sea level is about $1.2\,kg\,m^{-3}$. Natural variations of the barometric pressure occur at any one altitude as a consequence of weather. The atmospheric density decreases as the altitude increases. This variation can be approximately modelled using the barometric formula. More sophisticated models are used by meteorologists and space agencies to predict weather and orbital decay of satellites.

Solar radiations while passing through the Earth's atmosphere are subjected to the mechanisms of atmospheric absorption and scattering. The X-rays and extreme ultraviolet radiations of the Sun are highly absorbed in the ionosphere by nitrogen, oxygen and other atmospheric gases. The ozone and water vapours largely absorb ultraviolet ($\lambda < 0.40$ μm) and infrared radiations ($\lambda > 2.3$ μm). There is almost complete absorption of short wave radiations ($\lambda < 0.29$ μm) in the atmosphere. Hence, the energy incident on the Earth's surface in wavelength radiation below 0.29 μm and above 2.3 μm of the spectra of the solar radiation is negligible.

The Earth's atmosphere has the following unique properties:[1]

(a) It absorbs the ultraviolet (UV) and far infrared radiation and allows only radiation having wavelength ranging between 0.29 μm and 2.3 μm, known as short wavelength radiation.
(b) It also does not allow radiation having wavelength $\lambda > 2.3$ μm, known as long wavelength radiation.

The phenomenon of blocking of UV radiation, referred to as the *global greenhouse effect*, occurred some 420 million years ago and this allowed plants to grow on the Earth. Fossils (remains of blue-green algae and bacteria) from at least 3×10^9 years ago have been found in rocks and water. Without the greenhouse effect, the Earth would be a frozen planet with an average temperature of about −18 °C (about 0 °F). For survival of living plants on the Earth, there should be a favourable environment (global environment) in the terrestrial region controlled by short wavelength radiation transmitted by the atmosphere.

However, a similar effect is observed by having transparent material over any surface because the transparent material also behaves as the atmosphere with respect to short wavelength radiation. The concept of trapping short wavelength radiation (thermal energy) in an enclosure has many applications, *e.g.*:

(a) *Flat-plate air collector*: A device having an insulated blackened flat surface with a transparent glass window above it that works with the micro greenhouse effect.
(b) *Solar dryer:* A device that uses solar energy for drying applications.
(c) *Greenhouse*: A microclimate, which can be created by using the transparent glass/plastic house similar to the global greenhouse concept. It can be used for optimum growth of living plants (*e.g.* flowers, vegetables, *etc.*) for maximum crop production during season as well as off-season (post-harvest and pre-harvest period) and is generally known as greenhouse technology. The greenhouse can also be used for crop drying for storage purposes.
(d) *Photovoltaic (PV device)*: A device used to convert short wavelength radiation into direct current (dc) electricity *etc.*
(e) *Solar still:* Used for desalination of saline water.

Thus, for optimum design of the above systems, elementary knowledge of solar radiation becomes necessary, which is briefly described as follows.

1.2 Measurement of Solar Radiation on Earth's Surface

The solar radiation reaching the Earth's surface through the atmosphere can be classified into two components: *beam* and *diffuse radiation.*

Beam radiation (I_b): The solar radiation propagating along the line joining the receiving surface and the Sun. It is also referred to as direct radiation.
Diffuse radiation (I_d): The solar radiation scattered by aerosols, dust and molecules. It does not have any unique direction.
Total radiation (I_t)*:* The sum of the beam and diffuse radiation, sometimes known as global radiation.

The following instruments are commonly used for measurement of solar radiation on Earth's surface.

1.2.1 Pyrheliometer

The pyrheliometer is a broadband instrument that measures the direct (or beam) component of solar radiation at normal incidence. This means the instrument is always aimed directly at the Sun, *via* a tracking mechanism that continuously follows the Sun. It is sensitive to wavelengths in the band from 280 to 3000 nm (0.284 μm to 0.3 μm). Solar irradiance enters the instrument through a sealed crystal-quartz window and the sunlight is directed onto a thermopile which converts heat to an electrical signal that can be recorded. A calibration factor is applied when converting the mV signal to an equivalent radiant energy flux, measured in watts per square metre.

In this instrument, two identical blackened manganin strips are arranged so that either one can be exposed to radiation at the base of collimating tubes by moving a reversible shutter. Each strip can be electrically heated and each is fitted with a thermocouple. With one strip shaded and one strip exposed to radiation, a current is passed through the shaded strip to heat it to the same temperature as the exposed strip. When there is no difference in temperature, the electrical energy to the shaded strip must equal the solar radiation absorbed by the exposed strip. Solar radiation is then determined by equating the electrical energy to the product of incident solar radiation, strip area and absorptance. Then the position of the shutter is reversed, interchanging the electrical and radiation heating, and the second value is determined. Alternating the shade and the functions of the two strips compensates for minor differences in the strips, such as edge effects and lack of uniformity of electrical heating.

1.2.2 Pyranometer

A pyranometer is a type of actinometer used to measure broadband solar irradiance on a planar surface and is a sensor that is designed to measure the solar radiation flux density (in watts per metre square) from a field of view of 180°.

The working principle of a pyranometer is the same as a pyrheliometer except for the fact that a sensitive surface is exposed to the total beam, diffused and reflected from Earth and surrounding radiation. The sensitive surface consists of a circular, blackened (hot junction) multijunction thermopile whose cold junctions are electrically insulated from the basement. The temperature difference between the hot and cold junctions is a function of the radiation falling on the surface. The sensitive surface is covered by two concentric hemispherical glass domes to shield it from wind and rain. This also reduces the convection currents. A pyranometer, when provided with an occulting disc, measures the diffuse radiation. This disc, or band, blocks the beam radiation from the surface. The standard distance between the glass dome and the shading ring is 0.3 m. It may be noted that pyranometers are calibrated so as to measure the solar radiation on a horizontal surface. Therefore, when tilted, the change in free convection regime within the glass dome may introduce an error in measurement. A photograph of a typical pyranometer is shown in Figure 1.2.

A pyranometer produces voltage, as a function of the incident solar radiation, from the thermopile detectors. A potentiometer is required to detect and record this output. Radiation data usually must be integrated over some period of time, such as an hour or a day. Integration can be done by means of planimetry or an electronic integrator. Pyranometers have also been based on

Figure 1.2 Photograph of a typical pyranometer.

photovoltaic (solar cell) detectors. Silicon cells are the most common for solar energy measurement, although cadmium sulfide and selenium cells have also been used. Silicon solar cells have the property that their light current (approximately equal to the short-circuit current at normal radiation levels) is a linear function of the incident solar radiation. They have the disadvantage that their spectral response is not linear, so instrument calibration is a function of the spectral distribution of the incident radiation.

A typical pyranometer does not require any power to operate and they are frequently used in meteorology, climatology, solar energy studies and building physics. They can be seen in many meteorological stations, often installed horizontally and next to solar panels, and the sensor is mounted in the surface plane of the panel. Pyranometers are standardized according to the ISO 9060 standard, which is also adopted by the World Meteorological Organization (WMO). Calibration is typically done relative to the World Radiometric Reference (WRR). This reference is maintained by World Radiation Centre (WRC) in Davos, Switzerland.

1.2.3 Sunshine Recorder

Sunshine recorders are used to indicate the amount of sunshine at a given location. The results are used to provide information on the climate of an area and some of the fields it is of importance to are science, agriculture and tourism.

Traditionally, sunshine recorders are divided into two groups. In the first group the time of the occurrence of the event is provided by the Sun itself and in the second a clock-type device is used to provide the time scale. The older type of recorder required the interpretation of the results by an observer and these may have differed from one person to another. Today, with the use of electronics and computers, it is possible to record the sunshine duration that does not rely on an observer's interpretation. At the same time the newer recorders can also measure the global and diffuse radiation.

A sunshine recorder consists of a glass sphere mounted in a section of a spherical brass bowl with grooves for holding the recorder cards. The sphere burns a trace on the card when exposed to the Sun, the length of the trace being a direct measure of the duration of bright sunshine. There are sets of grooves for taking three sets of cards: long curved for summer, short curved for winter and straight cards at equinoxes.

1.3 Sun–Earth Angles

The energy flux of beam radiation on a surface with arbitrary orientation can be obtained by the flux either on a surface perpendicular to the Sun rays or on a horizontal surface. The various Sun–Earth angles required to understand the solar energy received are as follows.

1.3.1 Zenith Angle (θ_z)

Let P be a point on the surface of the Earth referred to as the position of the observer and PN normal to the horizontal plane as shown in Figure 1.3. The direction PN is known as the zenith direction. The zenith angle (θ_z) is the angle of the Sun's ray (SP) away from the zenith direction, which varies from 0° to 90°. When the Sun is either rising or setting the zenith angle is near 90° whereas

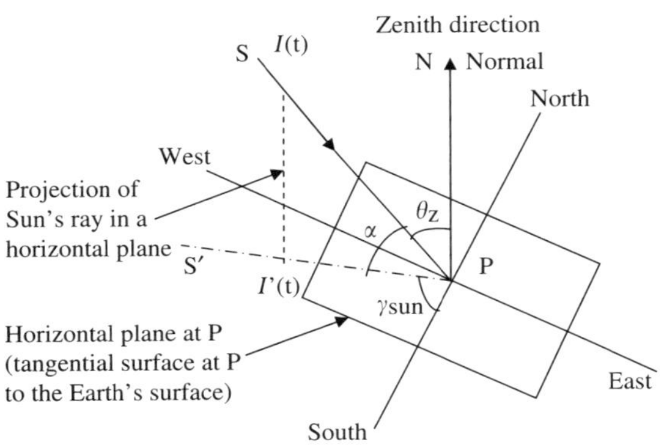

Figure 1.3 Zenith, solar altitude and solar azimuth angles.

at noon it is equal to or very near to zero. The zenith angle varies throughout the day with the movement of the Sun.

1.3.2 Solar Altitude (α)

The solar altitude (α) is the angle between the rays of the Sun (SP) and the horizontal plane under consideration. PS' is the projection of the Sun's rays on a horizontal surface. Thus, PS′ represents the horizontal surface. The angle S′PS is the solar altitude, as shown in Figure 1.3. Hence

$$\alpha + \theta_z = 90^\circ$$

The altitude angle is zero at sunrise and sunset, whereas at noon it is near to 90°. The altitude angle also varies throughout the day with the movement of the Sun.

1.3.3 Solar Azimuth Angle (γ_{Sun})

This angle is measured with respect to the south direction (the directions pointed to by a compass are magnetic south and north). We must consider geographic south, which is different from magnetic south. A person standing vertically at noon (noon is the moment at which shadows are shortest) makes their shortest shadow on the Earth pointing towards geographic south and north. If the person is facing the Sun then that direction is geographic south, whereas the direction of the back of the person will be geographical north. Considering Figure 1.3, the angle between the south direction and the projection of the rays of the Sun on a horizontal plane is known as the solar azimuth angle γ_{Sun}.

1.3.4 Wall Azimuth Angle (γ_{wall})

'Wall' does not mean any vertical surface. It can also mean an inclined surface. The angle that the projection of normal at the inclined surface on the horizontal surface makes with the south direction is known as the wall azimuth angle or surface azimuth angle γ_{wall}, as shown in Figure 1.4.

The following are the main points:

(i) γ_{wall} for the surface facing north will be $\pm 180^\circ$.
(ii) γ_{wall} for the surface facing south will be 0°.
(iii) If the wall or surface has $\beta = 90^\circ$, the wall is vertical.
(iv) If $\beta = 0^\circ$, the surface is horizontal.
(v) If N′ falls on the west side of the south direction, then γ_{wall} is taken as positive.
(vi) If N′ falls on the east side of the south direction, then γ_{wall} is negative.
(vii) The angle of incidence θ_i is the angle between the beam radiation on a surface and the normal to that surface.

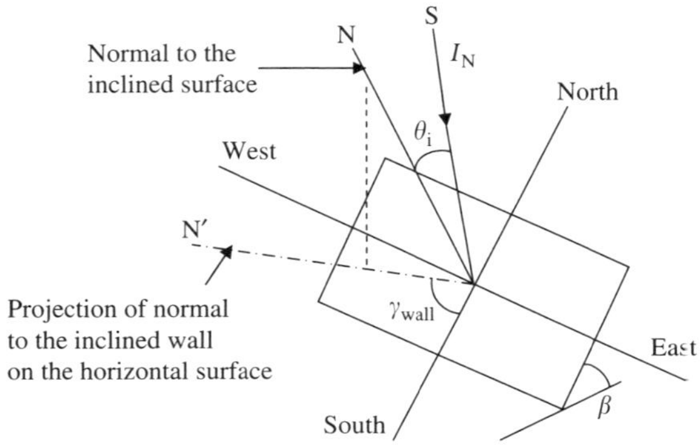

Figure 1.4 Wall azimuth angle for an inclined surface.

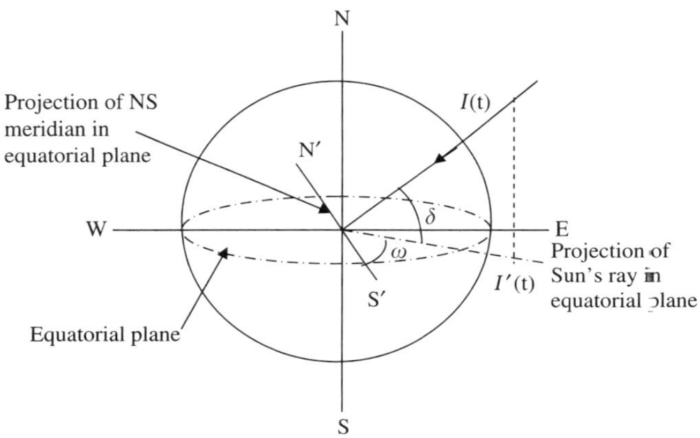

Figure 1.5 Solar declination angle.

1.3.5 Solar Declination (δ)

The angle that the Sun's rays make with the equatorial plane is known as the declination angle (Figure 1.5). In other words, the solar rays hit our planet at a certain angle with respect to the equator; this angle is the solar declination.

On any day, δ is taken as a constant which changes on the next day. Cooper's empirical relation for calculating the solar declination angle (in degrees) is[2]

$$\delta = 23.45 \sin\left[(284 + n) \times \frac{360}{365}\right] \tag{1.1}$$

where $n =$ day of the year ($1 \leq n \leq 365$).

Solar declination can also be defined as the angle between the line joining the centres of the Sun and the Earth and its projection on the equatorial plane. The solar declination changes mainly due to the rotation of Earth about an axis. Its maximum value is 23.45° on 21 December and the minimum is – 23.45° on 21 June.

Key Points

(i) The line joining the centre of the Sun and the Earth is important for δ.
(ii) The axis of rotation remains pointed in the same direction. It is never perpendicular to the orbital plane.
(iii) The equinox dates when solar declination is zero are 22 March and 22 September, *i.e.* when night is equal to day.
(iv) The longest day is 22 June and the shortest is 22 December.

Example 1.1

Calculate the δ on July 20, 2008.

Solution

For July 20, 2008,

$$n = \text{day of the year} = 201$$

Therefore, using eqn (1.1) we get

$$\delta = 23.45 \sin\left[(284 + n) \times \frac{360}{365}\right] = 20.63^\circ$$

1.3.6 Latitude (ϕ) and Longitude (L_t)

We can describe a location on Earth using latitude and longitude. Consider P to be a place under consideration on Earth's surface (Figure 1.6). Angle ϕ

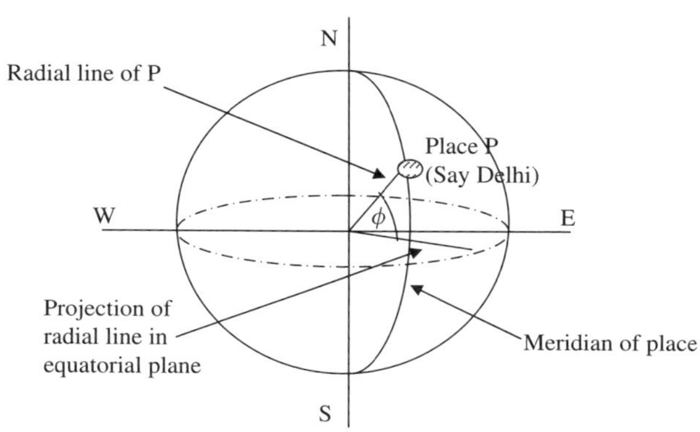

Figure 1.6 Latitude angle.

represents the latitude of the place *P*. Normally we need three coordinates to define any point in space (radius, 1st angle and 2nd angle). For describing any place on Earth's surface, the radius of the Earth is fixed. Now, there are only two angles, *i.e.* the 1st and 2nd angles, which are to be understood to describe any place. These angles are *latitude* and *longitude*.

The latitude of a location is the angle made by the radial line joining the given location to the centre of the Earth with its projection on the equatorial plane. If an observer at any point on the surface of the Earth is represented by point *P* (Figure 1.6), then ϕ represents the latitude of the place where the observer is standing. This angle indicates how far we are from the equatorial plane. The higher is ϕ, the further we are away from the equator and nearer to either of the poles. At the poles we receive much less solar radiation. The imaginary sphere of radius equal to the average Earth–Sun distance that envelops the Earth is known as the *celestial sphere*.

Latitude ϕ gives the location of a place on Earth, *i.e.* north or south of the equator. Latitude is an angular measurement ranging from 0° at the equator to 90° at the poles (90°N or 90°S) for the north and south poles, respectively. It is essential to mention here that the equator is an imaginary circle drawn around a planet at a distance halfway between the poles. The equator divides the planet into two halves, *viz*. a northern hemisphere and a southern hemisphere. The latitude of the equator is, by definition, 0°. The length of Earth's equator is about 40,075.0 km, or 24,901.5 miles. Thus, it is well understood that latitude shows us how far we are from north or south, *i.e.* it is zero at the equator, 90° N at the north pole and 90° S at the south pole. Lines of latitude run parallel to the equator.

The equator is one of the five main circles of latitude based on the relationship of the Earth's rotation and plane of orbit around the Sun. Additionally, the equator is the only line of latitude which is also a great circle. On the Earth, a *circle of latitude* is an imaginary east–west circle that connects all the locations with a given fixed latitude. The position of any place on the circle of latitude is given by the longitude. Each is perpendicular to all meridians at the intersection points. Those parallels closer to the poles are smaller than those at or near the equator.

The five major circles of latitude are (Figure 1.7):

(i) Arctic Circle (66° 33′ 38″ N)
(ii) Tropic of Cancer (23° 26′ 22″ N)
(iii) Equator (0° N)
(iv) Tropic of Capricorn (Sagittarius) (23° 26′ 22″ S)
(v) Antarctic Circle (66° 33′ 38″ S)

The Arctic Circle and Antarctic Circle represent the southernmost and northernmost locations where it is possible to have a day without a sunrise.

The Tropic of Cancer and Tropic of Capricorn represent the northernmost and southernmost locations where the Sun may be seen directly overhead (midsummer and midwinter, respectively).

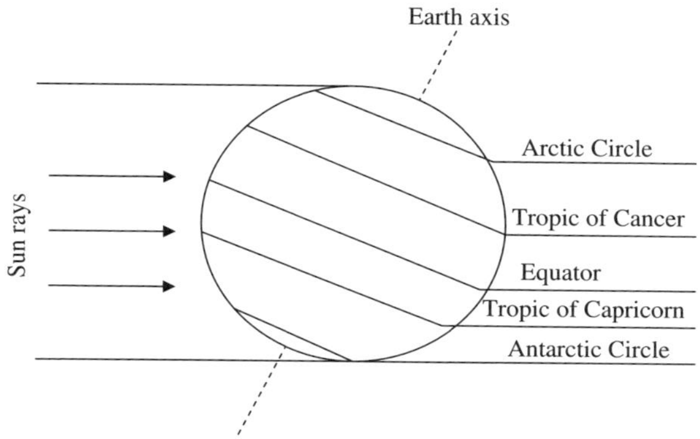

Figure 1.7 Sunrays falling on the Earth.

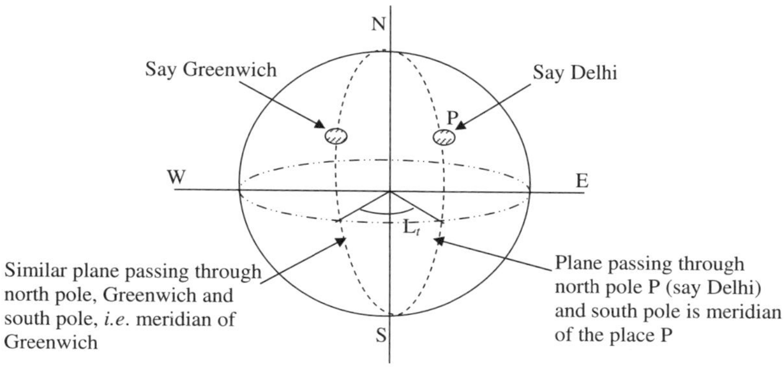

Figure 1.8 Longitude angle.

On the Earth, a *meridian* is an imaginary north–south line between the north pole and the south pole that connects all locations with a given longitude. The position on the meridian is given by the latitude, each being perpendicular to all circles of latitude at the intersection points.

The meridian that passes through Greenwich (England) is considered as the prime meridian, *i.e.* zero degrees of longitude. Any other meridian is referred to from the prime meridian, and has a fixed angular distance from the prime meridian known as the longitude of that meridian (Figure 1.8). All the places on that meridian have the same longitude. The Earth can be divided in two parts with reference to the prime meridian, *viz.* eastern and western hemispheres. The maximum distant meridian on both sides can be at 0 to 180° from the principal meridian.

Key Points

(i) The latitude is taken as positive for the northern hemisphere and negative for the southern hemisphere.

(ii) In the case of Delhi, the longitude is 77.2° E, also known as –77.2°, *i.e.* towards east from the Greenwich meridian and is considered as negative.

1.3.7 Hour Angle (ω)

The hour angle is the angle through which the Earth has to rotate to bring the meridian plane of any place or location under the Sun. This angle continuously decreases from sunrise to noon, becomes zero at noon and then starts increasing when its value becomes positive. At sunset the hour angle is maximum positive and at sunrise it is maximum negative for any place. In other words, the hour angle is the measure of the angular displacement of the Sun through which the Earth has to rotate to bring the meridian of the place directly under the Sun. Thus, it is very clear that ω will vary with the time of the day as shown in Figure 1.9.

The angle between an observer's (at a particular place on Earth) meridian and the hour circle on which some celestial body lies is known as the *hour angle*. This angle is conventionally expressed in units of time (hours, minutes and seconds), which gives the time elapsed since the celestial body's last transit at the observer's meridian (for a positive hour angle), or the time expected for the next transit (for a negative hour angle) (1 hour = 15°).

At sunrise, the value of ω will be maximum, then it will slowly and steadily reduce and keep reducing with time until solar noon. At this point ω becomes zero. It starts increasing the moment after solar noon and will be maximum at sunset. The values at sunrise and sunset are numerically the same but have opposite signs.

An expression for the hour angle, ω (in degrees), is given by

$$\omega = (ST - 12) \times 15 \tag{1.2}$$

where ST is local solar time.

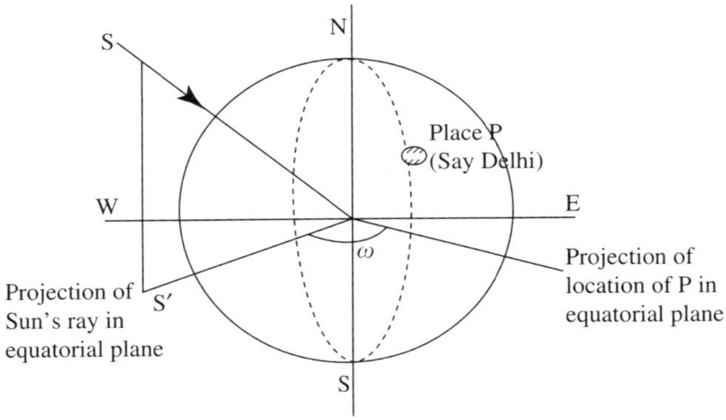

Figure 1.9 Hour angle.

Example 1.2

Calculate the hour angle at 10 a.m. and 2 p.m.

Solution

Using eqn (1.2),

(i) for 10 a.m., $ST = 10$,
hence, the hour angle $\omega = (10 - 12) \times 15 = -30°$
(ii) for 2 p.m., $ST = 14$,
hence, $\omega = (14 - 12) \times 15 = 30°$.

1.3.8 Solar Time

Solar time is based on the idea that when the Sun reaches its highest point in the sky, it is noon.

There are two different times to be understood: solar time and *clock time*. Clocks and watches show *clock time*. Neither kind of time is intrinsically 'better' than the other. Both are useful and interesting for their separate purposes.

Solar time is recognized when the Sun reaches its highest point (when it just crosses the meridian), at noon. The next day, when the Sun again crosses the meridian, it is again noon. The time that elapses between successive noons is important. Sometimes it is more and sometimes less than 24 hours of clock time. In the middle months of the year, the day length is close to 24 hours, but around 15 September the days are only some 23 hours, 59 minutes and 40 seconds long. Around Christmas the days are 24 hours and 20 seconds long.

Clock time recognized each day is exactly 24 hours long, which is not actually true. But it is obviously much more convenient to have a clock time which takes exactly 24 hours for each day because mechanical clocks and watches (and more recently electronics) can be made to measure these exactly equal time intervals.

Obviously, these small differences in the lengths of days produce larger differences between solar and clock time. These differences reach a peak of just over 14 minutes in mid February, when solar time is slower relative to clock time. It reaches just over 16 minutes at the beginning of November when solar time is fast relative to clock time. There are also two minor peaks: first in mid May, when solar time is nearly 4 minutes fast, and second in late July, when solar time is just over 6 minutes slow. These minor peaks contribute towards the fortunate effect in the northern hemisphere. The differences are relatively small during most of the months when there is a reasonable amount of sunshine.

The differences do not cumulate across the years as the clock time is arranged in such a way that over the course of a four-year cycle, including a leap year, these two times come back very nearly to the same time they started. The 'very nearly' is because 'clock time' still has to be adjusted by not having a leap year at the turn of each century, except when the year is exactly divisible by 400, *e.g.* 1900 was not a leap year but 2000 was. Even with this correction, we had an extra second added to 'clock time' recently.

The reasons for these differences are discussed below, followed by some information on what the differences are at given times of the year.

Days of Different Lengths[3] arise from two quite separate causes. First, the plane of the equator is not the same as the plane of the Earth's orbit around the Sun, but is offset from it by the *angle of obliquity*. Second, the orbit of the Earth around the Sun is an ellipse and not a circle, and the apparent motion of the Sun is, thus, not exactly equal throughout the year. The Sun appears to be moving fastest when the Earth is closest to the Sun.

The sum of the two effects is the *equation of time* and the magnitude of that is shown by the curve in Figure 1.10. The characteristic twin peaks discussed above are also shown.

The standard meridian place (time zone) of any country is the place with reference to which the 'clock time' of that country is decided. The time shown by a clock is clock time. For India it is IST (Indian Standard Time) If you are an observer at the standard meridian place of the country, *e.g.* Allahabad for India, and the Sun is passing through the meridian of that place, *i.e.* Allahabad, and at that instance our watch shows 12 noon, then it is solar time for that day and that place. It can also be called *solar noon* for that standard meridian place (time zone).

Solar noon is that moment of the day that divides the daylight hours for that day exactly into half. It is the time, at a specific location, when the Sun reaches its highest apparent point in the sky. To determine solar noon, calculate the length of the day from the times of sunset and sunrise and divide by two. Solar noon may be quite a bit different from 'clock' noon. However, this solar noon will be the same as clock noon for the place of standard meridian of the country (time zone). The shadow at any place will be shortest if the Sun is passing through the meridian of that place. In other words, if at any place in India our watch shows 12 noon, it means it is the time when the Sun is passing through the standard

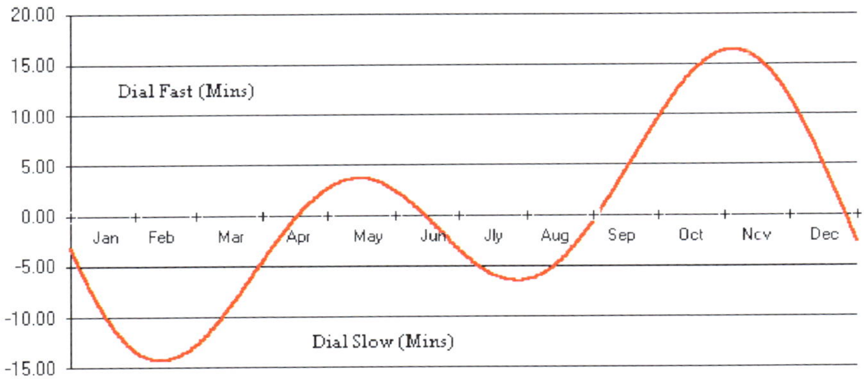

Figure 1.10 Equation of time.[9]

meridian of Allahabad or it is the solar noon at Allahabad. Thus, it is very easy to understand that at the occurrence of solar noon at the standard meridian of the time zone, in those places which are not on the standard meridian (say Allahabad), the solar noon either has occurred or is yet to occur, depending on the longitude of the place under consideration. We can calculate 'solar time' for any place with reference to standard time (clock time) by the relation:

$$\text{solar time} = \text{clock time} + \text{longitude correction} + \text{equation of time } (E) \quad (1.3a)$$

or,

$$\text{solar time} - \text{clock time} = 4(\mathrm{L_{st}} - \mathrm{L_{loc}}) + E \quad (1.3b)$$

where $\mathrm{L_{st}}$ is the standard meridian for the local time zone. $\mathrm{L_{st}}$ for India has the value 81° 44′. $\mathrm{L_{loc}}$ is the longitude of the location in question (in degrees west) (Table 1.1) and E is the equation of time (in minutes) (Table 1.2) and is given by the expression:[1]

$$E = 229.2(0.000075 + 0.001868\cos\mathrm{B} - 0.032077\sin\mathrm{B} - 0.014615\cos 2\mathrm{B} - 0.04089\sin 2\mathrm{B}) \quad (1.3c)$$

where $\mathrm{B} = (n - 1)360/365$, n = day of the year.

The equation of time (minutes: seconds) for typical days for different months for Delhi (Longitude 77° 12′ E) has been given in Table 1.2.

1.3.9 Angle of Incidence

The angle of incidence is the angle between a beam incident on a surface and the line perpendicular to the surface at the point of incidence called the *normal* (Figure 1.11).

An expression for $\cos\theta_i$ is given by

$$\cos\theta_i = (\cos\phi\cos\beta + \sin\phi\sin\beta\cos\gamma)\cos\delta\cos\omega + \cos\delta\sin\omega\sin\beta\sin\gamma + (\sin\phi\cos\beta - \cos\phi\sin\beta\cos\gamma) \times \sin\delta \quad (1.4)$$

Table 1.1 Latitude, longitude and elevation for different places in India.[1]

Place	*Latitude* (ϕ)	*Longitude* (L_{loc})	*Elevation* (E_0)
Bangalore	12° 58′ N	77° 35′ E	921 m above msl
Bombay	18° 54′ N	72° 49′ E	11 m above msl
Jodhpur	26° 18′ N	73° 01′ E	224 m above msl
Mt. Abu	24° 36′ N	72° 43′ E	1195 m above msl
New Delhi	28° 35′ N	77° 12′ E	216 m above msl
Simla	31° 06′ N	77° 10′ E	2202 m above msl
Srinagar	34° 05′ N	74° 50′ E	1586 m above msl
Calcutta	22° 32′ N	88° 20′ E	6 m above msl
Madras	13° 00′ N	80° 11′ E	16 m above msl

Table 1.2 The Sun's equation of time (E) (minutes: seconds).[1]

Month	*1*	*8*	*15*	*22*
January	–(3:16)	–(6:26)	–(9:12)	–(11:27)
February	–(13:34)	–(14:14)	–(14:15)	–(13:41)
March	–(12:36)	–(11:04)	–(9:14)	–(7:12)
April	–(4:11)	–(2:07)	–(0:15)	(1:19)
May	2:50	3:31	3:44	3:30
June	2:25	1:15	–(0:09)	–(1:40)
July	–(3:33)	–(4:48)	–(5:45)	–(6:19)
August	–(6:17)	–(5:40)	–(4:35)	–(3:04)
September	–(0:15)	2:03	4:29	6:58
October	10:02	12:11	13:59	15:20
November	16:20	16:16	15:29	14:02
December	11:14	8:26	5:13	1:47

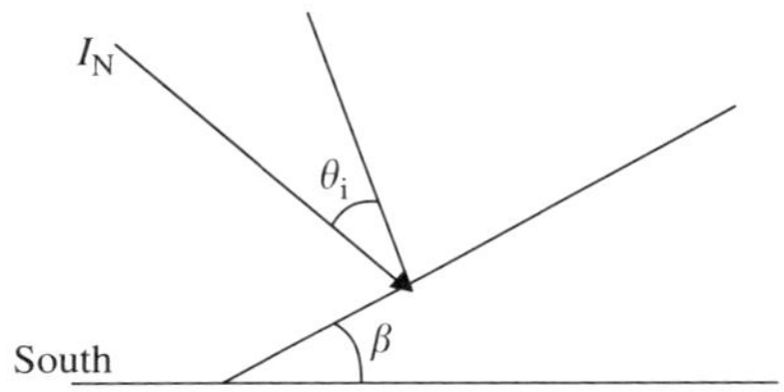

Figure 1.11 View of an inclined surface.

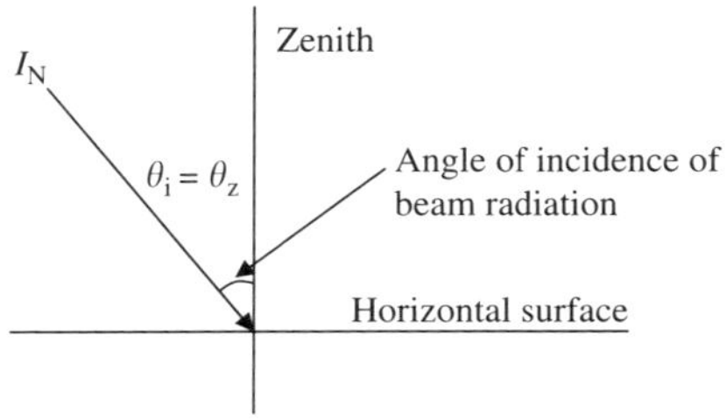

Figure 1.12 View of a horizontal surface.

where β is the inclination of the plane (a surface on which beam radiation is falling) with the horizontal surface and γ is the wall azimuth angle (due south) that specifies the orientation of the surface. This angle decides the distance of a tilted plane from the south orientation. If its value is 0° then the surface is facing towards south.

If the plane under consideration is horizontal, *i.e.* $\beta = 0$ and also $\gamma = 0$, then the angle of incidence θ_i becomes equal to the zenith angle (Figure 1.12).

The following expression for cos θ_z is obtained from eqn (1.4):

$$\cos\theta_z = \cos\phi\cos\delta\cos\omega + \sin\phi\sin\delta \tag{1.5}$$

Example 1.3

Calculate the solar zenith angle for Examples 1.1 and 1.2 for New Delhi ($\phi = 28^\circ\ 35'$).

Solution

$1^\circ = 60'$ (angle),
hence, $\phi = 28^\circ\ 35' = 28^\circ + \left(\frac{35}{60}\right)^0 = 28.58^\circ$.
For 10 a.m., $ST = 10$, then using eqn (1.2),
hour angle $\omega = (10 - 12) \times 15 = -30^\circ$.
From Example 1.1, $\delta = 20.63^\circ$.
Now from eqn (1.5), at 10 a.m., we get

$$\cos\theta_z = \cos(28.58) \times \cos(20.63) \times \cos(-30) + \sin(28.58) \times \sin(20.63)$$
$$= 0.880$$
$$\theta_z = \cos^{-1}(0.880) = 28.32^\circ.$$

Similarly, $\theta_z = 28.32^\circ$ at 2 p.m. ($ST = 14$ and $\omega = 30^\circ$).

1.4 Solar Radiation on a Horizontal Surface

The combination of both forms of solar energy (beam and diffuse) incident on a horizontal plane at the Earth's surface is referred to as global solar energy and these three quantities (specifically their rate or irradiance) are linked mathematically as

$$I_G = I_N \cos\theta_z + I_d \tag{1.6}$$

where I_G is the global irradiance on a horizontal surface, I_d the diffuse irradiance, I_N the direct beam irradiance on a surface perpendicular to the direct beam and θ_z the Sun's zenith angle (eqn (1.5)). By measuring the three components separately, a useful quality assurance test is immediately available by comparing the measured quantity with that calculated from the other two. Thus global solar irradiance is a measure of the rate of total incoming solar energy (both direct and diffuse) on a horizontal plane at the Earth's surface. A pyranometer sensor can be used to measure this quantity with limited accuracy. The most accurate measurements are obtained by summing the diffuse and horizontal component of the direct irradiance.

The radiant energy flux received per second by a surface of unit area held normal to the direction of the Sun's rays at the mean Earth–Sun distance, outside the atmosphere (extra-terrestrial region), is practically constant throughout the year. This value is termed the solar constant I_{SC}, and its value is now adopted to be 1367 W m^{-2}. However, this extraterrestrial radiation suffers variation due to the fact that the Earth revolves around the Sun not in a circular orbit but follows an elliptic path, with the Sun at one of the foci. The intensity of extraterrestrial radiation measured on a plane normal to the radiation on the nth day of the year is given in terms of the solar constant I_{SC} as[4]

$$I_{ON} = I_{SC}\left[1 + 0.033\cos\left(\frac{360 \times n}{365}\right)\right] \tag{1.7}$$

Example 1.4

Calculate the solar intensity in the extraterrestrial region for Example 1.1.

Solution

From Example 1.1, $n = 201$.

From eqn (1.7), we get

$$I_{ON} = 1367\left[1 + 0.033\cos\left(\frac{360 \times 201}{365}\right)\right] = 1324\,\mathrm{W\,m^{-2}}$$

The range of wavelength radiation emitted from the Sun, the attenuation of its amplitude during propagation from the Sun to the atmosphere and further attenuation of radiation in the atmosphere, as well as the long wavelength radiation emitted from Earth, is shown in Figure 1.13. Thus, from the view of terrestrial applications of solar energy, only radiation of wavelength between 0.29 and 2.3 μm is significant.

Following Singh and Tiwari,[5] the rate of beam (direct) radiation reaching the terrestrial region can be written as:

$$I_N = I_{ON}\exp\left[-(m.\ \varepsilon.\ T_R + \alpha)\right] \tag{1.8}$$

where m, ε, T_R and α are the air mass, the integrated Rayleigh scattering optical thickness of the atmosphere, the Linke turbidity factor and a lumped atmospheric parameter for beam radiation, respectively.

The parameters m and ε are expressed in the following form:[6,7]

$$m = \left[\cos\theta_z + 0.15 \times (93.885 - \theta_z)^{-1.235}\right]^{-1} \tag{1.9a}$$

and

$$\varepsilon = 4.529 \times 10^{-4}m^2 - 9.66865 \times 10^{-3}m + 0.108014 \tag{1.9b}$$

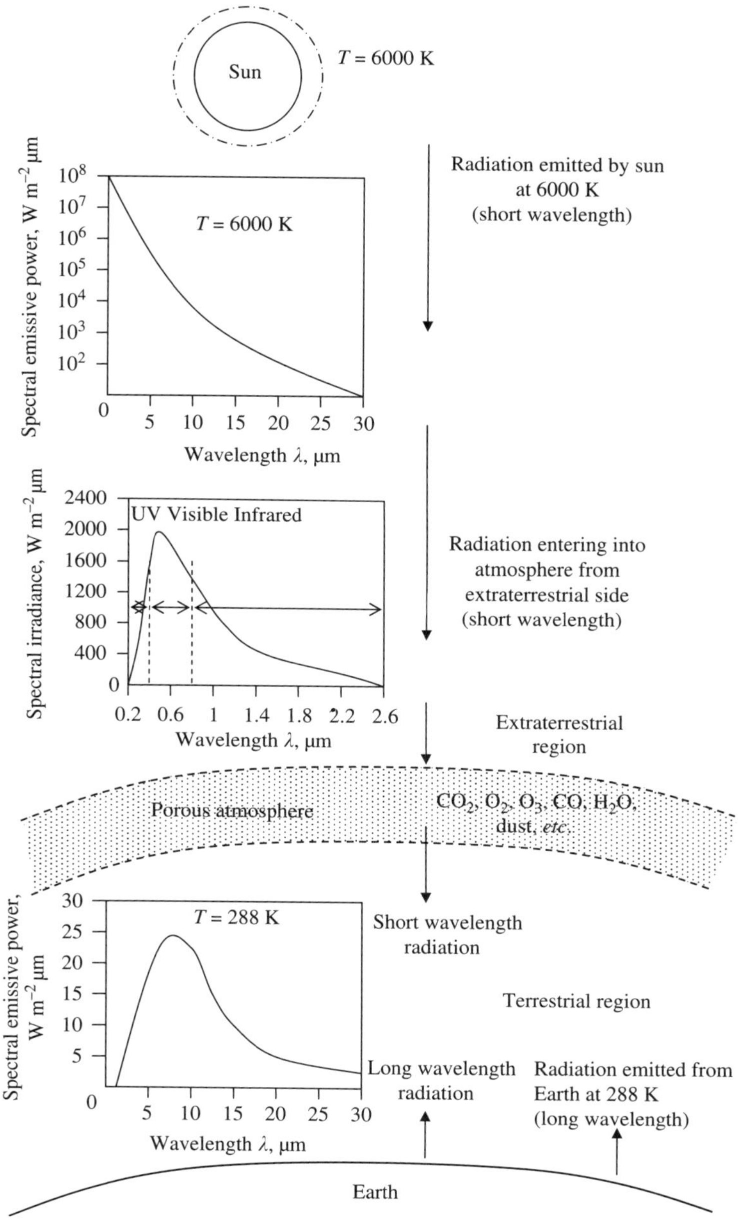

Figure 1.13 Natural flow of solar radiation and its absorption on Earth's surface.

The terrestrial beam radiation received on the horizontal surface is expressed by the classical equation as

$$I_b = I_N \cos\theta_z = I_{ON} \exp\left[-(m.\ \varepsilon.\ T_R + \alpha)\right] \cos\theta_z \tag{1.10}$$

where θ_z is the solar zenith angle for a given time (eqn (1.5)).[1] The term α determines the additional depletion in direct normal irradiance in the terrestrial region due to cloudiness/haziness and transient and unpredictable changes. The diffuse radiation on the horizontal surface can be rewritten in terms of constants K_1 (dimensionless) and K_2 (W m^{-2}) as

$$I_d = K_1(I_{ON} - I_N)\cos\theta_z + K_2 \tag{1.11}$$

where I_N (W m^{-2}) is the normal terrestrial solar radiation at the ground level (eqn (1.10)). The constants K_1 and K_2 can be defined as lumped atmospheric parameters for diffuse radiation.[5] Further, the constant K_1 can be interpreted as the 'perturbation factor' for describing scattering out of beam traversing the lumped atmosphere and K_2 can be referred to as 'background diffuse radiation'.

The values of T_R, α, K_1 and K_2 will be different for the following four weather conditions of New Delhi (composite climate) and those for other stations are given in Appendix II.

Type a (clear day; blue sky): In this case, the ratio of daily diffuse to daily global radiation has been considered as less than or equal to 0.25. The number of hours of sunshine is greater than or equal to 9 h.

Type b (hazy day; fully): In this case, the ratio of daily diffuse to daily global radiation has been considered as between 0.25 and 0.50. The sunshine hours lie between 7 and 9 h.

Type c (hazy and cloudy; partially): The ratio of daily diffuse to daily global radiation lies between 0.50 and 0.75 and sunshine hours falls between 5 and 7 h.

Type d (cloudy day; fully): The ratio of daily diffuse to daily global radiation is greater than or equal to 0.75. The number of sunshine hours is less than or equal to 5 h.

It has been well understood that at sunrise and sunset the zenith angle is equal to 90°. If this value of θ_z is substituted in the expression of the zenith angle the expression reduces to

$$\cos 90 = \cos\phi \cos\delta \cos\omega_s + \sin\phi \sin\delta$$

$$\cos\omega_s = -\frac{\sin\phi \cdot \sin\delta}{\cos\phi \cdot \cos\delta} = -\tan\phi \cdot \tan\delta$$

$$\omega_s = \cos^{-1}[-\tan\phi \tan\delta]$$

where ω_s is the hour angle on sunrise and sunset. It is important to evaluate the total sunshine hours for any particular day for any particular place on Earth. In 24 hours the Sun rotates (actually the Earth rotates) 360°, *i.e.* in 1 hour the Sun rotates 15° or, in other words, it takes 4 minutes for 1° movement of the Sun. Hence, for ω_s rotation the time taken can be calculated, which will be helpful to calculate the total sunshine hours. Since ω_s is equal on both sides of solar noon for any day, the total number of sunshine hours is given by

$$Day\ length = 2 \cdot \frac{\omega_s}{15} = \frac{2}{15} \cdot \cos^{-1}[-\tan\phi \cdot \tan\delta]$$

On the day of equinox, since the solar declination δ is supposed to be zero

$$\begin{aligned} Day\ length &= 2 \cdot \frac{\omega_s}{15} = \frac{2}{15} \cdot \cos^{-1}[-\tan\phi \cdot \tan 0] = \frac{2}{15} \cdot \cos^{-1}(0) = \frac{2}{15} \cdot 90 \\ &= 12\,\mathrm{h} \end{aligned}$$

Thus, an interesting conclusion is arrived at, which is that the day and night are equal only on the equinox day, otherwise they are either longer or shorter depending on the place and day under consideration.

Example 1.5

Calculate the air mass and optical thickness of the atmosphere for Example 1.3.

Solution

From eqn (1.9a), the air mass m is given by

$$\begin{aligned} m &= [\cos\theta_z + 0.15 \times (93.885 - \theta_z) - 1.253]^{-1} \\ &= [0.88 + 0.15 \times (93.885 - 28.32) - 1.253]^{-1} \\ &= 1.135. \end{aligned}$$

From eqn (1.9b), the optical thickness of the atmosphere (ε) is

$$\begin{aligned} \varepsilon &= 4.529 \times 10^{-4} m^2 - 9.66865 \times 10^{-3} m + 0.108014 \\ &= 4.529 \times 10^{-4} \times (1.135)^2 - 9.66865 \times 10^{-3} \times (1.135) + 0.108014 \\ &= 0.0976. \end{aligned}$$

1.5 Solar Radiation on an Inclined Surface

Knowing the hourly beam and diffuse radiation by eqns (1.10) and (1.11) on a horizontal surface, the total radiation for any inclined (inclination $= \beta$) with any orientation of solar thermal device (for east, south, west and north) $\gamma = -90°, 0°, +90°$ and $\pm 180°$ for a given latitude ϕ can be evaluated using the Liu and Jordan[8] formula

$$I_t = I_b R_b + I_d R_d + \rho R_r (I_b + I_d) \tag{1.12}$$

Example 1.6

Calculate the beam I_b and diffuse radiation on a horizontal surface for Example 1.2 to 1.5 for clear sky conditions.

Solution

From Example 1.2 to 1.5, we get

$$I_{ON} = 1324\,\mathrm{W\,m^{-2}},\ m = 1.135,\ \varepsilon = 0.0976 \text{ and } \cos\theta_z = 0.88$$

For clear sky conditions and the month of July, the values of the constants of eqns (1.10) and (1.11) are

$$T_R = 2.40,\ \alpha = 0.24,\ K_1 = 0.49 \text{ and } K_2 = -69.12.$$

From eqn (1.10), the beam radiation on a horizontal surface I_b is

$$I_b = I_N \cos\theta_z = I_{ON}\exp[-(m.\ \varepsilon.\ T_R + \alpha)]\cos\theta_z$$
$$I_b = 1324\exp[-(1.135 \times 0.0976 \times 2.4 + 0.24)] \times 0.88$$
$$= 892.4\,\mathrm{W\,m^{-2}}.$$

From eqn (1.11), the diffuse radiation on a horizontal surface I_d is

$$I_d = K_1[I_{ON} - I_N]\cos\theta_z + K_2$$
$$I_d = K_1 I_{ON}[1 - \exp\{-(m.\ \varepsilon.\ T_R + \alpha)\}]\cos\theta_z + K_2$$
$$I_d = 0.49 \times 1324 \times [1 - \exp\{-(1.135 \times 0.0976 \times 2.4 + 0.24)\}] \times 0.88 - 69.12$$
$$= 157.54\,\mathrm{W\,m^{-2}}.$$

where R_b, R_d and R_r are known as conversion factors for beam, diffuse and reflected components, respectively, and ρ is the reflection coefficient of the ground ($=0.2$ and 0.6 for non-snow-covered and snow-covered ground, respectively).

Expressions for these conversion factors are:

(i) R_b is defined as the ratio of flux of beam radiation incident on an inclined surface to that on a horizontal surface.

$$I_b R_b = I_N \cos\theta_i$$

$$R_b = \frac{I_N \cos\theta_i}{I_b} = \frac{I_N \cos\theta_i}{I_N \cos\theta_z} = \frac{\cos\theta_i}{\cos\theta_z}$$

(ii) R_d is defined as the ratio of the flux of diffuse radiation falling on the tilted surface to that on the horizontal surface.

This conversion factor depends on the distribution of diffuse radiation over the sky and on the portion of sky seen by the surface. But a satisfactory method of estimating the distribution of diffuse radiation over the sky is yet to be found. It is, however, widely accepted that the sky is an isotropic source of diffuse radiation.

$$R_d = \frac{1 + \cos\beta}{2} \tag{1.13a}$$

For $\beta = 0°$ $R_d = 1$ and $\beta = 90°$ $R_d = 1/2$.

(iii) R_r is the reflected component, which comes mainly from the ground and other surfaces, and is given by

$$R_r = \frac{1 - \cos\beta}{2} \tag{1.13b}$$

For $\beta = 0°$ $R_r = 0$ and $\beta = 90°$ $R_r = 1/2$.

Example 1.7

Calculate an angle of incidence θ_i of solar radiation for an inclined surface having inclination of 10° and facing east, south and west at 11 a.m. for New Delhi on July 20, 2008.

Solution

For New Delhi $\phi = 28.58°$ (Example 1.3).

$\beta = 10°$ (given), $\gamma = -90°$ (east), 0° (south) and +90° (west)

$\delta = 20.63°$ (Example 1.1), $\omega = 30°$ (Example 1.2).

From eqn (1.4), the angle of incidence for a surface is given by

$$\begin{aligned}\cos\theta_i =& (\cos\phi\cos\beta + \sin\phi\sin\beta\cos\gamma)\cos\delta\cos\omega \\ &+ \cos\delta\sin\omega\sin\beta\sin\gamma \\ &+ (\sin\phi\cos\beta - \cos\phi\sin\beta\cos\gamma) \times \sin\delta.\end{aligned}$$

(a) Now, the angle of incidence for a south-facing surface is

$$\begin{aligned}\cos\theta_i =& (\cos 28.58 \cos 10 + \sin 28.58 \sin 10 \cos 0) \times \cos 20.63 \cos 30 \\ &+ \cos 20.63 \sin 30 \sin 10 \sin 0 \\ &+ (\sin 28.58 \cos 10 - \cos 28.58 \sin 10 \cos 0) \times \sin 20.63 \\ =& 0.88\end{aligned}$$

$\theta_i = \cos^{-1}(0.88) = 28.35°$.

(b) For an east-facing surface

$$\begin{aligned}\cos\theta_i =& (\cos 28.58 \cos 10 + \sin 28.58 \sin 10 \cos(-90)) \times \cos 20.63 \cos 30\\ &+ \cos 20.63 \sin 30 \sin 10 \sin(-90)\\ &+ (\sin 28.58 \cos 10 - \cos 28.58 \sin 10 \cos(-90)) \times \sin 20.63\\ =& 0.785.\end{aligned}$$

$$\theta_i = \cos^{-1}(0.785) = 38.2°.$$

(c) For a west-facing surface

$$\begin{aligned}\cos\theta_i =& (\cos 28.58 \cos 10 + \sin 28.58 \sin 10 \cos(90)) \times \cos 20.63 \cos 30\\ &+ \cos 20.63 \sin 30 \sin 10 \sin(90)\\ &+ \sin 28.58 \cos 10 - \cos 28.58 \sin 10 \cos(90)) \times \sin 20.63\\ =& 0.948.\end{aligned}$$

$$\theta_i = \cos^{-1}(0.948) = 18.52°.$$

Example 1.8

Calculate the beam radiation on the inclined surfaces mentioned in Example 1.6 and 1.7.

Solution

(a) For a south-facing surface, the beam radiation is
$I_b = I_N \cos\theta_z$ (eqn (1.12)).
Here $I_N = 892.4$ (Example 1.6)
and $\cos\theta_i = 0.88$ (Example 1.7).
Then $I_b = 894.2 \times 0.88 = 786.8\ \mathrm{W\,m^{-2}}$.
(b) For an east-facing surface, $\cos\theta_i = 0.785$ (Example 1.7)

$$I_b = 894.2 \times 0.785 = 701.9\ \mathrm{W\,m^{-2}}.$$

(c) For a west-facing surface, $\cos\theta_i = 0.948$ (Example 1.7)

$$I_b = 894.2 \times 0.948 = 847.7\ \mathrm{W\,m^{-2}}$$

Example 1.9

Calculate the conversion factor for diffuse I_d and reflected radiation for the surfaces having an inclination of 30°.

Solution

From eqns (1.13(a)) and (1.13(b)), we have

$$R_{\rm d} = \frac{1 + \cos 30}{2} = 0.933$$

and

$$R_{\rm r} = \frac{1 - \cos 30}{2} = 0.066$$

It is important to note that the conversion factors are independent of surface orientation.

Example 1.10

Calculate the total solar radiation on the inclined surfaces mentioned in Example 1.6 and 1.7 ($\rho = 0.2$).

Solution

Given $\rho = 0.2$,

Using eqn (1.12), the total solar radiation on any surface is given by

$$I_{\rm t} = I_{\rm N}\cos\theta_{\rm i} + I_{\rm d}R_{\rm d} + \rho R_{\rm r}(I_{\rm b} + I_{\rm d})$$

$I_{\rm d} = 157.5\,{\rm W\,m^{-2}}$ (Example 1.6); $R_{\rm d} = 0.933$ and $R_{\rm r} = 0.066$ (Example 1.9).

(a) For a south-facing surface, $I_{\rm N}\cos\theta_{\rm i} = 786.8\,{\rm W\,m^{-2}}$ (Example 1.8). Therefore, the total solar radiation is

$$\begin{aligned} I_{\rm t} &= 786.8 + 0.933 \times 157.5 + 0.2 \times 0.066 \times (786.8 + 157.5) \\ &= 946.2\,{\rm W\,m^{-2}}. \end{aligned}$$

(b) For an east-facing surface, $I_{\rm N}\cos\theta_{\rm i} = 701.9\,{\rm W\,m^{-2}}$ (Example 1.8). Therefore, the total solar radiation is

$$\begin{aligned} I_{\rm t} &= 701.9 + 0.933 \times 157.5 + 0.2 \times 0.066 \times (701.9 + 157.5) \\ &= 860.2\,{\rm W\,m^{-2}}. \end{aligned}$$

(c) For a west-facing surface, $I_{\rm N}\cos\theta_{\rm i} = 847.7\,{\rm W\,m^{-2}}$ (Example 1.8). Therefore, the total solar radiation is

$$\begin{aligned} I_{\rm t} &= 847.7 + 0.933 \times 157.5 + 0.2 \times 0.066 \times (847.7 + 157.5) \\ &= 1007.9\,{\rm W\,m^{-2}}. \end{aligned}$$

Problems

1.1 Plot the variation of solar declination angle (δ) with the nth day of the year. Hint: use eqn (1.1).

1.2 Plot the variation of intensity of extraterrestrial radiation with the n th day of the year. Hint: use eqn (1.7) and vary n from 1 to 365.

1.3 Plot the variation of zenith angle (θ_z) with hour angle for New Delhi ($\phi = 28^\circ\ 35'$) for February 10, 2007, and June 06, 2008. Hint: use eqn (1.5).

1.4 Write down the hour angle for all sunshine hours from 6 a.m. to 6 p.m. Hint: use eqn (1.2).

1.5 Plot the hourly variation of direct radiation (I_N) in the terrestrial region for New Delhi for clear sky conditions for the month of January 2007. Hint: use eqn (1.8).

1.6 Calculate the declination angle (δ) for March 31 in a leap year and hour angle (ω) at 2.00 p.m. Hint: use eqn (1.1) and $\omega = 15\,(ST - 12\ \text{hours})$, ST is in hours.

1.7 Calculate the number of daylight hours at Delhi on December 21 and June 21 in a leap year. Hint: use eqn $Day\ length = \frac{2}{15} \cdot \cos^{-1}[-\tan\phi \cdot \tan\delta]$.

1.8 Calculate the total solar radiation for the east and west surfaces of a greenhouse dryer. Hint: for east $\gamma = -90^\circ$, for west $\gamma = +90^\circ$ and use eqn (1.12).

1.9 Repeat Problems 1.6 to 1.8 for all months and weather conditions of New Delhi.

1.10 Calculate the hourly variation of beam (I_b) and diffuse (I_d) radiations on a horizontal surface for Problem 1.5. Hint: use eqns (1.10) and (1.11).

1.11 Calculate the total solar radiation on the south-facing surface of a flat-plate collector for New Delhi conditions for Problem 1.10. Hint: use eqn (1.12) and $\gamma = 0$.

References

1. G. N. Tiwari, *Solar Energy, Fundamentals, Design, Modeling and Applications*, Narosa Publishing House, New Delhi, India, 2004.
2. P. I. Cooper, *Sol. Energ.*, 1969, **12**(3), 333–346.
3. G. N. Tiwari and P. Barnwal, *Fundamentals of Solar Dryers*, Anamaya Publishers, New Delhi, India, 2008.
4. J. A. Duffie and W. A. Beckman, *Solar Engineering of Thermal Processes*, John Wiley and Sons Inc., New York, 1991.
5. H. N. Singh and G. N. Tiwari, *Energy*, 2005, **30**, 1589–1601.
6. F. Kasten, *Arch. Meteor. Geophys. Bioclim., Series B*, 1965, **14**, 206–223.
7. F. Kasten and A. T. Young, *Applied Optics*, 1989, **28**(22), 4735–4738.
8. B. Y. H. Liu and R. C. Jordan, *ASHRAE Journal*, 1962, **3**(10), 53–59.
9. Equation of Time, http://ourworld.compuserve.com/homepages/patrick_powers/sundials.htm, accessed 1 October 2008.

CHAPTER 2

History of PV-integrated Systems

2.1 Introduction

Renewable energy (RE) resources have enormous potential and can meet the present world energy demand. They can enhance diversity in energy supply markets, secure long-term sustainable energy supply and reduce local and global atmospheric emissions. They can also provide commercially attractive options to meet specific needs for energy services (particularly in developing countries and rural areas), and offer possibilities for local manufacturing of equipment. In addition, the uses of RE resources have been charted specifically in many of the roadmaps of the developed countries. One of the most promising RE technologies is photovoltaic (PV) technology. Photovoltaic systems are popularly configured as stand-alone, grid-connected and hybrid systems. They are developing rapidly in the world, in both developed and developing nations. The performance of the PV system depends upon several factors, especially the meteorological conditions such as solar radiation, ambient temperature and wind speed.

Since the nineteenth century solar thermal collectors have been in commercial production. During the 1960s, R&D was mainly concentrated on the space industry due to the higher cost of solar cells. In 1973–1974, after the OPEC oil embargo, oil prices considerably increased and many governments were strongly stimulated to undertake research into renewable energy. A PV-Thermal (PV/T) collector is a module in which the PV not only produces electricity but also serves as a thermal absorber. In this way, heat and power are produced simultaneously. Since the demand for solar heat and solar electricity are often supplementary, it seems a logical idea to develop a device that can comply with both demands. Over the years, a large amount of PV/T research has been carried out, originating from several independent developments that all resulted in the idea of integrating PV and thermal into one module. In PV/T system applications the production of electricity is the main

RSC Energy Series No. 2
Fundamentals of Photovoltaic Modules and Their Applications
By G. N. Tiwari and Swapnil Dubey

Published by the Royal Society of Chemistry, www.rsc.org

priority, therefore it is necessary to operate the PV modules at low temperature in order to keep PV cell electrical efficiency at a sufficient level. This requirement limits the effective operation range of the PV/T unit for low temperatures, thus the extracted heat can be used mainly for low-temperature applications such as space heating, water or air preheating and natural ventilation in buildings. Water-cooled PV/T systems are practical systems for water heating in domestic buildings but their application has been limited up to now. Air-cooled PV/T systems have already been applied in buildings, usually integrated on their inclined roofs or façades. These systems keep the electrical output at a sufficient level, covering building space heating needs during winter and ventilation needs during summer, also avoiding building overheating.

2.2 History of PV/T Air Heating

Air-type PV/T collectors are distinguished according to the air-flow pattern. These are differentiated with respect to the flow of air above the absorber, below the absorber and on both sides of the absorber in single and in double pass. The applications of PV/T air heating are classified as follows.

2.2.1 PV Integrated with Air Collector

The first PV/T air collector integrated in a house called 'Solar One' was built in 1973/1974 at the University of Delaware by Professor Böer.[1] At that time Professor Böer had done a large amount of work on PV. In the roof and facade of this house, air collectors were integrated, and 4 of the 24 roof collectors were equipped with CdS/Cu2S cells.[2] After the pioneering work of Professor Böer, in the late 1970s and early 1980s the main research in PV/T air was carried out in the group of Hendrie[3–6] and also at Sandia and Brown University. In 1978, MIT Lincoln laboratory and Sandia laboratories acquired jointly two full-size flat-plate prototype PV/T air collectors manufactured by ARCO and Spectrolab[3] and the insufficient performance of this first generation of PV/T collectors motivated the development of a second generation, for which a number of novel concepts were developed. In Japan, Ito and Miura[7] did measurements on partially transparent photovoltaic modules as the top cover of an unglazed air collector. This design was chosen over the design in which the air was flowing between the PV and the top cover, because of the higher PV temperatures involved in the latter design. Thermal efficiencies were found in the order of 40%, strongly depending on the wind speed. In the early 1990s, in Israel, an unglazed PV/T collector was developed and commercialized with both liquid and air heat extraction.[8] However, the main purpose of the hot-air option is to provide additional cooling of the PV. In the air-type collector investigated by Raghuraman,[6] air flows between the upper absorber consisting of PV-cells and the lower absorber consisting of a black thermal absorber. He finds a thermal efficiency of 42%. Cox and Raghuraman[9] performed simulations on the collector design of Raghuraman as described above. They concluded that for sufficiently large cell coverage the secondary absorber underneath the silicon cells should not be spectrally selective since the reduced energy loss due to low emission is offset by the reflection of long-wavelength radiation that is emitted by the hot upper absorber. Zondag *et al.*[10] calculated the thermal efficiency of a PV/T-liquid channel collector, with either the channel underneath opaque PV, or the channel underneath transparent PV with a secondary absorber at the rear. It was found that for the case with the additional rear absorber, the thermal efficiency was 63% instead of 60% for the opaque PV case.

Air has a thermal conductivity that is 24 times lower than for water. Since $h = Nu \times k/D$, this reduces the heat transfer. This leads to the fact that for air collectors the channel height has a large influence on the thermal efficiency. Due to the much lower heat capacity the flow rate in an air collector is necessarily much larger than in a liquid collector. Loferski *et al.*[11] report on a PV/T-air system in which a fin is connected to the back of each PV-cell. The fins increase the surface area available for heat-exchange by a factor of four and the thermal yield of the cells by a factor of two over non-finned cells. The fins are connected by means of a Dow Corning RTV silicone, which is UV-resistant and can withstand temperatures of over 120 °C. Prakash[12] modelled a channel-type PV/T collector for the cases of both air (100–300 $kg\,h^{-1}$) and water (40–120 $kg\,h^{-1}$). He finds that decreasing the duct depth from 3 to 1 cm increases the thermal performance from 17% to 34% for an air heater (100 $kg\,h^{-1}$) and from 50% to 64% for a water heater (40 $kg\,h^{-1}$). For the case of 0.01 m duct depth, increasing the flow rate from 100 to 300 $kg\,h^{-1}$ increased the thermal performance of the air heater from 34% to 51%, while for the water heater an increase of 40–120 $kg\,h^{-1}$ increased the efficiency from 64% to 67%. Obviously, the heat transfer is much more critical for an air collector than for a liquid collector. The importance of the heat transfer to the air is further underscored by measurements made by Hendrie,[3] which indicated that for the first generation Spectrolab PV/T collector at an average fluid temperature of 28 °C the cell temperature was 74 °C. The heat transfer was inhibited by a badly applied encapsulant layer below the cells, which resulted in a wrinkly sheet that caused recirculation zones in the airflow through the collector, thereby reducing the area available for heat transfer. This effect was also reported by Raghuraman.[6] Because of the critical heat transfer to the air, it is very important to model the heat transfer properly. First of all, one should be aware that for a sufficiently wide channel, the hydraulic diameter is twice the channel height. Next, for laminar flow, the entrance length is often substantial. Tripanagnostopoulos *et al.*[13–18] have improved the heat transfer in his PV/T air collector by inserting a blackened metal sheet at half height along the full length of the air channel. The metal sheet gets heated due to thermal radiance from the PV, and thereby adds to the effective heat transfer area, increasing his thermal efficiency from 35% to 40% at zero reduced temperature. Eicker[19] presents an overview of entrance-effect heat transfer relations for air collectors, showing a variation of about 10% in average Nusselt number

Figure 2.1 Photograph of a PV/T solar air heater.

when integrated over the entrance length. For fully developed laminar flow she recommends to use the fixed Nusselt value of 5, 4. An indoor test procedure for PV/T air collectors has been developed by Solanki *et al.*[20] A photograph of the test simulator is shown in Figure 2.1. They have compared experimental and theoretical results and found that the thermal, electrical and overall efficiency of the solar heater obtained in indoor conditions are 42%, 8.4% and 50%, respectively. In the case of indoor simulation, the effect of mass flow rate on thermal, electrical and overall efficiency at constant solar radiation $600\,\mathrm{W\,m^{-2}}$ and $T_{fi} = 38\,°\mathrm{C}$ and the variation of instantaneous efficiency and electrical efficiency with $(T_{fi}–T_a)/I(t)$ is shown in Figures 2.2 and 2.3, respectively. Younger *et al.*[5] did measurements on a PV/T-air system in which the upper side of the air channel consisted of a PV-laminate. In order to increase the heat transport from the PV laminate to the fluid he used a PV-laminate whose backside consisted of roughened Teflon with a roughness of 60 mm. Garg and Adhikari[21] did a parametric study for a PV/T air collector. They concluded that the reduction in heat loss due to the addition of an extra cover does not justify the increased transmission loss.

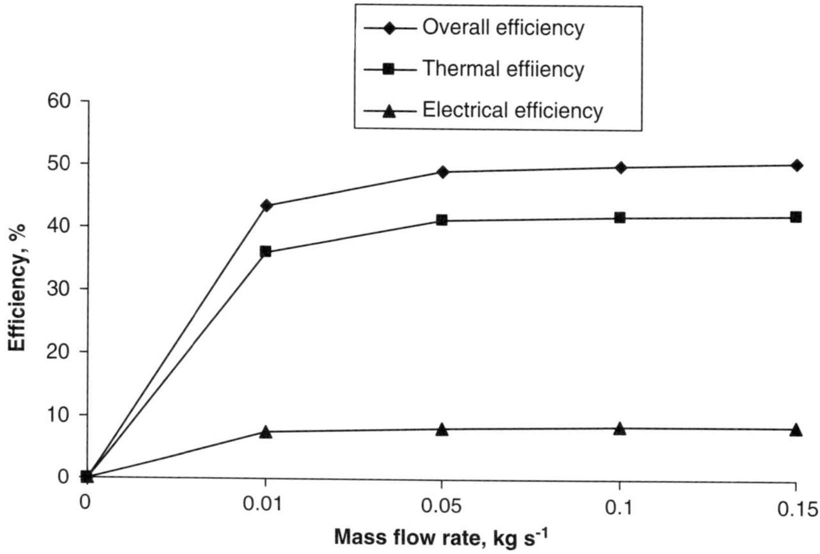

Figure 2.2 Effect of mass flow rate on thermal, electrical and overall efficiency at solar radiation 600 W m^{-2} and $T_{fi} = 38$ °C.

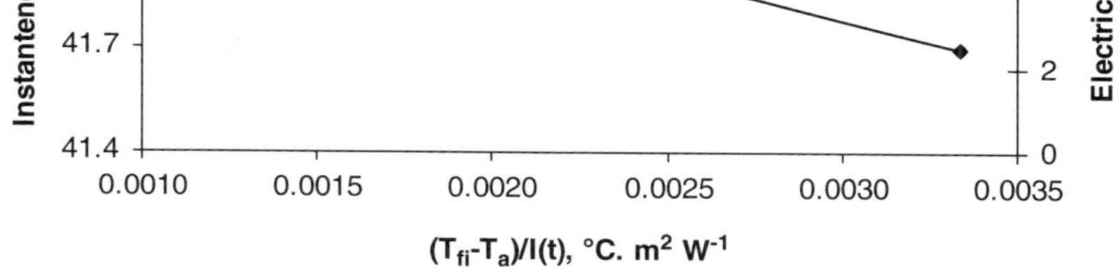

Figure 2.3 Variation of instantaneous efficiency and electrical efficiency with $(T_{fi}-T_a)/I(t)$.

2.2.2 Ventilated BIPV System

Building integrated photovoltaic (BIPV) systems are photovoltaic materials that are used to replace conventional building materials in parts of the building envelope such as the roof, skylights or facades. They are increasingly being incorporated into the construction of new buildings as a principal or ancillary source of electrical power, although existing buildings may be retrofitted with BIPV modules as well. The advantage of integrated photovoltaics over more common non-integrated systems is that the initial cost can be offset by reducing the amount spent on building materials and labour that would normally be used to construct the part of the building that the BIPV modules replace. In addition, since BIPV systems are an integral part of the design, they generally blend in better and are more aesthetically appealing than other solar options. These advantages make BIPV one of the fastest growing segments of the photovoltaic industry. A photograph of the CIS Tower in Manchester, England, which was clad in PV panels at a cost of £5.5 million, is shown in Figure 2.4. Transparent solar panels have also been used to replace conventional window glass and take advantage of the combined functions of power

Figure 2.4 The CIS Tower, Manchester, England, was clad in PV panels at a cost of £5.5 million (courtesy: http://en.wikipedia.org/wiki/File:CIS_Tower.jpg).

Figure 2.5 Photograph of south roof integrated transparent solar panels (courtesy: www.cmhc-schl.gc.ca/ . . . /enefcosa_003.cfm).

generation, lighting and temperature control. Transparent solar panels use a tin oxide coating on the inner surface of the glass panes to conduct current out of the cell. The cell contains titanium oxide that is coated with a photoelectric dye. Transparent solar cells use ultraviolet light to generate electricity but allow visible light to pass through them. Most conventional solar cells use visible and infrared light to generate electricity. In contrast, the innovative new solar cell uses ultraviolet radiation. A photograph of south roof integrated transparent solar panels is shown in Figure 2.5.

In the case of ventilated BIPV systems, air collector modules are constructed as part of the building shell, through applying an air gap between the wall and the PV-laminate. Similar to PV/T air collectors, in BIPV/T the heat transfer to the air is crucial to the thermal performance. Widely different thermal efficiencies are reported for air systems: from 14% up to 60%. This is due to the fact that the heat transfer can be increased strongly by increasing the air velocity. The air flow along the PV is driven either in natural mode or in mixed and forced mode. Bollo *et al.*[22] examined a solar chimney (duct depth 0.4 m) for different configurations. Among others, he investigated: (a) PV on the front surface and (b) low-emittance glass on the front surface and PV within the air channel, 0.05 m below the front surface, in both cases for natural convection and the chimney below 37 °C. It was found that for these cases, in configuration (b), a substantially higher temperature increase was found over the height of the solar chimney, but also that the electrical efficiency was reduced by 35% due to the reduced transmittance of the low-emissivity glazing and the high PV temperature that reached a maximum of about 100 °C. Experimental and numerical work on a PV-clad wall has been described by Brinkworth *et al.*[23] He calculated a base case with a duct width of 0.12 m and a duct height of 5 m, for facade integration for a sunny day in Cardiff (peak irradiance for vertical surface 620 $W\,m^{-2}$). Temperature rise increases with increasing height of the duct and thereby the buoyancy induced flow rate. Flow rate increases from 0.2 $m\,s^{-1}$ at a height of 1 m, to a fixed value of 0.4 $m\,s^{-1}$ for heights over 30 m. In order to increase the accuracy of the measurements, experiments on heating foils was carried out instead of solar irradiance and thermal results have been calculated by Wouters and Vandaele.[24] In order to facilitate the design of PV/T facades, a model for the thermal performance of the PV/T facade was developed,[25–30] based on the conventional U- and g-values of glazing.

In most PVT/air systems the air circulates through a channel formed between the rear PV surface and the system thermal insulation, and in some other systems through channels on both PV module sides, in series or in parallel flow. The usual heat extraction mode is the direct air heating from the PV module rear surface by natural or forced convection and the thermal efficiency depends on channel depth, air-flow mode and air-flow rate. Small channel depth and high flow rate increase heat extraction, but also increase the pressure drop, which reduces the system net electrical output in the case of forced air flow, because of the increased power for the fan. In applications with natural air circulation, the small channel depth reduces air flow and this results in an increase of PV module temperature. In these systems a large depth of air channel (minimum 0.1 m) is necessary.[31] Nagano *et al.*[32,33] did experiments with six different PV/T modules of 1.4 m^2 each, which were tested as hybrid wallboards under an inclination of 80°. Experimental results on a ventilated BIPV test site in Sydney, Australia, containing 20 solar tile modules has been presented by Bazilian,[34,35] of which 20% is fitted with a heat recovery unit. Through an air gap of 0.15 m, the module is cooled by means of forced convection (flow rate 0.35 $m\,s^{-1}$), which results in a thermal efficiency of about 30%. A Direct Numerical Simulation (DNS) study was carried out on the airflow distribution within a facade, where it has observed that there was a strong localized turbulence due to the inlet and an inhomogeneous heat transfer (Gandini *et al.*[36]). Experiments on a forced ventilation PV-air system with a channel of 0.15 m width has been carried out and it was found that the efficiency was improved by inserting metal fins of 1.5 and 4 cm to the rear surface, Tripanagnostopoulos *et al.*[17] and Crick *et al.*[37] glued aluminium fins to the back of a PV module in their naturally ventilated facade, which they found to increase the heat transfer considerably. Physical implementation of a BIPV system on the south wall of a room has been studied by Dubey *et al.*[38] They have found that the differences between room air temperature and ambient temperature during summer and winter conditions of Srinagar are 6.5 °C and 2.8 °C, respectively. A schematic diagram of a PV/T air duct integrated on a south wall is shown in Figure 2.6.

Chow *et al.*[39] calculate the electrical performance for integrated BIPV, ventilated PV and PV/T, with monocrystalline cells, for a hotel in Macao. They find that the electrical yield is largest for the ventilated PV and lowest for the BIPV, but the differences are very small. Guiavarch and Peuportier[40] compare the electrical yield of non-integrated PV, integrated PV without air gap,

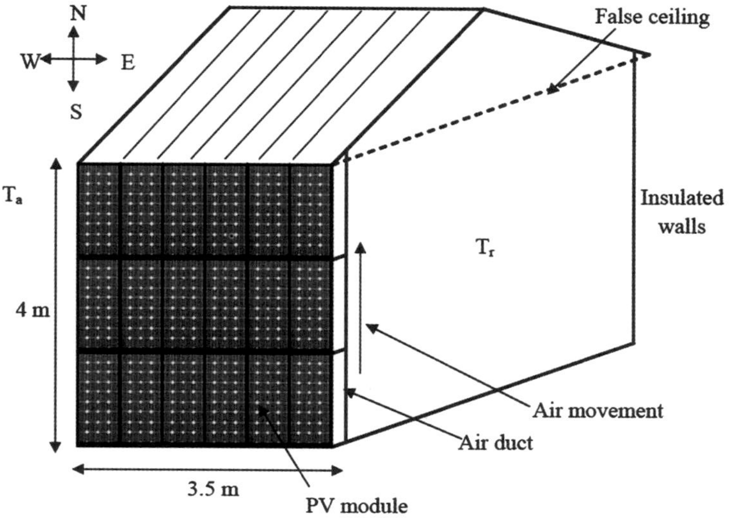

Figure 2.6 Schematic diagram of a PV/T air duct integrated on a south wall.[38]

ventilated PV and a ventilated PV with heat recovery. The lowest performance was obtained from PV without air gap, of about 7% less than the ventilated PV for Paris, and 8% less for Nice. An interesting building application of solar energy systems is to use linear Fresnel lenses as transparent material of atria, sunspaces, *etc.*, to control lighting and temperature of these spaces, also providing electricity and heat to cover building energy needs. In buildings, shading devices (Tsangrassoulis *et al.*[41]) and double-glazed windows with motorized reflective blinds (Athienitis and Tzempelikos[42]) aim to reduce the absorbed solar energy and to keep the average temperature of the interior space at a comfortable level. Flat or curved (CPC) reflectors are suggested as lightguides to provide sunlight to the building interior spaces (Molteni *et al.*[43], Scartezzini and Courret[44]). Fresnel lenses (optical devices) are of practical interest for solar radiation concentration, because of their low volume and weight and also their smaller focal length and lower cost compared to thick ordinary lenses. The advantage of linear Fresnel lenses to separate the direct from the diffuse solar radiation makes them suitable for illumination control in the building interior space, providing light of suitable intensity level without sharp contrasts and achieving shading by absorbing a great part of the incoming solar radiation.

The concentration of the direct part of the incident solar radiation on a thermal absorber of small width located at the focal position has been suggested by Jirka *et al.*[45] to achieve a lower illumination level, to avoid space overheating and to contribute to the thermal needs of the building. An effective combination of Fresnel lenses can be the use of hybrid PV/T small width absorbers to extract the concentrated solar radiation in the form of electricity and heat (Tripanagnostopoulos *et al.*[46]). This compound system can be also used to achieve illumination control of buildings during the day, storing the surplus energy for space heating during the night. This system can contribute to the ventilation needs during the day and also to cover other building electrical loads. In low intensity solar radiation, the absorbers can be out of focus, leaving the light to come into the interior space and to keep the illumination at an acceptable level. Laboratory scale experimental results give an idea about the application of this new optical system for lighting (reduction by about 60–80%) and cooling control (reduction by 3–10 °C) of building interior spaces (Tripanagnostopoulos *et al.*[46]), estimating that the system is promising for building applications and effectively combined with PV/T type absorbers. In BIPV/T applications, flow velocities are generally low and buoyancy and wind have significant effects. Due to the large effect of flow rate and channel design, a substantial variation in thermal module efficiencies is reported. However, for practical flow rates, the thermal efficiency for unglazed modules is rather low. Effective and low-cost methods to increase the heat transfer need to be investigated and implemented. The electrical performance is enhanced by about 10% as compared to non-ventilated PV.

The expression for the thermal efficiency of a BIPV/T system, based on the transmittance–absorptance product of BIPV/T accounting for the packing factor can be written as[47]

$$\eta_{th} = F_R((S \times \tau\alpha_{PV}) + (1 - S)\tau\alpha_T) - F_R U_{loss} \times \frac{T_{fi} - T_a}{I(t)} \quad (2.1)$$

where S = packing factor and F_R = heat removal factor.

The expression for the top loss coefficient using Klein's empirical equation can be given as

$$U_{top} = \left[\frac{N}{\frac{C}{T_{pm}}\left(\frac{T_{pm}-T_a}{N-f}\right)^e} + \frac{1}{h_w} \right]^{-1} + \frac{\sigma\left(T_{pm} + T_a\right)\left(T_{pm}^2 + T_a^2\right)}{\left(\varepsilon_p + 0.000591Nh_w\right)^{-1} + \frac{2N + f - 1 + 0.133\varepsilon_p}{\varepsilon_g} - N} \quad (2.2)$$

where

$$C = (520 - 0.000051\beta^2)$$
$$f = (1 + 0.089h_w - 0.116h_w\varepsilon_p)(1 + 0.07866N)$$
$$e = 0.430\left(1 - \frac{100}{T_{pm}}\right), \qquad T_{pm} = T_i + \frac{Q/A_{collector}}{F_R U_{loss}}(1 - F_R)$$

where β is the collector mounting, σ is the Stefan–Boltzmann constant, N is the number of covers or glazing layers, ε_g is the emittance of the cover or glazing, ε_p is the emittance of the plate, h_w is the convection heat transfer due to the wind and F_R is the heat removal factor.

Photographs of a test setup of PV/T air collectors are shown in Figures 2.7 and 2.8. The collector sucks fresh air through a number of small holes and is suitable for preheating of air for ventilation or drying. An innovative technological system for building integration, integrated solar roof (TIS), of hybrid PV/T air collectors has been studied by Aste *et al.*[48] The performance of PV/T air collectors has been evaluated for Milan climatic conditions. The airflow into the gap between the sandwich and the absorber plate can be achieved by the forced flow using a fan or natural flow through buoyancy effect. A photograph of the system is shown in Figure 2.9. They have found that the daily thermal efficiency varies on average from 20% to 40%. However, the daily average electrical efficiency obtained was around 9–10%. The expression for the PV thermal–spectral actual efficiency can be given as[48]

$$\eta_{\mathrm{th,sp}} = \eta_{\mathrm{o}} \times \frac{100 - \gamma(T_{\mathrm{PV}} - 25)}{100 \times SCF} \tag{2.3}$$

where γ = temperature power coefficient of the PV cells and SCF = spectrum correction factor of PV efficiency.

Thermal Exchange Coefficients of PV/T Air Collector[48]

The external convective and radiative heat transfer coefficients depend on the average temperatures. The external radiative coefficient between the PV sandwich and the sky can be expressed as

$$h_{\mathrm{r,PV-sky}} = F \times 4 \times \varepsilon_{\mathrm{PV}} \times \sigma \times T^3_{\mathrm{PV-sky}} \tag{2.4a}$$

Figure 2.7 PV/T air collector having small holes to suck fresh air (courtesy: Ivan Katic, Denmark).

Figure 2.8 Photograph of test setup of PV/T air collectors (courtesy: Ivan Katic, Denmark).

Figure 2.9 Photograph of a prototype PV/T air collector[48] (courtesy: Niccolò Aste, Italy).

The glass–absorber radiative coefficient is calculated as

$$h_{r,G-P} = 4 \times \varepsilon_{G-P} \times \sigma \times T_{G-P}^3 \tag{2.4b}$$

and the PV cells–absorber radiative coefficient is calculated as

$$h_{r,PV-P} = 4 \times \varepsilon_{PV-P} \times \sigma \times T_{PV-P}^3 \tag{2.4c}$$

$$\text{Here, } \varepsilon_{PV-P} = \frac{1}{1/\varepsilon_{PV} + 1/\varepsilon_P - 1}$$

where F = sky view factor of collector, σ = Stefan Boltzmann constant and ε = emissivity.

Example 2.1

Calculate the thermal efficiency of a BIPV/T system, when the packing factor is 89%, the heat removal factor is 0.942, transmittance–absorptance of PV and tedlar are 0.85 and 0.5, respectively, $U_{loss} = 3.4\ \text{W m}^{-2}\,°\text{C}^{-1}$, ambient temperature = 30 °C, inlet temperature = 40 °C and $I(t) = 700\ \text{W m}^{-2}$.

Solution

Using eqn (2.1), we get

$$\begin{aligned}\eta_{th} &= 0.942((0.89 \times 0.85) + (1 - 0.89)0.5) - 0.942 \times 3.4\frac{40 - 30}{700} \\ &= 0.7186 \\ &= 71.8\%\end{aligned}$$

Example 2.2

Calculate the PV thermal–spectral efficiency in the case of an air heating system when the standard efficiency = 12%, temperature power coefficient = 0.0045, spectrum correction factor = 1 and PV cell temperature = 65 °C.

Solution

Using eqn (2.3), we get

$$\begin{aligned}\eta_{th,sp} &= 0.12 \times \frac{100 - 0.0045(65 - 25)}{100 \times 1} \\ &= 0.1197 \\ &= 11.97\%\end{aligned}$$

Example 2.3

Calculate the PV cells–absorber radiative coefficient for Example 2.2, when the emissivity of the PV and the plate are 0.95 and 0.9, respectively. The Stefan–Boltzmann constant = $5.67 \times 10^{-8}\,\mathrm{W\,m^{-2}\,K^{4}}$.

Solution

Using eqn (2.4c), we get

$$\varepsilon_{PV-P} = \frac{1}{1/0.95 + 1/0.9 - 1} = 0.859$$

$$h_{r,PV-P} = 4 \times 0.859 \times 5.67 \times 10^{-8} \times (65 + 273)^3 = 7.52\,\mathrm{W/m^2}$$

2.3 History of PV/T Water Heating

Martin Wolf[49] analysed a silicon solar array mounted inside a stationary non-concentrating thermal collector, using a lead-acid battery as the storage element for residential heating; this was the first work carried out on flat-plate PV/T water collectors. He concluded that the system was technically feasible and cost effective. As a demonstration project, Professor Böer applied 13 PVT-liquid collectors at his own home 'Solar Knoll' in about 1978. After the pioneering study of Martin Wolf in 1976, the subject of PV/T liquid was quickly taken on by other groups and research has started. Research and modelling on PV concentrators and actively cooled PV/T concentrators was carried out using the TRNSYS-application at the Arizona State University during 1974–1978.[50] This work was extended to include PV/T flat-plate collectors as well[50–52] and was the basis for the PV/T model TYPE 50 that is presently available in TRNSYS. MIT Lincoln laboratory and Sandia jointly acquired three full-size flat-plate prototype PV/T collectors in 1978.[3] These collectors were manufactured by ARCO (both an air-type and a liquid-type) and Spectrolab (air-type). In the subsequent testing of these collectors at MIT, it was found that the performance of these collectors turned out to be below the initial specifications of 6.5% electrical and 40% thermal efficiency. For that reason, a second generation of PV/T collectors was developed, consisting of two production-ready PV/T liquid designs, two experimental advanced PV/T air designs and three new PV/T liquid concepts (a dual flow concept, an advanced unglazed concept and a two-phase Freon concept in which the PV/T functioned as the evaporator of a heat pump). Out of the two designs, one was developed by MIT and the PV-manufacturer Spire Corporation (a concept for mounting on top of an existing roof) and the other by Solar Design Associates and Spire Corporation under the auspices of MIT (a roof-integrated collector replacing roofing material). However, due to the termination of the funding program, not all concepts could be built. The results of the work have been published in a number of papers and a final report.[3,5,9] Research on PV/T systems was also carried out at MIT.[53–55] Research on the effect of the thermal gradient on the electrical performance was carried out at Sandia Labs.[56,57] In 1980, PVT research was started at JPL and Brookhaven laboratories[58,59]. A large amount of PV/T module research involving comparative experimental studies on glazed and unglazed PV/T collectors, with and without booster reflectors, was carried out at the University of Patras.[14,16,60,61] An economic study was also carried out.[62] In Cyprus, a numerical study comprising a literature review was presented and carried out for thermosyphon PV/T systems.[63–65] Further modelling work on the PV/T thermosyphon studies was carried out in cooperation with the University of Patras.[66,67]

Sheet-and-tube is a conventional design which is used for solar collectors. The thermal efficiency of a sheet-and-tube collector depends on its ratio of W/D, in which W is the distance between the tubes and D is the tube diameter. The W/D ratio used in practice is a conciliation between optimized heat transfer and economic aspects. However, the optimum for a PV/T system is different for a conventional solar thermal collector. In addition, there are two effects in a reduction of the W/D ratio; one is the increase of the fin efficiency due to the shorter fin length, while the other is a decrease of the flow velocity in the case of parallel risers (due to the increased flow area, assuming the flow rate is kept constant) or an increase in pressure drop in the case of a spiral tube. Efforts have been undertaken to improve the heat transfer from the absorber to the liquid. The best heat transfer is obtained by leading the heat-collecting medium through a thin channel over the full width of the absorber. Huang *et al.*[68,69] built 2 unglazed PV/T prototypes based on a sheet-and-tube construction. They used W/D ratios of 10 (copper tube to aluminium plate) and 6.2 (extruded tube-in-sheet aluminium). Since they found that the thermal performance of sheet-and-tube construction was not satisfactory, it was decided to build a polycarbonate multi-channel structure (W/D = 1). A temperature difference of 4 °C was found between the PV and the water in the tank. For an M/A ratio of $82\,\mathrm{l\,m^{-2}}$, 9% electrical efficiency together with 38% characteristic daily efficiency was found. Tiwari and Sodha[70] did a simulation study based on the collector system of Huang *et al.*[68,69] (for $M/A = 87\,\mathrm{kg\,m^{-2}}$), for which 35% thermal efficiency together with 9% electrical efficiency was calculated. A thermosyphon PV/T system with a PV/T module based on an extruded aluminium channel absorber with a W/D ratio of 1 to obtain an optimal heat transfer to the fluid was built by Chow *et al.*[71] For an M/A-ratio of $65.2\,\mathrm{kg\,m^{-2}}$, He *et al.*[72] measured an average daily module electrical efficiency of about 5%. For an improved prototype in which the collector area was fully covered with PV cells, Ji *et al.*[73] present over 45% average daily efficiency while the electrical efficiency was about 10%. In addition, calculations on the efficiency of a channel-type PV/T as a function of channel width, in which also entrance effects are taken into account, was presented by Ji *et al.*[74].

Cristofari *et al.*[75] presented a simulation model of finite differences of a water heating system using a hybrid PV/T collector manufactured in a copolymer material and running in low flow-rate conditions. The main favourable properties of polymeric material (polycarbonate) when applied to solar collectors are: low density, mechanical strength, no special surface treatment required, no corrosion, processing techniques adapted to mass production; but there are some disadvantages such as low thermal conductivity, large thermal expansion and limited service temperature. Hybrid PV/T systems are developed to generate electricity and hot water/air simultaneously. During the operation, a heat carrier fluid removes heat from the absorber and PV cells. The collected heat can be used as preheated water (Figure 2.10). The main advantages of such solar collectors are (Zondag *et al.*[76], Sandnes and Rekstad[77]):

- economical order compared to a combination of separate thermal and photovoltaic panels;
- the area covered with a hybrid solar collector produces more electrical and thermal energy than a corresponding area half covered with standard PV panels and half with a conventional thermal collector. This is particularly useful because the space on the roof of a house is often reduced;
- the average temperature of operation for a hybrid collector being generally lower than for a standard PV module, its electrical production will be increased;
- a hybrid collector provides architectural uniformity on a roof in contrast to an association of two separate solar collectors.

The PV/T collector studied by Cristofari *et al.*[75] was composed of a polycrystalline module pasted to an absorber–exchanger, which transforms the solar radiation to heat. This 'absorber–exchanger' has back and side insulations (expanded polyurethane), which are inserted in the body of the collector and allow good mechanical behaviour of the collector structure shown in Figure 2.11. The absorber–exchanger in the copolymer material must satisfy the following constraints: UV protected, high thermal conductivity, water-resistant and glycol-resistant, good thermal range of utilization (–10 to + 150 °C), good mechanical strength and chemically stable. The complete layout of a solar water heating system is shown in Figure 2.12. It was found that the daily solar irradiation of the

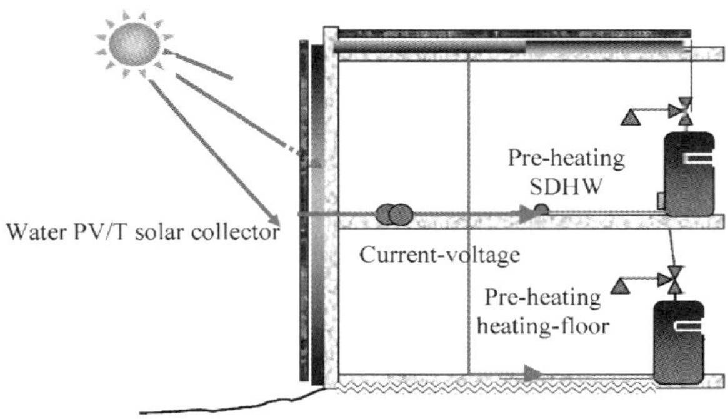

Figure 2.10 Utilization of a PV/T water solar collector[75] (courtesy: Gilles Notton, France).

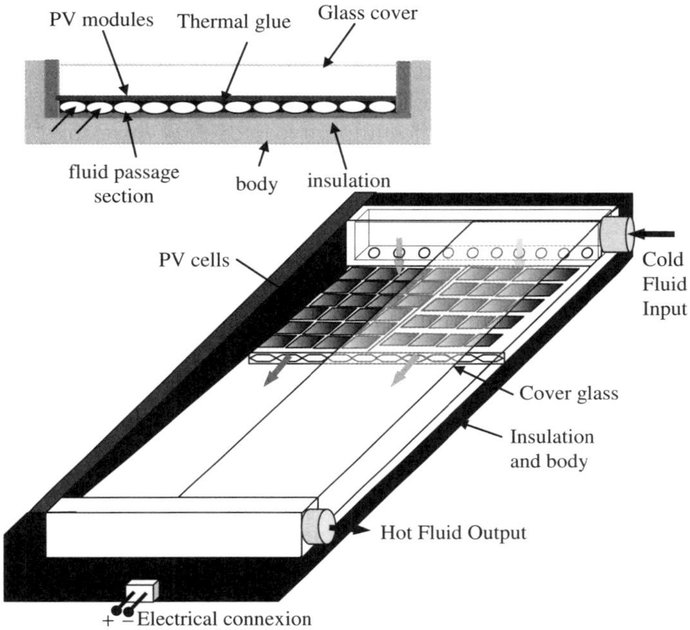

Figure 2.11 The photovoltaic/thermal solar collector[75] (courtesy: Gilles Notton, France).

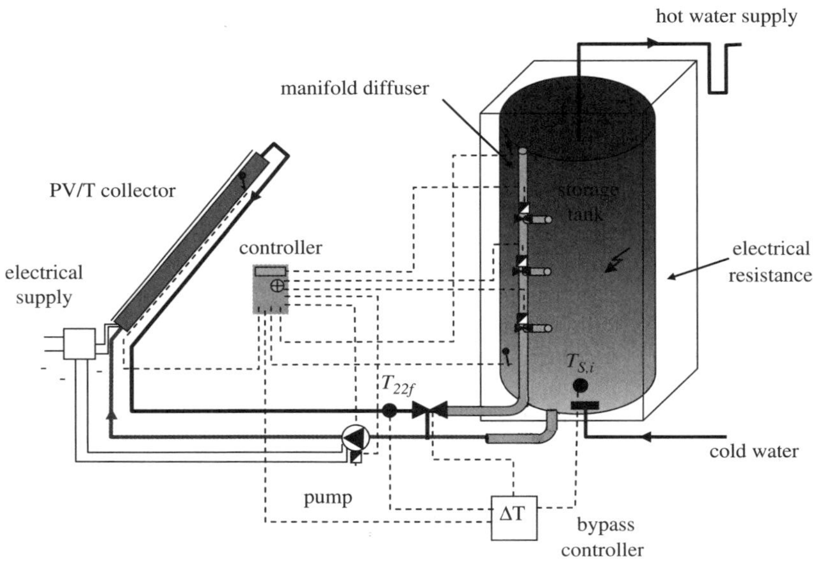

Figure 2.12 Layout of the solar system installation[75] (courtesy: Gilles Notton, France).

solar collector was 8.89 kWh; daily produced thermal energy was 4.93 kWh; daily electricity production was 1.13 kWh.

On an annual basis, the average efficiencies are: 55.5% from a thermal point of view and 12.7% from an electrical point of view. The optical coefficient and heat loss coefficient of the collector were 0.61 and 9.04 respectively.

A novel concept of a dual-flow PV/T-liquid collector, in which the incoming water flow through the collector flows directly underneath the PV laminate, whereas the outgoing water flows directly over the PV-laminate, was discussed by Hendrie.[3] De Vries[78] proposed a dual-flow PV/T-collector like that of Hendrie,[3] but with a reversed water flow (water inlet above the PV, water exit below the PV). In addition, he proposed an additional insulating air layer between the PV and the lower channel. He found that the yearly yield of a PV/T system could be raised by 2% by using a water channel underneath the cells instead of a sheet-and-tube construction. The yearly yield could be raised by another 6% for a water layer flowing over the PV/T laminate instead of underneath (the design is equivalent to a double-covered collector in which water is flowing through the lower channel). However, due to the additional use of a glass layer the annually averaged electrical efficiency was reduced from 6.6% to 6.2%. Fraisse *et al.*[79] presented energy performance of water hybrid PV/T collectors applied to combisystems of Direct Solar Floor type by using poly-crystalline photovoltaic modules for the Mácon area in France. A photograph of a glass-covered water PV/T prototype is shown in Figure 2.13. They have studied four different cases of PV and PV/T and found that the annual

Figure 2.13 Photograph of glass-covered water PV/T prototype (polycrystalline PV)[79] (courtesy: G. Fraisse, France).

photovoltaic cell efficiency was 6.8% less than the conventional PV module due to an increase in temperature related to the additional glass cover; without a glass cover the efficiency is 10%.

Energy and exergy analysis of glazed and unglazed PV/T collectors has been studied by Chow *et al.*[80] Experiments have been conducted for outdoor conditions in Hong Kong. The thermal efficiency and solar cell conversion efficiency for the glazed collector were 50.3% and 9.3%, respectively, and for the unglazed collector the values were 40.8% and 12.1%, respectively. A photograph of PV/T collectors with and without a glass cover is shown in Figure 2.14. The energy balance equations of the individual components of a PV/T water heating system considering heat capacity is given in Table 2.1. In this dynamic model, each constituent layer was represented by a single node. While the edge loss of the thin PV module is considered negligible, the working temperature of each layer can be taken as uniform and the heat flow as uni-directional (Chow *et al.*[81]).

The expression for the first law efficiency of thermodynamics for the time period from t_1 to t_2 can be given as[80]

$$\eta_{\mathrm{pv/t}} = \frac{\int_{t_1}^{t_2} \left(A_{\mathrm{c}}\dot{E}_{\mathrm{t}} + A_{\mathrm{pv}}\dot{E}_{\mathrm{pv}}\right) dt}{A_{\mathrm{c}} \int_{t_1}^{t_2} I(t)\, dt} = \eta_{\mathrm{t}} + \zeta\eta_{\mathrm{pv}} \tag{2.5}$$

where $\dot{E}$ = energy rate per unit area, W m^{-2}, and ζ = packing factor.

Figure 2.14 PV/T collectors with and without glass cover[80] (courtesy: T. T. Chow, China).

Table 2.1 Energy balance equations of individual components of a PV/T water heating system considering heat capacity (Chow *et al.*[81]).

Front glazing	$\rho_g\delta_g C_g \frac{dT_g}{dt} = \alpha_g I(t) + (h_{wind} + h_{r,g-a})(T_a - T_g) + \frac{A_c}{A_g} U_{c-g}(T_c - T_g) + \frac{A_p}{A_g} U_{p-g}(T_p - T_g)$
PV encapsulation	$\rho_c\delta_c C_c \frac{dT_c}{dt} = \alpha_c I(t) - E + U_{c-g}(T_g - T_c) + U_{c-p}(T_p - T_c)$
Thermal absorber	$\rho_p\delta_p C_p \frac{dT_p}{dt} = \frac{A_g}{A_p}\left(\alpha_p I(t) + U_{p-g}(T_g - T_p)\right) + \frac{A_f}{A_p} h_{p-f}(T_f - T_p) + \frac{A_c}{A_p} U_{c-p}(T_c - T_p) + U_{ins-p}(T_{ins} - T_p)$
Water in channels	$\rho_f C_f \frac{dT_f}{dt} = \frac{A_p}{A_f} h_{p-f}(T_p - T_f) - u_f \rho_f C_f \frac{dT_f}{dy}$
Water layer in storage tank (middle layer)	$\frac{1}{3}\rho_f V_{tk} C_f \frac{dT_{tk,mid}}{dt} = \dot{m}_f C_f(T_{tk,up} - T_{tk,mid}) + 3k_f A_s \frac{(T_{tk,up} - T_{tk,mid})}{\delta_{tk}} - 3k_f A_s \frac{(T_{tk,low} - T_{tk,mid})}{\delta_{tk}} \frac{1}{3} h_{tk} A_{tk}(T_a - T_{tk,mid})$
Water segment in connecting pipe	$\rho_f D_{pi} C_f \frac{dT_{ws}}{dt} = 4\frac{A_{pe}}{A_{pi}} U_{a-ws}(T_a - T_{ws}) - D_{pi}\rho_f u_f C_f \frac{dT_{ws}}{dy}$
Thermal insulation	$\rho_{ins}\delta_{ins} C_{ins} \frac{dT_{ins}}{dt} = (h_{wind} + h_{r,a-ins})(T_a - T_{ins}) + U_{ins-p}(T_p - T_{ins})$

Here, E = Electrical power output, ins = insulation, up = upper layer, mid = middle layer, low = lower layer, tk = tank, ws = water segment in connecting pipe, pi = pipe interior, pe – pipe exterior.

Chow *et al.*[80] concluded that, for both glazed and unglazed conditions, the increase of cell efficiency, M/A_c, and ambient temperature would lead to an increase of $\eta_{pv/t}$. On the contrary, higher radiation and wind velocity would lead to a decrease of $\eta_{pv/t}$. The increase of the packing factor would lead to an increase of $\eta_{pv/t}$ for the unglazed condition, but a decrease for the glazed condition. From the viewpoint of the first law of thermodynamics, the glazed condition would be a better choice for the PV/T collector system than the unglazed condition for maximizing the overall energy output.

The expression for the second law efficiency of thermodynamics for the time period from t_1 to t_2 can be given as[80]

$$\varepsilon_{pv/t} = \varepsilon_{pv} + \varepsilon_t = \eta_{pv} + \left(1 - \frac{T_a}{T_{final}}\right)\eta_t \tag{2.6}$$

$$\text{or,}\ \varepsilon_{pv/t} = \frac{\int_{t1}^{t2} \left(A_c \dot{E}x_t + A_{pv}\dot{E}x_{pv}\right) dt}{A_c \int_{t1}^{t2} \dot{E}x_{sun} dt} = \varepsilon_t + \zeta\varepsilon_{pv} \tag{2.7}$$

where $\dot{E}x$ = energy rate per unit area, W m^{-2}.

The expression for the exergy of solar radiation can be given as

$$\dot{E}x_{sun} = \left[1 + \frac{1}{3}\left(\frac{T_a}{T_{sun}}\right)^4 - \frac{4T_a}{3T_{sun}}\right] I(t),\ \text{Petela}^{82} \tag{2.8a}$$

$$\dot{E}x_{sun} = \left[1 - \frac{4T_a}{3T_{sun}}\right] I(t),\ \text{Spaner}^{83} \tag{2.8b}$$

$$\dot{E}x_{sun} = \left[1 - \frac{T_a}{T_{sun}}\right] I(t),\ \text{Jeter}^{84} \tag{2.8c}$$

where T_a = ambient temperature, K, and T_{Sun} = solar radiation temperature = 6000 K. Normally, the differences in the results coming from these three calculation methods are less than 2%.

In the case of second law efficiency of thermodynamics,[80] ε_{pv} has a favourable factor from the photovoltaic viewpoint. But it has an unfavourable factor from the photothermal viewpoint since ε_t will decrease when more irradiation is converted into electricity. For both the glazed and unglazed conditions, when the PV cell efficiency improves, $\varepsilon_{pv/t}$ increases along with the increase of cell efficiency. Similarly, the packing factor is a favourable factor for PV but an unfavourable factor for thermal energy. In either glazed or unglazed conditions, an increase of M/A_c leads to the increase of both electrical gain and thermal gain and results in an increased $\eta_{pv/t}$. T_a is a favourable factor for ε_t but an unfavourable factor for ε_{pv}. Within the normal range of wind velocity, $\varepsilon_{pv/t}$ of the unglazed condition is always better than for the glazed condition.

Zondag *et al.*[85] and Jong[86] made a comparison between various types of PV/T designs (covered and uncovered) and various types of thermal systems (domestic hot water or heating with or without heat pump) in the Dutch climate. In this study, it was concluded that an uncovered PV/T system only gives a good efficiency for the case in which the PV/T is used for low-temperature ground storage combined with a heat pump. This is due to the increase in thermal efficiency and running hours for such low inflow temperatures. But due to the energy consumption by the heat pump, the net electrical efficiency of the system becomes negative. Monitoring results from the 54 m^2 glazed PV/T array installed at the head office of Renewable Energy Systems were presented by Zondag *et al.*[87] The PV/T heats an underground seasonal storage tank and stored heat is used to provide space heating during the winter. Kalogirou and Tripanagnostopoulos[88] calculated the yield of a 4-m^2 PV/T thermosyphon system for different climates. For their crystalline silicon PV/T module they found a useful thermal gain of 5.7 GJ, 5.0 GJ and 3.8 GJ for Nicosia, Athens and Madison, respectively, while the electrical performance ranged from 532 to 499 kWh. For the same module using a-Si, the thermal performance was slightly higher while the electrical performance was halved.

Concentrating photovoltaic (CPV) systems are used to reduce the area of the expensive photovoltaic cells. CPVT collectors may operate at temperatures above 100 °C, and the thermal energy can drive processes such as refrigeration, desalination and steam production. These systems require dish, trough and Fresnel-lens concentrators and are usually on the 100–200 m^2 scale. These relatively large devices are suitable for utility scale power plants in open areas, but are difficult to fit on rooftops and in an urban environment; much smaller units are needed for such applications. A novel miniature concentrating PV (MCPV) system producing both electrical and thermal energy was designed and analysed by Kirbus *et al.*[89] A photograph of the system is shown in Figure 2.15. The MCPV collector is not limited to low-temperature applications as in the case of PV/T collectors. The concentrator is a simple on-axis parabolic dish. The target is placed at the focal point and its aperture is perpendicular to the optical axis. The reflector is made of a single piece of glass, thermally bent to shape and then back-coated with silver to produce the reflective surface. An external protective coating prevents exposure of the silver to the environment. The thickness of the glass will ensure that the dish is self-supporting. The light to electricity conversion efficiency is given as[89]

$$\eta_{ele} = \eta_{opt} \times \eta_{pv}\left(1 - \frac{Q_{PAR}}{Q_{GRO}}\right)\eta_{inv} \tag{2.9}$$

η_{opt}, η_{pv} and η_{inv} are the efficiencies of the optics, the PV module and the inverter subsystems. Q_{PAR} and Q_{GRO} are the parasitic power consumption (for tracking motors and coolant pump) and the gross electrical power produced by the module. For the thermal energy product, the conversion efficiency is given as

$$\eta_{th} = \eta_{opt} \times (1 - \eta_{pv})\eta_{erc} \tag{2.10}$$

Figure 2.15 Photograph of miniature concentrating PV (MCPV) system[89] (courtesy: Abraham Kribus, Israel).

η_{rec} is the receiver efficiency, accounting for thermal losses from the receiver module to the environment.

Kirbus *et al.*[89] have found that electrical efficiency is about 20% at low temperature and is gradually reduced at elevated temperature. The cost of electricity produced from MCPV was \$2.5 Wp^{-1}. The performance and cost of a CPVT system with single effect absorption cooling are investigated by Mittelman *et al.*[90] The overall electrical conversion efficiency of a CPVT plant decreases about 23% at 50 °C coolant outlet temperature and about 20% at 150 °C coolant outlet temperature. The overall system electrical efficiency was lower than the cell efficiency due to the optics, module and inverter losses. The COP of the system lies between 0.6 and 0.75. The collector installation cost was \$2 Wp^{-1} and the cost of the cooling from the CPVT system was \$2.22 k Wh^{-1}.

A study of a PV/T water heating system operated in natural mode was presented by Ji *et al.*[73] The collector was fully covered with single-crystalline silicon cells. The test was performed in outdoor climatic conditions at the University of Science and Technology, China. A photograph of the system is shown in Figure 2.16. The system was composed of 144 black single-crystalline silicon cells connected in series, one converter, four accumulator batteries (12

Figure 2.16 Photograph of fully covered PV/T collector[73] (courtesy: T. T. Chow, China).

V, 100 Ah), and the associated switches and wiring. Each solar cell was 0.0625 × 0.125 m, with 14.5% photoelectric conversion efficiency in the standard testing conditions. The expression for the characteristic daily total primary energy (E_f) saving is defined as[73]

$$E_f = E_f' - U_L \frac{T_{fi} - \bar{T}_a}{I(t)} \tag{2.11}$$

Test results (simulation results) showed that as the hot-water load per unit heat-collecting area exceeded 80 kg m^{-2}, the daily electrical efficiency was about 10.2%, the characteristic daily thermal efficiency exceeded 45%, the characteristic daily total efficiency was above 52% and the characteristic daily primary-energy saving was up to 65% for this system with a PV cell covering factor of 0.63 and front-glazing transmissivity of 0.83. The effect on thermal and electrical performance by varying the covering factor of a PV cell shows that as the covering factor increased from 0.5 to 0.9, the water heat gain decreased from 4 to 3.6 kWh day^{-1}, electrical gain increased from 0.43 to 0.77 kWh day^{-1}, thermal efficiency decreased from 48.3 to 44.0%, electrical and overall efficiencies were nearly the same and primary energy saving increased from 66 to

68.7%. When the transmissivity increased from 07 to 0.9, the water heat gain increased from 3.6 to 3.9 kWh day^{-1}, electrical gain increased from 0.46 to 0.59 kWh day^{-1}, thermal efficiency increased from 44.4 to 48.2%, electrical efficiency increased from 8.8 to 11.3%, overall efficiency increased from 50 to 55.3% and primary energy saving increased from 59 to 66.9%.

Chow *et al.*[91] presented annual performance of a building integrated photovoltaic/water-heating system for a warm climate application in Hong Kong. The PV/T collectors are integrated on the south wall of an air-conditioned building. The PV-wall (PVW) wall was composed of six PVW collectors mounted on a 100-mm brick wall with plastering on both interior and exterior wall surfaces. The PVW collector adopted the flat-box thermal absorber design and was provided with polycrystalline silicon PV cells. A photograph and constituent layers of the BIPVW system are shown in Figures 2.17 and 2.18, respectively. The system thermal performance under natural water circulation was found to be better than the pump-circulation mode. For a specific BIPVW system at a vertical wall of a fully air-conditioned building and with collectors equipped with flat-box-type thermal absorber and polycrystalline silicon cells, the year-round thermal and cell conversion efficiencies were found to be 37.5% and 9.39% respectively under typical Hong Kong weather conditions. The overall heat transmission through the PVW wall is reduced to 38% of the normal building facade. When serving as a water pre-heating system, the economical pay back period was estimated to be around 14 years. The dynamic behaviour of the BIPVW system has also been evaluated by Chow *et al.*[80] using the finite-difference control volume approach. The daily thermal and electrical

Figure 2.17 Experimental setup of building integrated photovoltaic/water-heating system at City University of Hong Kong[91] (courtesy: T. T. Chow, China).

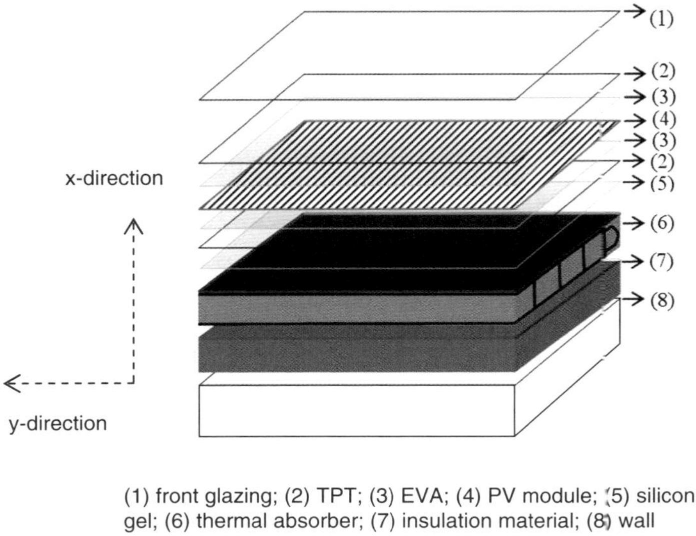

Figure 2.18 Constituent layers of the BIPVW system[80] (courtesy: T. T. Chow, China).

efficiency of the system under the thermosyphon test was 26% and 9.4% respectively and under the pump-operated test it was 25.5% and 9.7%, respectively. The effect of fluid flow and the packing factor on the energy performance of a wall-mounted hybrid PV/T water heating collector system has been presented by Ji *et al.*[92] A schematic diagram of the flow circuit of the PV/T collector system and thermal resistance circuit diagram of heat flow network are shown in Figures 2.19 and 2.20. Ji *et al.*[92] have found that as the mass flow rate increased from 0.07 to 0.09 kg s^{-1} at a pipe diameter of 0.025 m, the required pumping power increased by 33%, which is beneficial for PV cooling.

Performance analysis of a photovoltaic solar-assisted heat pump (PV-SAHP) system in terms of the coefficient of performance has been studied by Ji *et al.*[93] They have found that the system has a superior COP to the conventional heat pump at the same time and gives higher electrical output. The COP of the heat pump was able to reach 10.4 and the average value was about 5.4. The average photovoltaic efficiency was around 13.4%. The expression for the comprehensive coefficient of thermal-and-electrical performance ($COP_{p/t}$) is defined by taking into consideration that the output power of the PV cells is transformed into the equivalent thermal power through the use of the average electricity-generation efficiency (η_{power}) of a coal-fired power plant as[94]

$$COP_{p/t} = \frac{Q_c + \left(W_p/\eta_{power}\right)}{W_{com}} \tag{2.12}$$

where Q_c = condenser power, W, W_p = output power of the PV cell, W, W_{com} = compressor power, W, and η_{power} = 38%.

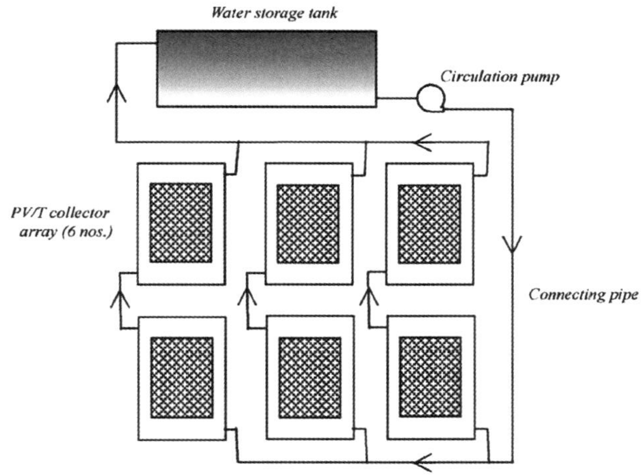

Figure 2.19 Schematic diagram of flow circuit of the PV/T collector system.[92]

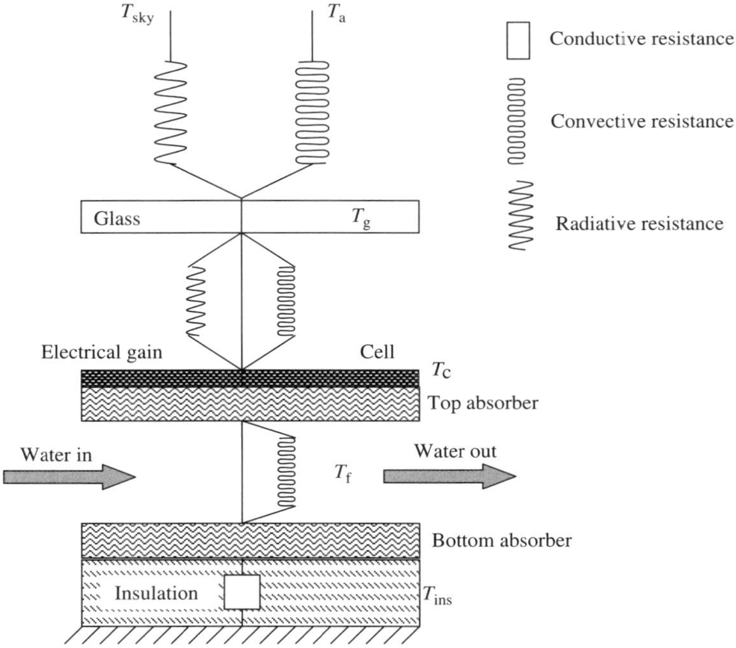

Figure 2.20 The thermal resistance circuit diagram of heat flow network.[92]

The highest overall coefficient of performance ($COP_{p/t}$), bringing into consideration both the photovoltaic and the thermal efficiency, was about 16.1.

An experimental study on energy generation with a photovoltaic solar thermal hybrid system for the geographic location of Cyprus was presented by Erdil *et al.*[94] In this system, the cooling medium is applied in front rather than at the rear of a solar module. The cooling fluid (*i.e.* water) was circulated between the glazing and the module and stored in a storage tank. For this, two collectors connected in parallel are used and it was found that the daily 2.8 kWh thermal energy can be stored as pre-heated water. A steady-state two-dimensional mathematical model of a PV/T bi-fluid (air and water) collector with a metal absorber was presented by Assoa *et al.*[95] The cross section of the PV/T hybrid bi-fluid collector is shown in Figure 2.21. The system consists of a ribbed sheet steel absorber on which a PV plate (0.24 m × 1.98 m; polycrystalline) is fixed through a thin layer of tedlar. An air gap between the absorber and an insulation layer in the rib has been provided. This was originally used for the mechanical rigidity of the sheet steel. This rib included an insulation layer of polystyrene covered by a thin reflective layer as well as a water circulation pipe. These small-diameter tubes are insulated by a cellular rubber half-cylinder. The thermal efficiency of the system has reached up to 80%. Robles-Ocampo *et al.*[96] studied a hybrid system with bifacial PV module and transparent plane collector experimentally. They have designed a water heating planner collector with reflecting planes. A transparent solar plane collector was placed above the surface of the PV module and filled with water, as shown in Figure 2.22. To collect solar radiation onto the rear face of the bifacial PV module, reflecting planes made of stainless steel have been set in the position corresponding to the equinox time, when the module has the inclination to a horizontal plane approximately equal to the latitude of the particular geographic point as shown in Figure 2.23. Robles-Ocampo *et al.*[96] have concluded that the estimated overall solar energy utilization efficiency for the system related to the direct radiation flux was of the order of 60%, with an electrical efficiency of 16.4%.

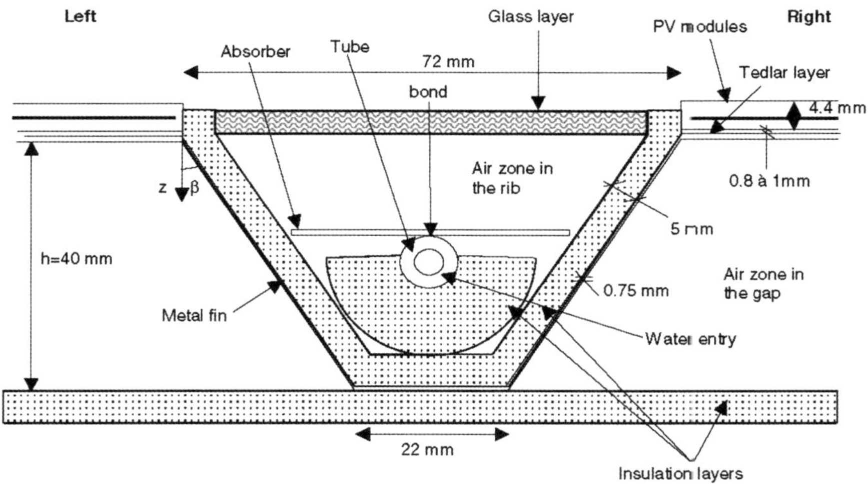

Figure 2.21 Cross section of the PV/T hybrid bi-fluid collector[95] (courtesy: Y. B. Assoa, France).

Figure 2.22 Photograph of the hybrid system (the collector is being filled with water)[96] (courtesy: B. Robles-Ocampo, Mexico).

Example 2.4

Compare the exergy of solar radiation obtained from the expressions given by the researchers, ambient temperature = 25 °C, solar radiation temperature = 6000 K, solar intensity = 750 W m^{-2}.

Solution

Using eqns (2.8a–2.8c), we get

$$\dot{E}x_{\text{sun}} = \left[1 + \frac{1}{3}\left(\frac{25+273}{6000}\right)^4 - \frac{4(25+273)}{3\times 6000}\right]750 = 700.3\ \text{W m}^{-2}$$

$$\dot{E}x_{\text{sun}} = \left[1 - \frac{4\times 298}{3\times 6000}\right]750 = 700.3\ \text{W m}^{-2}$$

Figure 2.23 Photograph of the reflecting planes set[96] (courtesy: B. Robles-Ocampo, Mexico).

$$\dot{E}x_{\text{sun}} = \left[1 - \frac{298}{6000}\right] 750 = 712.7\,\text{W m}^{-2}$$

Here, we can say that the differences in the results coming from these three calculation methods are less than 2%.

Example 2.5

Calculate the comprehensive coefficient of thermal-and-electrical performance of a photovoltaic solar-assisted heat pump when the water inlet temperature at the condenser is 30 °C, condenser power = 2400 W, compressor power = 325 W and PV power = 540 W.

Solution

Using eqn (2.12), we get

$$COP_{\text{p/t}} = \frac{2400 + (540/0.38)}{325} = 11.75$$

2.4 Temperature-dependent Electrical Performance of PV Module

The operating temperature plays a central role in the photovoltaic conversion process. Both the electrical efficiency and the power output of a PV module depend linearly on the operating temperature. The various correlations proposed in the literature represent simplified working equations which can be applied to PV modules or PV arrays mounted on free-standing frames, PV/Thermal collectors and BIPV arrays, respectively.

The electrical performance is primarily influenced by the type of PV used. In practice, only a-Si and crystalline Si have been found in the literature on PV/T. The higher efficiency of crystalline Si will result in a higher electrical efficiency and a higher electrical-to-thermal ratio of the PV/T than in the case of a-Si. Tripanagnostopoulos *et al.*[16] present experimental measurements on PV/T-liquid and PV/T-air collectors for both a-Si and c-Si. They find that at zero reduced temperature, for the PV/T liquid collector, the efficiency of the c-Si prototype is 55% and the a-Si prototype 60%, while for the PV/T air collector the c-Si prototype is 38% and the a-Si prototype 45%. However, the electrical performance for the c-Si modules is 12% and for the a-Si it is 6%. A higher thermal yield was also found for a-Si by Ji *et al.*[97]. However, in other experiments a lower thermal efficiency was found for a-Si than for c-Si (Affolter *et al.*[98,99], Platz *et al.*[100]). Zondag *et al.*[10] compared a conventional PV module, an unglazed PV/T module and a glazed PV/T module. The average annual electrical efficiency was found to be 7.2%, 7.6% and 6.6% respectively. Since glass with a transparency of 92% was used in the calculations, the reduction in electrical performance for the glazed PV/T as compared to the conventional PV laminate is exactly what one would expect from the additional reflection losses, which means that for the glazed PV/T the additional temperature effect cancelled over the year, while for the unglazed PV/T the temperature effect was positive. Chow[101] calculated the electrical performance of a thermosyphon PV/T collector with the PV at the high end and at the low end of the absorber. For the colder low end, he found a 3% higher electrical efficiency. Naveed *et al.*[102] examined a PV/T air system in which PV was connected to an unglazed transpired collector. It was found that a temperature reduction of 3–9 °C resulted in an improved electrical performance, allowing a reduction in PV area from 25 to 23 m^2. Krauter and Ochs[103] and Krauter[104,105] have been developed an unglazed integrated solar home system, in which a PV laminate is connected to a triangular water tank. The tank serves to cool the PV by means of an 'extended heat capacity'. Typically, at high irradiance, a PV temperature reduction of about 20 °C is reported relative to a conventional solar home system, which leads to a 9–12% increase in electrical yield, depending on the stratification. The stratification in the tank causes a temperature difference of about 6 °C between the upper and the lower PV module.

2.4.1 PV Module Efficiency as a Function of the Operating Temperature

The correlations expressing the PV cell temperature (T_c) as a function of weather variables such as the ambient temperature (T_a), local wind speed (V_w), solar radiation (I(t)), material- and system-dependent properties such as glazing-cover transmittance (τ), plate absorptance (α), *etc.* have been discussed in this section (Skopalki and Palyvos[106]). The effect of temperature on the electrical efficiency of a PV cell/module can be obtained by using the fundamental equation

$$P_m = I_m V_m = (FF) I_{sc} V_{oc} \tag{2.13}$$

In this equation FF is the fill factor, I_{sc} is the short circuit current, V_{oc} is the open circuit voltage and the subscript m refers to the maximum power point in the module's I–V curve. Both the open circuit voltage and the fill factor decrease substantially with temperature (as the thermally excited electrons begin to dominate the electrical properties of the semi-conductor), while the short-circuit current increases, but only slightly (Zondag[107]). Thus, the net effect leads to a linear relation in the form

$$\eta_c = \eta_{Tref}[1 - \beta_{ref}(T_c - T_{ref}) + \gamma \log_{10} I(t)] \tag{2.14}$$

in which η_{Tref} is the module's electrical efficiency at the reference temperature, T_{ref}, and at solar radiation of 1000 W m^{-2}. The temperature coefficient, β_{ref}, and the solar radiation coefficient, γ, are mainly material properties, having values of about 0.004 K^{-1} and 0.12, respectively, for crystalline silicon modules (Notton *et al.*[108]). The latter, however, is usually taken as zero (Evans[109]), and eqn (2.14) reduces to

$$\eta_c = \eta_{Tref}[1 - \beta_{ref}(T_c - T_{ref})] \tag{2.15}$$

which represents the traditional linear expression for the PV electrical efficiency (Evans and Florschuetz[110]). The quantities η_{Tref} and β_{ref} are normally given by the PV manufacturer. However, they can be obtained from flash tests in which the module's electrical output is measured at two different temperatures for a given solar radiation flux (Hart and Raghuraman[111]). The actual value of the temperature coefficient, in particular, depends not only on the PV material but on T_{ref} as well. It is given by the ratio

$$\beta_{ref} = \frac{1}{T_o - T_{ref}} \tag{2.16}$$

in which T_0 is the (high) temperature at which the PV module's electrical efficiency drops to zero (Garg and Agarwal[112]). For crystalline silicon solar cells this temperature is 270 °C (Evans and Florschuetz[113]). In a number of correlations, the cell/module temperature – which is not readily available – has been replaced by T_{NOCT}, *i.e.* by the nominal operating cell temperature. One such expression is

$$\eta = \eta_{ref}\left[1 - \beta\left[T_a - T_{ref} + (T_{NOCT} - T_a)\frac{I(t)}{I(t)_{NOCT}}\right]\right] \tag{2.17}$$

The quantities labelled as NOCT are measured under open-circuit conditions (*i.e.* with no load attached) while operating in the so-called nominal terrestrial environment (NTE), which is defined as follows (Stultz and Wen[114]):

Global solar flux: 800 W m^{-2}
Air temperature: 293.16 K (20 °C),
Average wind speed: 1 m s^{-1}
Mounting: open rack, tilted normally to the solar noon Sun

In addition to the 'instantaneous' values for the PV electrical efficiency, expressions for the monthly average efficiency can be written. The monthly electrical energy output of a PV array can be estimated on the basis of the following equation:

$$\bar{\eta} = \eta_{Tref}\left[1 - \beta_{ref}(\bar{T}_a - T_{ref}) - \frac{\beta_{ref}(\overline{\tau\alpha})\overline{VH_T}}{nU_L}\right] \tag{2.18}$$

in which the over-bar denotes monthly average quantities, n is the number of hours per day, U_L is the overall thermal loss coefficient, H_T is the monthly average daily insolation on the plane of the array and V is a dimensionless function of such quantities as the sunset angle, the monthly average clearness index and the ratio of the monthly total radiation on the array to that on a horizontal surface (Siegel *et al.*[115]). A number of equations found in the literature for the efficiency of PV cells/modules are shown in Tables 2.2 and 2.3. The first table contains values for the parameters of eqn (2.15), as reported by a number of authors, and the second presents additional forms for η_c, including pertinent comments for each correlation. On the basis of data listed in Table 2.2 for $T_{ref} = 25$ °C, average $\eta_{ref} \approx 0.12$ and average $\beta_{ref} \approx 0.0045$ °C^{-1}.

2.4.2 PV Power Output Dependence on Module Operating Temperature

The prediction of PV module performance in terms of electrical power output in the field condition, that is, the deviation from the standard test conditions reported by the manufacturer of the module. For example, a recently proposed correlation for PV power, similar to eqn (2.15), is

$$P = \eta_{Tref}\tau_{pv}AI(t)[1 - 0.0045(T_c - 25)] \tag{2.19}$$

Table 2.2 Evans–Florschuetz PV efficiency correlation coefficients, $\eta_c = \eta_{Tref}[1-\beta_{ref}(T_c-T_{ref})]$.

T_{ref} (°C)	η_{Tref}	β_{ref}	*Comments*	*References*
25	0.15	0.0041	Mono-Si	Evans and Florschuetz[110]
28	0.117 (average) (0.104–0.124)	0.0038 (average) (0.0032–0.0046)	Average of Sandia and commercial cells	OTA[117]
25	0.11	0.003	Mono-Si	Truncellito and Sattolo[118]
25	0.13	0.0041	PV/T system	Mertens[119]
		0.005		Barra and Coiante[120]
20	0.10	0.004	PV/T system	Prakash[121]
25	0.10	0.0041	PV/T system	Garg and Agarwal[112]
20	0.125	0.004	PV/T system	Hegazy[122]
25		0.0026	a-Si	Yamawaki *et al.*[123]
25	0.13	0.004	Mono-Si	RETScreen[124]
	0.11	0.004	Poly-Si	
	0.05	0.0011	a-Si	
25	0.178	0.00375	PV/T system	Nagano *et al.*[125]
25	0.12	0.0045	Mono-Si	Chow[101]
25	0.097	0.0045	PV/T system	Zondag *et al.*[10]
25	0.09	0.0045	PV/T system	Tiwari and Sodha[70]
25	0.12	0.0045	PV/T system	Tiwari and Sodha[126]
25	0.12	0.0045	PV/T system	Assoa *et al.*[95]
25	0.127	0.0063	PV/T system	Tonui and Tripanagnostopoulos[127]
25	0.127 unglazed 0.117 glazed	0.006	PV/T system	Tonui and Tripanagnostopoulos[128]
25		0.0054	PV/T system	Othman e*t al.*[129]

Table 2.4 lists a number of correlations found in the literature for PV electrical power as a function of cell/module operating temperature and basic environmental variables. Many of them are linear and similar to eqns (2.15) or (2.19). Correlations involved basic environmental variables, while the numerical parameters are not only material dependent but also system dependent. Thus, care should be exercised in applying a particular expression for the electrical efficiency or the power output of a PV module or array, as each equation has been developed for a specific mounting frame geometry or level of building integration. The same holds for choosing a PV module rating method, the details and limitations of which should be very clear to the prospective user. The reader, therefore, should consult the original sources and try to make intelligent decisions when seeking a correlation or a rating procedure to suit his/her needs.

From the PV-system designer's point of view, the ultimate interest is the proper sizing of the installation for a given service and, thus, the actual energy yield of the relevant array. In order to estimate, the designer starts with the PV module manufacturer's reported performance of his modules at standard test conditions. But such energy/power figures are only useful for comparing the peak performance of different module makes and types. That is, the STC rating is not capable of predicting exactly how much energy a module will produce in the field, *i.e.* when operating under real conditions. For this, there are several proposals for a PV module's energy rating procedure which would attempt to account for the varying operating conditions encountered in the field. In most cases, actual field measurements lead to a regression equation for power (or energy) that is based on a particular model and, having calculated the regression coefficients, a straightforward application to standard conditions gives the true power rating for the module (Taylor[116]).

2.5 Artificial Intelligence Techniques for PV systems

Artificial intelligence (AI) techniques are becoming useful as alternate approaches to conventional techniques or as components of integrated systems. AI techniques have been used to solve complicated practical problems in various areas and are becoming more and more popular nowadays. AI-based systems are being developed and deployed worldwide in a myriad of applications, mainly because of their symbolic reasoning, flexibility and explanation capabilities. AI has been used and applied in different sectors, such as engineering, economics, medicine, military, marine, *etc.* It has also been applied for modelling, identification, optimization, prediction, forecasting and control of complex systems. Many of the researchers used AI techniques as a design tool for sizing photovoltaic (PV) systems: stand-alone PVs, grid-connected PV systems, PV-wind hybrid systems, *etc.* Additionally, the advantage of using an AI-based sizing of PV systems is that it provides good optimization, especially in isolated areas, where the weather data are not always available.[169] The term artificial intelligence (AI) has been applied to computer systems and programs

Table 2.3 PV array efficiency as a function of temperature.

Correlation	*Comments*	*References*
$\eta(I(t),\ T_c) = \eta(I(t),\ 25\,°C)[1 + c_3(T_c - 25)]$	$c_3 = -0.5$ (% loss per C) for c-Si, $-0.02, \ldots, -0.41$ for thin film cells	Mohring *et al.*[130]
$\eta_T = \eta_0 - K(T^{1/4} - T_0^{1/4})$	$T_0 = 273$ K, $K = 22.4$	Ravindra and Srivastava[131]
$\eta_a = \eta_n \times k_\gamma \times k_\theta \times k_\alpha \times k_\lambda$, with $k_\gamma = 1 - \gamma(T_c - 25)/100$	k_γ = power temperature coefficient, k_j, $j = \theta, \alpha, \lambda$ optical, absorption, spectrum correction factors	Aste *et al.*[132]
$\eta = \eta_{Tref}\left[1 - \beta_{ref}(T_a - T_{ref}) - \frac{\beta_{ref}\tau\alpha I(t)}{U_L}\right]$	5% low predictions, $\beta_{ref} \sim 0.004\,°C^{-1}$, $\eta_{ref} = 0.15$, $T_{ref} = 0\,°C$	Siegel *et al.*[115]
$\bar{\eta} = \eta_{Tref}\left[1 - \beta_{ref}(\bar{T}_a - T_{ref}) - \frac{\beta_{ref}\overline{(\tau\alpha)}VH_T}{nU_L}\right]$	$\bar{\eta}$ = Monthly average efficiency, V = dimensionless, $\beta_{ref} \sim 0.004\,°C^{-1}$	Siegel *et al.*[115]
$\eta_i = \eta_{Tref}[1 - \beta_{ref}(T_{c,i} - T_{ref}) + \gamma \log_{10} I_i]$	η_i = hourly efficiency, I_i = incident hourly insol, $\beta_{ref} \sim 0.0045°C^{-1}$, $\gamma \sim 0.12$	Evans[109] and Cristofari *et al.*[133]
$\eta = \eta_{Tref}\left[\begin{array}{l}1 - \beta_{ref}(T_c - T_{ref}) \\ + \gamma \log_{10} I(t)\end{array}\right]$	η = instantaneous efficiency, $\beta_{ref} \sim 0.0044°C^{-1}$, $\eta_{Tref} = 0.125$, $T_{ref} = 25\,°C$	Notton *et al.*[108]
$\bar{\eta} = \eta_{Tref}\left\{1 - \beta_{ref}\left[\begin{array}{c}(T_c - T_a) - (T_a - \overline{T_a}) \\ -(\overline{T_a} - T_{ref})\end{array}\right] + \gamma \log_{10} I\right\}$	$\bar{\eta}$ = monthly average efficiency, $\beta_{ref} \sim 0.0045\,°C^{-1}$, $v \sim 0.12$	Evans[109]
$\eta = \eta_{ref}[1 - a_1(T_c - T_{ref}) + a_2 \ln(I(t)/1000)]$	For Si $a_1 = 0.005$, $a_2 = 0.052$, omitting the ln term slightly overestimates η	Anis *et al.*[134]
$\eta = a + b\frac{T_{in} - T_a}{I(t)}$	PV/T collector, PV cover: 100% → a = 0.123, b = −0.464 50% → a = 0.121, b = −0.450	Chow *et al.*[71]
$\eta = 0.94 - 0.0043\left[\overline{T}_a + \frac{\overline{I(t)}}{(22.4 + 8.7\overline{V}_{w})} - 25\right] \pm 2.6\%$	Overbars denotes daily averages. $\bar{G}_r$ = Wh/m^2 received/length of day(h), $\bar{V}_w$ in m/s	CLEFS CEA[135]
$\eta_{MPP}(I(t),\ T) = \eta_{MPP}(I(t),\ 25\,°C)(1 + \alpha(T - 25))$ $eta_{MPP}(I(t),\ 25\,°C) = a_1 + a_2 I(t) + a_3 \ln(I(t))$	a_1–a_3 device specific parameters, MPP tracking system	Beyer *et al.*[136]
$\eta = \eta_{NOCT}[1 - MPTC(T_{NOCT} - T_C)]$		Perlman *et al.*[137]

$\eta_{T} = \eta_{Tref}[1 - \beta_{ref}(T - T_{ref})]$	MPCT = Maxmimun power temperature coefficient with $MPCT = -0.5\%$ loss per °C, the efficiency is $\eta = 11.523–0.0512T_c$ $T_{ref} = 25°C$, $\eta_{Tref} = 25\,°C$, $\eta_{Tref} = 0.15$, $\beta_{ref} = 0.0041°C^{-1}$ *c–Si*, T in °C	Evans and Florschuetz[110]
$\eta_{PV} = \eta_{ref} - \mu(T_c - T_{ref})$	μ = overall cell temperature coefficient	Bazilian and Prasad[138]
$\eta(XI(t), T) = \eta(I(t), T_{ref})[1 - \beta_{ref}(T - T_0)] \left(1 + \frac{k_B T}{q}\frac{\ln X}{V_\infty(I(t), T_0)}\right)$	X = concentration factor, for X = 1 it reduces to Eq. $\eta_c = \eta_{Tref}\,[\,1 - \beta_{ref}(T_c - T_{ref})\gamma \log_{10} I(t)]$	Lasnier and Ang[139]
$\eta = \eta_{ref}\left[1 - \beta\left[T_a - T_{ref} + (T_{NOCT} - T_a)\frac{I(t)}{I(t)_{NOCT}}\right]\right]$	The T_c expression from Kou *et al.* [140] is introduced into the η expression in Evans and Florechetz[110]	Kou *et al.*[140] and Evans and Florschuetz[110]
$\eta = \eta_{ref}\left\{1 - \beta\left[T_a - T_{ref} + \left(\frac{9.5}{5.7 + 3.8V_w}\right)(T_{NOCT} - T_a)\frac{I(t)}{I(t)_{NOCT}}\right]\right\}$	The T_c expression from Duffie and Beckman[141] is introduced into the η expression is Evans and Florschuetz[110]	Duffie and Beckman[141] and Evans and Florschuetz[110]
$\eta = \eta_{ref}\left[1 - 0.9\beta\frac{I(t)}{I(t)_{,NOCT}}(T_{c,NOCT} - T_{a,NOCT}) - \beta(T_a - T_{ref})\right]$	Assumes $\eta \approx 0.9(\tau\alpha)$	Hove[142]
$\eta_{nom} = -0.05T_{surface} + 13.75$ $\eta_{meas} = -0.053T_{back} + 12.62$	$T_{surface} = 1.06T_{back} + 22.6$ Nominal vs measured values	Yamaguchi *et al.*[143]
$\eta = a_0 + a_1\frac{T_c(x.t) - T_\infty}{T_\infty} + a_2\frac{I(t) - I(t)_{ref}}{I(t)_{ref}}$	Ak, k = 0,1 and 2 are empirical constants, $T\infty$ is the indoor ambient temperature	Zhu *et al.*[144]
$\eta = \eta_a - c(\overline{T} - T_a)$	$\bar{T}$ = mean solar cell temp, η_a = efficiency at T_a, c = temperature coefficient	Bergene and Løvvik[145]
$\eta = \eta_{25} + b(T_c - 25)$	$b = b(I(t))$, T in °C	Durisch *et al.*[146]

Table 2.4 PV array power as a function of temperature, $P = \eta_i A I(t)$.

Correlation	*Comments*	*References*
$P = P_0[1 + (\alpha - \beta_{ref})\Delta T]$	α:0.0005°C^{-1}, β:0.0005°C^{-1}	Patel[147]
$P = (\alpha T_c + \beta)I(t)$	α = temperature coefficient, β = calibration constant	Yang *et al.*[148]
$P = -4.0 + 0.053I(t) + 0.13T_c - 0.00026I(t)T_c$	MPPT racked 100kWp system	Risser and Fuentes[149]
$P = -0.4905 + 0.05089I(t) + 0.00753T_c - 0.000289I(t)T_a$	MPPT racked 100kWp system	Risser and Fuentes[149]
$P = \eta_{Tref} A I(t)\tau_{pv}[1 - 0.0045(T_c - 25)]$	η_{Tref} = 0.14, T_c in °C^{-1}, τ_{pv} = pv cell glazing transmittance	Jie *et al.*[150]
$P = \eta_{Tref} A I(t)[1 - \beta_{ref}(T_c - T_{ref}) + \gamma \log_{10} I(t)]$	β_{ref} = 0.0044°C^{-1} for pc-Si , γ is usually taken as 0	Cristofari *et al.*[133]
$P_T = P_{Tref}[1 - \beta_{ref}(T - T_{ref})]$	β_{ref} = 0.004–0.006°C^{-1} T in °C, T_{ref} = reference temperature	Buresch[151]
$P_T = P_{Tref}[1 - \beta_{ref}(T - T_{ref})]$	β_{ref} = 0.004	Twidell and Weir[152]
$P(T) = P(25)[1 - \gamma(T - 25)]$	γ = 0.0053°C^{-1} for c-Si range:0.004–0.006°C^{-1}	Parretta *et al.*[153]
$P_T = P_{25}[1 - 0.0026(T - 25)]$	a-Si, T in °C, power degrades to 0.82 P_{init}	Yamawaki *et al.*[123]
$-P_T = P_{25} + \frac{dP}{dT}(T - 25)$	$\frac{dP}{dT} = -0.00407,\ 0.00535$, Si space cells, T in°C	Osterwald[154]
$P(T) \approx I(t)[\eta_0 - c(T - T_a)]$	η_0 = efficiency at $T_{a,c}$ = temperature dependence factor	Bergene and Løvvik[145]
$P_{max} = P_{max,ref}[1 - D_f(T_c - 25)]$	Df = "deficiency factor" = 0.005°C^{-1}	Al-Sabounchi[155]
$P = A(0.128I(t) - 0.239 \times 10^{-3} T_a)$	P-Si, hybrid PV-fuel cells system G_T in kW/m^2, P in kW, T_a in °C	Zervas *et al.*[156]
$P = P_{ref} I(t) K_{pt} K_w K_e K_c$ with $K_{pt} = 1 + \alpha(T_c - 25)$	K_w, K_e, K_c loss coefficients due to mounting, dirt etc., AC conversation. semitransparent	Wong *et al.*[157]
$P = \eta_{Tref} A I(t)(\tau\alpha)[1 - \beta_{ref}(T_P - T_{ref})]$	T_p = plate temperature, η_{Tref} = 0.118 at 45°C–air coll, η_{Tref} = 0.108 at 28°C–water coll	Hendrie[158]

$P_{Tc} = \eta_{Tref} A I(t) K_f [1 + \alpha(T_C - 25)]$	$T_{ref} = 25\,°C$, $\eta_{Tref} = 0.13$, $\alpha = -0.004\,°C^{-1}$, K_f factor for rest, frame installation, T_c in °C	Nishioka *et al.*[159]
$P_{max} = P_{max,ref} \frac{I(t)}{I(t)_{,ref}} [1 + \gamma(T_C - T_{ref})]$	γ = temperature factor for power, $\gamma = -0.0035$ (range–0.005 °C^{-1} to –0.003 °C^{-1})T_c in °C	Menicucci and Fernandez[160]
$P_{max} = P_{max,ref} \frac{I(t)}{I(t)_{ref}} [1 + \gamma(T_c - 25)]$	$\gamma = -0.0035$(range–0.005 °C^{-1} to –0.003 °C^{-1})T_c in °C	Fuentes *et al.*[161]
$P_{max} = P_{max,ref} \frac{I(t)}{1000} [1 + \gamma(T_c - T_{ref})]$	γ = temperature factor for power, $T_{ref} = 25\,°C$, used in PVFORM	Marion[162]
$P_{mp,Tr} = I_{mp,T}[1 - \alpha(T - T_r)] \left[V_{mp,T} - \beta V_{mp}^{STP}(T - T_r)\right]$	STC refers to ASTM standard conditions (1000W/m^2, AMI = 1.5, $T_r = 25\,°C$)	King *et al.*[163]
$P_{max} = P_{max,ref} \frac{I(t)}{I(t)_{ref}} [1 + \alpha(T - T_{ref})][1 + \beta_{ref}(T - T_{ref})] \left[1 + \delta(T) \ln\left(\frac{I(t)}{I(t)_{ref}}\right)\right]$	Adapted from the MER model. Coefficientj δ evaluated at actual conditions	Kroposki *et al.*[164]
$P_T = -8.6415 + 0.076128 I(t) + 1.02318 \times I(t)^2 + 0.20178 T - 4.9886 \times 10^{-3} T^2$	T is the panel temperature (K)	Jie *et al.*[165]
$P = I(t)(b_1 + b_2 I(t) + b_3 T_a + b_4 V_f)$	EPTC model, b_j regression coefficient, V_w^f wind sped 10 m above ground	Farmer[166]
$P = c_1 + (c_2 + c_3 T_a) I(t) + (c_4 + c_5 V_w) I(t)^2$	c_j regression coefficients based on STC module tests	Taylor[116]
$P_{mp} = D_1 I(t) + D_2 T_c + D_3[\ln(I(t))]^m + D_4 T_c [\ln(I(t))]^m$	$D_j (j = 1–4)$, m parameters	Rosell and Ibáñez[167]
$P = V_c I_c \left[1 - \frac{I(t) - 500}{2.0 \times 10^{-4}} + \frac{C_{Tc}}{4 \times 10^4}(50 - T_c)^2\right]$	I_c = outputs current(A), V_c = output voltage(V), T_c in k, $C_{Tc} = 1$ if $T_c \leq 50°$ or 3 if $T_c \geq 50°$	Furushima *et al.*[168]
$P = \eta_e A I(t) \tau_g p [1 - \beta_{ref}(T_c - 25)]$	p = packing factor, T_c in °C, τ_g = glazing transmissitiy	Chow *et al.*[71]
$P = \eta_{Tref} A I(t)[1 - 0.0045(T_c - 298.15)]$	$\eta_{Tref} = 0.14$, T_c in K	Jie *et al.*[148]

that can perform tasks more complex than straightforward programming, although still far from the realm of actual thought. AI consists of many branches, such as expert systems (ES), artificial neural networks (ANN), genetic algorithms (GA), fuzzy logic (FL) and various hybrid systems, which are combinations of two or more of the branches mentioned previously.[170] AI technologies have a natural synergism that can be exploited to produce powerful computing systems. A theme that can be found in these alternatives is the attempt to make up for deficiencies in the conventional approaches. In some cases, the goal is to produce better, more efficient and effective computing systems. Sometimes this requires adding features associated with human intelligence such as learning and the ability to interpolate from current knowledge. The appropriate use of intelligent technologies leads to useful systems with improved performance or other characteristics that cannot be achieved through traditional methods.[171] AI techniques have been used in several domains and applications.[172–176] In order to size a PV system using AI techniques so that it can work properly, efficiently and economically to meet the desired load requirements under the local meteorological conditions, the characteristic performance of each component in the PV system is required. Normally, the information provided about the PV module and other components from the manufacturers is used for sizing the PV system by a rough estimation of the system output based on average values of daily meteorological data inputs.[177] An optimal and economic PV system is very important, particularly in isolated sites (Sahara regions, small island archipelagos, remote areas in developing nations, mountainous locations, rural regions, *etc.*).

In the design of stand-alone renewable energy power systems, the optimal sizing is an important and challenging task. A stand-alone photovoltaic power system consists of a photovoltaic array, a storage component and control and power processing components. The conventional methodology (empirical, analytical, numerical, hybrid, *etc.*) for sizing PV-systems has been used generally for locations where the required weather data (irradiation, temperature, humidity, clearness index, wind speed, *etc.*) and information are available. In this case, these methods present a good solution. However, these techniques could not be used for sizing PV systems in remote areas, where the required data are not available, especially solar radiation. In all of these, accuracy is achieved by using data from daily global irradiation series. Moreover, the majority of alternative approaches need long-term meteorological data such as total solar irradiation, air temperature, wind speed, *etc.* for their operation. In order to overcome this situation, methods have been developed for sizing the parameters for PV-systems based on AI techniques.[178]

2.5.1 Artificial Neural Networks

An artificial neural network (ANN) is a collection of small, individually interconnected processing units. Information is passed through these units along interconnections. An incoming connection has two values associated with it: an input value and a weight. The output of the unit is a function of the summed value. ANN's implemented on computers are trained with respect to data sets until they learn the patterns used as inputs. Once they are trained, new patterns may be presented to them for prediction or classification. ANNs can automatically learn to recognize patterns in data from real systems or from physical models, computer programs or other sources. An ANN can handle many inputs and produce answers that are in a form suitable for designers.[170] A typical ANN comprises several layers of interconnected neurons, each of which is connected to other neurons in the ensuing layer. Data are presented to the neural network *via* an input layer, while an output layer holds the response of the network to the input. One or more hidden layers may exist between the input layer and the output layer. All hidden and output neurons process their layer inputs by multiplying each input by its weight, summing the product and then processing the sum using a non-linear transfer function to generate a result.[174] Neural networks have the potential to provide some of the human characteristics of problem solving that are difficult to simulate using the logical, analytical techniques of expert system or standard software technologies. For example, neural networks can analyse large quantities of data to establish patterns and characteristics in situations where rules are not known and can in many cases make sense of incomplete or noisy data. These capabilities have thus far proven too difficult for traditional symbolic or logic-based approaches.[170] The immediate practical implication of neural computing is its emergence as an alternative or supplement to conventional computing systems and AI techniques. As an alternative, neural computing can offer the advantage of execution speed, once the network has been trained. The ability to train the system with data sets, rather than having to write programs, may be more cost effective and may be more convenient when changes become necessary. In applications where rules cannot be known, neural networks may be able to represent those rules implicitly as stored connection weights.[170] The greatest advantage of ANNs over other modelling techniques is their capability to model complex, non-linear processes without having to assume the form of the relationship between input and output variables. There are several ANN architectures used in the literature, such as multilayer perceptron (MLP), radial basis function network (RBF) and recurrent neural network (RNN).[179]

2.5.2 Fuzzy Logic

Fuzzy systems (FS) are based on fuzzy set theory and associated techniques pioneered by Lotfi Zadeh.[180,181] A goal of this approach is to mimic the aspect of human cognition that can be called approximate reasoning. Fuzzy systems may be less precise than conventional systems but are more like our everyday experiences as human decision-making. Fuzzy logic (FL) is used mainly in control engineering. It is based on fuzzy logic reasoning which employs linguistic rules in the form of 'if–then' statements. Fuzzy logic and fuzzy control feature a relative simplification of a control methodology description. This allows the application of a 'human language' to describe the problems and their fuzzy solutions. In many control applications, the model of the system is unknown or the input parameters are highly variable and unstable. In such cases, fuzzy controllers can be applied. Fuzzy logic is very useful in modelling complex and imprecise systems. Under the fuzzy set theory, elements of a fuzzy set are mapped to a universe of membership values using a function–theoretic form belonging to the close interval from 0 to 1. An important step in applying fuzzy methods is the assessment of the membership function of a variable, which parallels the estimation of probability in stochastic models.

2.5.3 Genetic Algorithm

Genetic algorithms (GAs) are inspired by the way living organisms are adapted to the harsh realities of life in a hostile world, *i.e.* by evolution and inheritance. The algorithm imitates in the process the evolution of population by selecting only fit individuals for reproduction. Therefore, a GA is an optimum search technique based on the concepts of natural selection and survival of the fittest. It works with a fixed-size population of possible solutions of a problem, called individuals, which are evolving in time. A GA utilizes three principal genetic operators: selection, crossover and mutation.[172,173] Genetic algorithms were envisaged by Holland[182] in the 1970s as an algorithmic concept based on a Darwinian-type survival-of-the-fittest strategy with sexual reproduction, where stronger individuals in the population have a higher chance of creating offspring. A genetic algorithm is implemented as a computerized search and optimization procedure that uses principles of natural genetics and natural selection. The basic approach is to model the possible solutions to the search problem as strings of ones and zeros. Various portions of these bit-strings represent parameters in the search problem. If a problem-solving mechanism can be represented in a reasonably compact form, then GA techniques can be applied using procedures to maintain a population of knowledge structure that represent candidate solutions, and then let that population evolve over time through competition (survival of the fittest and controlled variation) A GA will generally include the three fundamental genetic operations of selection, crossover and mutation. These operations are used to modify the chosen solutions and select the most appropriate offspring to pass on to succeeding generations.[183]

Genetic algorithm applications are appearing as alternatives to conventional approaches and in some cases are useful where other techniques have been completely unsuccessful. Genetic algorithms are also used with other intelligent technologies, such as neural networks, expert systems and case-based reasoning.

2.5.4 Wavelet

Wavelet transform (WT) is a novel signal-processing technique developed from the Fourier transform and has been widely used in signal processing. The main characteristic of wavelet transform is its time-frequency localization. Wavelet transformation (WT) has versatile basis functions, which are selected based on the type of the signal analysed. Wavelets have generated a tremendous interest in both theoretical and applied areas, especially over the past few years. The number of researchers applying wavelets is already large and continues to grow, so progress is being made at a rapid pace. In fact, advancements in the area are occurring at such a rate that the very meaning of 'wavelet analysis' keeps changing to incorporate new ideas. In a rapidly developing field, overview papers are particularly useful and several good ones concerning wavelets are already available.[184]

2.5.5 Hybrid Systems

The increased popularity of hybrid intelligent systems (HIS) in recent years stems from the extensive success of these systems in many real-world complex problems. The main reason for this success seems to be the synergy derived by the computational intelligent components, such as machine learning, fuzzy logic, neural networks and genetic algorithms. Each of these methodologies provides hybrid systems with complementary reasoning and searching methods that allow the use of domain knowledge and empirical data to solve complex problems.[185,186] Hybrid systems combining fuzzy logic, neural networks, genetic algorithms and expert systems are proving their effectiveness in a wide variety of real-world problems.

2.6 Market Potential of PV/T Systems

Over the past five decades, as the demand for energy has escalated and the consumption of fossil fuels has accelerated, people have sought renewable sources as an alternative way to meet growing energy requirements. PV is an increasingly important energy technology. Deriving energy from the Sun offers numerous environmental benefits. It is an extremely clean energy source, and few other power-generating technologies have as little environmental impact as photovoltaics. As it quietly generates electricity from light, PV produces no air pollution or hazardous waste. Moreover, it does not require liquid or gaseous fuels to be transported or combusted. Also, because its energy source, sunlight, is free and abundant, PV systems can offer virtually guaranteed access to electric power.

However, this technology faces several large obstacles, most notably the costs relating to power generation and transmission as well as difficulties in obtaining funding for the development of advanced technology. Research is under way for development of so-called second generation – or thin-film – PV technologies to bring down the costs associated with PV energy. The largest market potential is seen for liquid-type PV/T for domestic hot water, combined with space heating. At the end of 2007, according to preliminary data, cumulative global production of solar PV systems was 12,400 megawatts. Roughly

90% of this generating capacity consists of grid-tied electrical systems. Such installations may be ground mounted (and sometimes integrated with farming and grazing) or building integrated. At present, about 90% of the European conventional solar collector market is residential, consisting of 80% domestic hot-water systems and 10% space-heating systems, which are normally called combi systems.[187] Although most collectors are installed on single-family houses, the share of large systems for collective applications is expected to increase. In the classification of PV/T systems, water-heating systems for the residential market are indicated as the main market for glazed PV/T systems, while public pool systems and large hot-water systems (both for collective applications and for utility applications such as hospitals, campgrounds and homes for the elderly) are presented as interesting niche markets. At present, for the glazed PV/T collectors required for this application, there is potential for further improvements with respect to the issues associated with high stagnation temperature, as well as the relatively large collector losses, both due to reflection and thermal losses. In addition, non-technical issues such as certification and building integration, and the development of plug-and-play installation, are also important and should receive more attention.

The market for conventional unglazed liquid collectors consists primarily of pool-heating applications. The potential of unglazed pool heating collectors in Europe is small; after the modest peak in the early 1990s, the market has declined in European countries such as the Netherlands, Austria and France, while in Germany and Sweden the amount of newly installed unglazed collectors has been more or less constant over the last decade.[187] However, in the USA or Australia, where the pool collector market is much larger,[188] a larger potential exists. Finally, a large new market for unglazed PV/T collectors is opened if these collectors can be combined successfully with a heat pump.

The commercially available PV/T modules are mainly air type with unglazed PV. In this application, the PV is effectively cooled; thereby increasing the electrical yield and conventional PV modules can be applied. A problem is the limited application of air-heating systems in the domestic market, as indicated by the fact that air collectors have a market share of less than 1% of the worldwide solar collector market.[188] However, the market for air-heating systems might well grow in the future, due to the reduction in domestic heating demand and the increasing application of ventilation systems with heat recovery, allowing for easy integration of air collectors. Particularly in the passive houses, in which the entire heating demand can in principle be met by heated ventilation air, this will gain increased attention and may be a standard for the future. Also the utility market is very interesting for air collectors, due to the requirements for air conditioning and often high required ventilation rates (*e.g.* schools), as well as a better match between solar supply and heating demand. The market for PV facades is expected to show a substantial growth in the future, due to increasing experience with PV facade integration and due to decreasing PV prices. The market for ventilated PV facades with heat recovery is expected to follow. The potential of this application is mainly seen in utility buildings. A strong point for this market is direct heating, as heating demand

and irradiance both peak during working hours. A problem might be that, in buildings with a large share of direct solar gain, the heat from the PV facade is in competition with the heat generated by passive means. Also, since the temperature level that can be provided will be low, due to the low thermal efficiency of PV facades which limits the thermal contribution of this system, this will demand a carefully optimized design. As concluded in the PV-HYBRIDPAS project,[24] an optimal integration with the HVAC design and an evaluation of the hybrid PV for each specific case is essential. In this case, a stack effect to boost the ventilation or preheating air for solar cooling can be applied to control the heat generated during the summer.

Problems

2.1 Repeat Examples 2.1 and 2.2 for $U_{loss} = 2.1\ \mathrm{W\,m^{-2}}$, ambient temperature = 25 °C, inlet temperature = 50 °C, $I(\mathrm{t}) = 800\ \mathrm{W\,m^{-2}}$ and PV cell temperature = 60 °C, 70 °C and 80 °C. Hint: use eqns (2.1) and (2.3).

2.2 Compare the exergy of solar radiation obtained from the expressions given by the researchers, ambient temperature = 20 °C, 30 °C, solar radiation temperature = 6000 K and solar intensity = $850\ \mathrm{W\,m^{-2}}$. Hint: use eqns (2.8a–2.8c).

2.3 Repeat Examples 2.5 for condenser power = 2200 W, compressor power = 395 W and PV power = 440 W. Hint: use eqn (2.12).

2.4 Calculate the temperature-dependent efficiency of a solar cell obtained at different temperature coefficients (β_{ref}) = $0.0032\ \mathrm{K^{-1}}$, $0.0045\ \mathrm{K^{-1}}$ and $0.006\ \mathrm{K^{-1}}$, when standard efficiency = 12%, PV cell temperature = 70 °C, reference temperature = 25 °C and the solar radiation coefficient (γ) = 0.12. Hint: use eqns (2.14) and (2.15).

2.5 Calculate the power generated by a PV module, when area = $0.605\ \mathrm{m^2}$, solar intensity = $700\ \mathrm{W\,m^{-2}}$, PV cell temperature = 70 °C, when standard efficiency = 12%. Hint: use eqn (2.19).

References

1. K. W. Boer and G. Tamm, *Sol. Energ.*, 2003, **74**, 525–528.
2. M. A. S. Malik, in *Solar One; Solar Energy Applications in Buildings*, ed. A. A. M. Sayigh, Academic Press, New York, 1979.
3. S. D. Hendrie, *Final Report, Report, MIT,* 1982.
4. S. D. Hendrie and P. Raghuraman, in *14th IEEE, San Diego*, 1980.
5. P. R. Younger, W. S. Kreisman, M. J. Nowlan, J. S. Solomon and J. S. Strong, in *15th IEEE, Orlando*, 1981.
6. P. Raghuraman, *J. Sol. Energ. Eng.*, 1981, **103**, 291–298.
7. S. Ito and N. Miura, in *ISES Solar World Congress, Budapest*, 1993.
8. A. Elazari, in *2nd WCPEC, Vienna, Austria*, 1998.
9. C. H. Cox and P. Raghuraman, *Sol. Energ.*, 1985, **35**(3), 227–241.

10. H. A. Zondag, D. W. De Vries, W. G. J. Van Helden, R. J. C. Van Zolingen and A. A. Van Steenhoven, *Sol. Energ.*, 2003, **74**, 253–269.
11. J. J. Loferski, C. Case, G. Doodlesack, B. Roessler, R. Dobbins and T. Russell, in *16th IEEE, San Diego*, 1982.
12. J. Prakash, *Energ. Convers. Manag.*, 1994, **35**(11), 967–972.
13. Y. Tripanagnostopoulos, in *17th EPSEC*, Munich, 2001.
14. Y. Tripanagnostopoulos, D. Tzavellas, I. Zoulia and M. Chortatou, in *17th EPSEC*, Munich, 2001.
15. Y. Tripanagnostopoulos, M. Bazilian, I. Zoulia and R. Battisti, in *PV in Europe, Rome*, 2002.
16. Y. Tripanagnostopoulos, T. Nousia, M. Souliotis and P. Yianoulis, *Sol. Energ.*, 2002, **72**(3), 217–234.
17. Y. Tripanagnostopoulos, T. Nousia and M. Souliotis, in *16th EPSEC, Glasgow*, 2000.
18. Y. Tripanagnostopoulos, M. Souliotis, R. Battisti and A. Corrado, *PIP*, 2006, **14**, 65–76.
19. U. Eicker, *Solar Technologies for Buildings*, Wiley, New York, 2003.
20. S. C. Solanki, S. Dubey and A. Tiwari, *Applied Energy*, 2009, in press, doi:10.1016/j.apenergy.2009.03.013.
21. H. P. Garg and R. S. Adhikari, *Renew. Energ.*, 1997, **11**(3), 363–385.
22. C. Bollo, A. Pappalardo and G. Marletta , in *13th EPSEC, Nice*, 1995.
23. B. J. Brinkworth, B. M. Cross, R. H. Marshall and Y. Hongxing, *Sol. Energ.*, 1997, **61**(3), 169–178.
24. P. Wouters and L. Vandaele, *PV-Hybrid Final Report*, 1998.
25. D. Infield, L. Mei, U. Eicker and V. Fux, in *ISES Solar World Congress, Jerusalem*, 1999.
26. D. Infield, L. Mei, U. Eicker and V. Fux, in *16th EPSEC, Glasgow*, 2000.
27. D. Infield, L. Mei and U. Eicker, *Sol. Energ.*, 2004, **76**, 93–98.
28. D. Infield, L. Mei, W. M. Lee and D. L. Loveday, in *Third WCPEC, Osaka*, 2003.
29. D. Infield, *Final Report for Project JOR3-CT97-0185*, 2000.
30. D. Infield, L. Mei, P. Rowley and U. Eicker, in *ISES Solar World Congress*, 2001.
31. A. K. Bhargava, H. P. Garg and R. K. Agarwal, *Energ. Convers. Manag.*, 1991, **31**, 471–479.
32. K. Nagano, T. Mochida, K. Shimakura and M. Murashita, in *ISES Solar World Congress, Adelaide*, 2001.
33. K. Nagano, T. Mochida, K. Shimakura, M. Murashita and S. Takeda, *SEMS*, 2003, **77**, 265–82.
34. M. D. Bazilian, D. Prasad, P. Erling and F. Leenders, in *PLEA 2001, Florianopolis*, 2001.
35. M. D. Bazilian and D. Prasad, in *ISES Solar World Congress, Adelaide, Australia*, 2001.
36. A. Gandini, L. Mazzarella and J. J. Bloem, *DAME-BC, Ispra*, 2003.
37. F. J. Crick, A. Wilshaw, N. M. Pearsall, K. Hynes, M. Shaw and G. Young, in *14th EPSEC, Barcelona*, 1997.

38. S. Dubey, S. C. Solanki and A. Tiwari, *Energ. Build.*, 2009, in press, doi:10.1016/j.enbuild.2009.03.010.
39. T. T. Chow, J. W. Hand and P. A. Strachan, *Applied Thermal Energy*, 2003, **23**, 2035–2049.
40. A. Guiavarch and B. Peuportier, *Sol. Energ.*, 2006, **80**(1), 65–77.
41. A. Tsangrassoulis, M. Santamouris and D. Asimakopoulos, *Energ. Build.*, **24**, 223–230.
42. A. K. Athienitis and A. Tzempelikos, *Sol. Energ.*, 2002, **72**, 271–281.
43. S. C. Molteni, G. Gourret, B. Paule, L. Michel and J. L. Scartezzini, *Sol. Energ.*, 2001, **69**, 117–129.
44. J. L. Scartezzini and G. Courret, *Sol. Energ.*, 2002, **73**, 123–135.
45. V. Jirka, V. Kuceravy, M. Maly, J. Pokorny and E. Rehor, in *Proceedings of World Renewable Energy Congress V, Part III*, 1998, 1595–1598.
46. Y. Tripanagnostopoulos, C. Siabekou and J. K. Tonui, *Sol. Energ.*, 2007, **81**, 661–675.
47. T. N. Anderson, M. Duke, G. L. Morrison and J. K. Carson, *Sol. Energ.*, 2009, **83**, 445–455.
48. N. Aste, G. Chiesa and F. Verri, *Renew. Energ.*, 2008, **33**, 914–927.
49. M. Wolf, *Energ. Convers. Manag.*, 1976, **16**, 79–90.
50. D. L. Evans and W. A. Facinelli, *Report ASU ERC-T-78017*, 1978.
51. L. W. Florschuetz, in *Sharing the Sun Joint Conference, ISES, Winnipeg*, 1976.
52. L. W. Florschuetz, *Sol. Energ.*, 1979, **22**, 361–366.
53. E. C. Kern and M. C. Russell, in *13th IEEE, Washington*, 1978.
54. E. C. Kern and M. C. Russell, *Report, MIT*, 1978.
55. D. B. Sheldon and M. C. Russel, in *Fifth Annual Heat Pump Technology Conference, Stillwater*, 1980.
56. D. R. Smith, K. L. Biringer and D. A. Pritchard, in *13th IEEE, Washington*, 1978.
57. T. J. Lambarski, in *17th IEEE, Kissimmee*, 1984.
58. J. W. Andrews, *Report, Brookhaven*, 1981.
59. S. Gasner and L. Wen, in *ASME Conference, Albuquerque*, 1982.
60. Y. Tripanagnostopoulos, S. Tselepis, M. Souliotis and J. K. Tonui, in *19th EPSEC, Paris*, 2004.
61. Y. Tripanagnostopoulos, T. Nousia and M. Soulioti, in *Fifth WREC, Florence*, 1998.
62. S. Tselepis and Y. Tripanagnostopoulos, in *PV in Europe, Rome*, 2002.
63. S. A. Kalogirou, *Renew. Energ.*, 2001, **23**, 247–260.
64. P. G. Charalambous, S. A. Kalogirou, G. Maidment and T. G. Karayiannis, *Photovoltaic Thermal (PV/T) Collectors: A Review*, HPC, Cyprus, 2004.
65. P. G. Charalambous, G. Maidment, S. A. Kalogirou and K. Yiakoumetti, *Applied Thermal Energy*, 2007, **27**, 275–86.
66. Y. Tripanagnostopoulos and S. Kalogirou, in *Eighth WREC, Denver*, 2004.
67. S. Kalogirou and Y. Tripanagnostopoulos, *WREC, Aberdeen*, 2005.

68. B. J. Huang, T. H. Lin, W. C. Hung and F. S. Sun, in *ISES Solar World Congress*, Jerusalem, 1999.
69. B. J. Huang, T. H. Lin, W. C. Hung and F. S. Sun, *Sol. Energ.*, 2001, **70**(5), 443–448.
70. A. Tiwari and M. S. Sodha, *Sol. Energ.*, 2006, **80**, 751–759.
71. T. T. Chow, W. He and J. Ji, *Sol. Energ.*, 2006, **80**(3), 298–306.
72. W. He, T. T. Chow, J. Ji, J. P. Lu, G. Pei and L. Chan, *Appl. Energ.*, 2006, **83**, 199–210.
73. J. Ji, J. P. Lu, T. T. Chow, W. He and G. Pei, *Appl. Energ.*, 2007, **84**, 222–237.
74. J. Ji, J. Han, T. T. Chow, C. Han, J. Lu and W. He, *J. Power Energ.*, 2006, **220**, 681–688.
75. C. Cristofari, G. Notton and J. L. Canalett, *Sol. Energ.*, 2009, in press, doi:10.1016/j.solener.2009.01.008.
76. H. A. Zondag, D. W. De Vries, W. G. J. Van Helden, R. J. C. Van Zolingen and A. A. Van Steenhoven, *Sol. Energ.*, 2002, **72**, 113–128.
77. B. Sandnes and J. Rekstad, *Sol. Energ.*, 2002, **72**, 63–73.
78. D. W De Vries, PhD Report, EUT, 1998.
79. G. Fraisse, C. Ménézo and K. Johannes, *Sol. Energ.*, 2007, **81**, 1426–1438.
80. T. T. Chow, W. He, A. L. S. Chan, K. F. Fong, Z. Lin and J. Ji, *Appl. Therm. Eng.*, 2008, **28**, 1356–1364.
81. T. T. Chow, W. He, J. Ji and A. L. S. Chan, *Sol. Energ.*, 2007, **81**, 123–130.
82. R. Petela, *J. Heat Trans.*, 1964, **86**, 187–192.
83. D. C. Spanner, *Introduction to Thermodynamics*, Academic Press, London, 1964.
84. S. M. Jeter, *Sol. Energ.*, 1981, **26**, 231–236.
85. H. A. Zondag, M. J. M. Jong and W. G. J. Van Helden, in *17th EPSEC 2001*, Munich, 2001.
86. M. J. M. Jong, in *11th Symposium Thermische Solarenergie*, Staffelstein, 2001.
87. H. A. Zondag, N. Bristow, E. Gramsbergen, W. G. J. Van Helden, D. Lloyd-Jones and S. Nelson, in *ESTEC 2005*, Freiburg, 2005.
88. S. Kalogirou and Y. Tripanagnostopoulos, *WREC*, Aberdeen, 2005.
89. A. Kribus, D. Kaftori, G. Mittelman, A. Hirshfeld, Y. Flitsanov and A. Dayan, *Energ. Convers. Manag.*, 2006, **47**, 3582–3590.
90. G. Mittelman, A. Kribus and A. Dayan, *Energ. Convers. Manag.*, 2007, **48**, 2481–2490.
91. T. T. Chow, A. L. S. Chan, K. F. Fong, Z. Lin, W. He and J. Ji, *Appl. Energ.*, 2009, **86**, 689–696.
92. J. Ji, J. Han, T. T. Chow, H. Yi, J. Lu and W. He, *Energ. Build.*, 2006, **38**, 1380–1387.
93. J. Ji, G. Pei, T. T. Chow, K. Liu, H. He, J. Lu and C. W. Han, *Sol. Energ.*, 2008, **82**, 43–52.
94. E. Erdil, M. Ilkan and F. Egelioglu, *Energy*, 2008, **33**, 1241–1245.

95. Y. B. Assoa, C. Ménézo, G. Fraisse, R. Yezou and J. Brau, *Sol. Energ.*, 2007, **81**, 1132–1143.
96. B. Robles-Ocampo, E. Ruýiz-Vasquez, H. Canseco-Sánchez, R. C. Cornejo-Meza, G. Trápaga-Martý′nez, F. J. García-Rodríguez, J. González-Hernández and Y. V. Vorobiev, *Sol. Energ. Mater. Sol. Cell.*, 2007, **91**, 1966–1971.
97. J. Ji, T. T. Chow and W. He, *Build. Environ.*, 2003, **38**, 1327–1334.
98. P. Affolter, A. Haller, D. Ruoss and P. Toggweiler, *16th EPSEC*, Glasgow, 2000.
99. New generation of hybrid solar PV/T collectors, Report DIS 56360/16868, 2000.
100. R. Platz, D. Fischer, M. A. Zufferey, J. A. Anna Selvan, A. Haller and A Shah, in *26th PVSC*, Anaheim, CA, 1997.
101. T. T. Chow, *Sol. Energ.*, 2003, **75**, 143–152.
102. A. T. Naveed, E. C. Kang and E. J. Lee, *J. Sol. Energ. Eng.*, 2006, **128**, 349–353.
103. S. C. W. Krauter and F. Ochs, in *Third WCPEC*, Osaka, 2003.
104. S. C. W. Krauter, *SEMS*, 2004, **82**, 119–130.
105. S. C. W. Krauter, in *19th EPSEC*, Paris, 2004.
106. E. Skoplaki and J. A. Palyvos, *Sol. Energ.*, 2009, **83**(5), 614–624.
107. H. A. Zondag, *Renew. Sustain. Energ. Rev.*, 2008, **12**(4), 891–959.
108. G. Notton, C. Cristofari, M. Mattei and P. Poggi, *Appl. Therm. Eng.*, 2005, **25**, 2854–2877.
109. D. L. Evans, *Sol. Energ.*, 1981, **27**, 555–560.
110. D. L. Evans and L. W. Florschuetz, *Sol. Energ.*, 1977, **19**, 255–262.
111. G. W. Hart and P. Raghuraman, *MIT Report*, 1982, DOE/ET/20279-202.
112. H. P. Garg and R. K. Agarwal, *Energ. Convers. Manag.*, 1995, **36**, 87–99.
113. D. L. Evans and L. W. Florschuetz, *Sol. Energ.*, 1978, **20**, 37–43.
114. J. W. Stultz and L. C. Wen, *LSA Task Report*, 1977, **5**, 101–131.
115. M. D. Siegel, S. A. Klein and W. A. Beckman, *Sol. Energ.*, 1981, **26**, 413–418.
116. R. W. Taylor, *Sol. Cell.*, 1986, **18**, 335–344.
117. OTA – Office of Technology Assessment, *Application of Solar Technology to Today's Energy Needs, Energy Conversion with Photovoltaics*, Princeton, 1978, vol. 10, p. 406.
118. N. T. Truncellito and A. J. Sattolo, *General Electric Advanced Energy Department*, 1979.
119. R. Mertens, in *Proceedings of UK-ISES Conference on C21 Photovoltaic Solar Energy Conversion*, September, 1979, p. 65.
120. L. Barra and D. Coiante, *Sol. Energ.*, 1993, **51**, 383–389.
121. J. Prakash, *Energ. Convers. Manag.*, 1994, **35**, 967–972.
122. A. A. Hegazy, *Energ. Convers. Manag.*, 2000, **41**, 861–881.
123. T. Yamawaki, S. Mizukami, T. Masui and H. Takahashi, *Sol. Energ. Mater. Sol. Cell.*, 2001, **67**, 369–377.
124. RETScreen, *International Photovoltaic Project Analysis*, 22, 2001.

125. K. Nagano, T. Mochida, K. Shimakura, K. Murashita and S. Takeda, *Sol. Energ. Mater. Sol. Cell.*, 2003, **77**, 265–282.
126. A. Tiwari and M. S. Sodha, *Renew. Energ.*, 2006, **31**, 2460–2474.
127. J. K. Tonui and Y. Tripanagnostopoulos, *Renew. Energ.*, 2007a, **32**, 623–637.
128. J. K. Tonui and Y. Tripanagnostopoulos, *Sol. Energ.*, 2007b, **81**, 498–511.
129. M. Y. Othman, B. Yatim, K. Sopian and M. N. A. Bakar, *Desalination*, 2007, **209**, 43–49.
130. H. D. Mohring, D. Stellbogen, R. Schäffler, S. Oelting, R. Gegenwart, P. Konttinen, T. Carlsson, M. Cendagorta and W. Hermann, in *Proceedings of 19th EC Photovoltaic Solar Energy Conference*, 2004, June 7–11, Paris, France, presentation 5CO.3.1.
131. N. M. Ravindra and V. K. Srivastava, *Sol. Cell.*, 1979/80, **1**, 107–109.
132. N. Aste, G. Chiesa and F. Verri, *Renew. Energ.*, 2008, **33**, 914–927.
133. C. Cristofari, P. Poggi, G. Notton and M. Muselli, in *Proceedings of Sixth IASTED International Conference on "Modeling, Simulation, and Optimization*", Gaborone, Botswana, 2006, September 11–13, 273–278.
134. W. R. Anis, R. P. Mertens and R. J. van Overstraeten, in *Proceedings of Fifth EC Photovoltaic Solar Energy Conference*, Athens, Greece, 1983, October 12–16, 520–524.
135. CLEFS CEA, *Influence of Temperature on Photovoltaic Module Efficiency*, 2004, **50/51**, 119.
136. H. G. Beyer, J. Bethke, A. Drews, D. Heinemann, E. Lorenz, G. Heilscher and S. Bofinger, in *19th EC Photovoltaic Solar Energy Conference*, Paris, France, June 7–11, 2004, http://www.pvsat.de/paper_beyer.pdf, accessed 18 December 2008.
137. J. Perlman, A. McNamara and D. Strobino, in *Proceedings of ISES Solar World Congress "Bringing Water to the World*", Orlando, FL, August 6–12 2005.
138. M. Bazilian and D. Prasad, *Renew. Energ.*, 2002, **27**, 57–68.
139. F. Lasnier and T. G. Ang, *Photovoltaic Engineering Handbook*, Adam Hilger, New York, NY, 1990, p. 80.
140. Q. Kou, S. A. Klein and W. A. Beckman, *Sol. Energ.*, 1998, **64**, 33–40.
141. J. A. Duffie and W. A. Beckman, *Solar Energy Thermal Processes*, Wiley, Hoboken, New Jersey, 2006.
142. T. Hove, *Renew. Energ.*, 2000, **21**, 207–229.
143. T. Yamaguchi, Y. Okamoto and M. Taberi, *Sol. Energ. Mater. Sol. Cell.*, 2003, **75**, 597–601.
144. Z. Zhu, H. Yang, R. Jiang and Q. Wu, *Heat Tran. Asian Res.*, 2004, **33**, 117–128.
145. T. Bergene and O. M. Lovvik, *Sol. Energ.*, 1995, **55**, 453–462.
146. W. Durisch, J. Urban and G. Smestad, *Renew. Energ.*, 1996, **8**, 359–366.
147. M. R. Patel, *Wind and Solar Power Systems*, CRC Press, Boca Raton, FL, 1999.
148. H. Yang, J. Burnett and J. Ji, *Energ. Build.*, 2000, **31**, 285–290.

149. V. V. Risser and M. K. Fuentes, in *Proceedings of Fifth EC Photovoltaic Solar Energy Conference*, Athens, Greece, 1983, October 12–16, pp. 623–627.
150. J. Jie, Y. Hua, H. Wei, P. Gang, L. Jianping and J. Bin, *Build. Environ.*, 2007, **42**, 1544–1552.
151. M. Buresch, *Photovoltaic Energy Systems*, McGraw-Hill, New York, 1983, p. 76.
152. J. Twidell and T. Weir, *Renewable Energy Resources*, E&FN Spon, London, UK, 1986, p. 160.
153. A. Parretta, A. Sarno and L. R. M. Vicari, *Optic. Comm.*, 1998, **153**, 153–163.
154. C. R. Osterwald, *Sol. Cell.*, 1986, **18**, 269–279.
155. A. M. Al-Sabounchi, *Renew. Energ.*, 1998, **14**, 149–155.
156. P. L. Zervas, H. Sarimveis, J. A. Palyvos and N. C. G. Markatos, *J. Power Sourc.*, doi:10.1016/j.jpowsour.2007.11.067.
157. P. W. Wong, Y. Shimoda, M. Nonaka, M. Inoue and M. Mizuno, *J. Asian Architect. Build. Eng.*, 2005, **4**, 549–556.
158. S. D. Hendrie, in *Proceedings of ISES Solar World Congress*, Atlanta, GA, 1979, May 28–June 1, pp. 1865–1869.
159. K. Nishioka, T. Hatayama, Y. Uraoka, T. Fuyuki, R. Hagihara and M. Watanabe, *Sol. Energ. Mater. Sol. Cell.*, 2003, **75**, 665–671.
160. D. Menicucci and J. P. Fernandez, *SAND85-0376*, Sandia National Laboratories, Albuquerque, NM, 1988.
161. M. Fuentes, G. Nofuentes, J. Aguilera, D. L. Talavera and M. Castro, *Sol. Energ.*, 2007, **81**, 1396–1408.
162. B. Marion, *Progress in Photovoltaics: Research and Applications*, 2002, **10**, 205–214.
163. D. L. King, J.A. Kratochvil and W. E. Boyson, in *Proceedings of 26th IEEE Photovoltaic Specialists Conference*, Anaheim, CA, 1997, September 29–October 3.
164. B. Kroposki, W. Marion, D. L. King, W. E. Boyson and J. A. Kratochvil, in *Proceedings of 28th IEEE Photovoltaic Specialists Conference*, Anchorage, AK, September 16–22, 2000, pp. 1407–1411.
165. J. Jie, H. Wei and H. N. Lam, *Sol. Energ. Mater. Sol. Cell.*, 2002, **71**, 435–448.
166. B. K. Farmer, *PVUSA Model Technical Specification for a Turnkey Photovoltaic Power Syste*m, 1992, Appendix C, p. 2.
167. J. I. Rosell and M. Ibáñez, *Energ. Convers. Manag.*, 2006, **47**, 2424–2430.
168. K. Furushima, Y. Nawata and M. Sadatomi, in *ASES Conference SOLAR – Renewable Energy Key to Climate Recovery*, Denver, Colorado, 2006, July 7–13.
169. A. Mellit, S. A. Kalogirou, L. Hontoria and S. Shaari, *Renew. Sustain. Energ. Rev.*, 2008, in press, doi:10.1016/j.rser.2008.01.006.
170. L. R. Medsker, *J. Netw. Comput. Appl.*, 1996, **19**, 213–234.
171. E. Rich and K. Knight, *Artificial Intelligence*, McGraw-Hill, New York, 1996.

172. S. A. Kalogirou, *Prog. Energ. Combust. Sci.*, 2003, **29**, 515–566.
173. S. A. Kalogirou, *Artificial Intelligence in Energy and Renewable Energy Systems*, Nova Publisher, 2007.
174. K. W. Chau, *J. Environ. Manag.*, 2006, **80**, 47–57.
175. K. W. Chau, *Mar. Pollut. Bull.*, 2006, **52**, 726–33.
176. S. P. Kyung and H. K. Soung, *Artif. Intell. Eng.*, 1998, **12**, 121–134.
177. Y. Sukamongkol, S. Chungpaibulpatana and W. Ongsakul, *Renew. Energ.*, 2002, **27**, 237–258.
178. L. Hontoria, J. Aguilera and P. Zufiria, *Sol. Energ.*, 2005, **78**, 313–319.
179. S. Haykin, *Neural Networks: A Comprehensive Foundation*, Macmillan, New York, 1994.
180. L. A. Zadeh, *Fuzzy Sets Inform. Control*, 1965, **8**, 338–353.
181. L. A. Zadeh, *Cybernet*, 1972, **2**, 4–34.
182. J. H. Holland, *Adaptation in Natural and Artificial Systems*, University of Michigan Press, Ann Arbor, 1975.
183. C. J. Lakhmi and N. M. Martin, *Fusion of Neural Networks, Fuzzy Systems and Genetic Algorithms: Industrial Applications*, CRC Press, LLC, 1998.
184. B. Jawerth and W. Sweldens, *Soc. Industrial Application Mathematics*, 1994, **36**(3), 377–412.
185. F. Robert, *Neural Fuzzy Systems*, Abo Akademi University, 1995.
186. A. Tsakonas and G. Dounias, *SETN LNAI*, 2002, **2308**, 494–511.
187. ESTIF, *Sun in Action II – a solar thermal strategy for Europe*, 2003.
188. W. Weiss, I. Bergmann and G. Faninger, *Solar heat worldwide – markets and contribution to the energy supply*, SHC report, 2006.

CHAPTER 3

Solar Cell Materials and Their Characteristics

3.1 Introduction

A solar cell or photovoltaic (PV) cell is a device that converts solar energy into electricity by the photovoltaic effect. Photovoltaics is the field of technology and research related to the application of solar cells as solar energy. Sometimes the term solar cell is reserved for devices intended specifically to capture energy from sunlight, while the term photovoltaic cell is used when the source is unspecified. Photovoltaic generation of power is caused by radiation that separates positive and negative charge carriers in an absorbing material. In the presence of an electric field, these charges can produce a current for use in an external circuit. Such fields exist permanently at junctions or inhomogeneities in materials as 'built-in' electric fields and provide the required e.m.f. for useful power production.

Junction devices are usually known as photovoltaic cells or solar cells, although the term is a misnomer in the sense that it is the current that is produced by the radiation photons and not the 'voltage'. The cell itself provides the source of electromagnetic force (e.m.f.). It is to be noted that photoelectric devices are electrical current sources driven by a flux of radiation. A majority of photovoltaic cells are silicon semi-conductor junction devices. Thus, in order to study the photovoltaic cells we should have an understanding of the basics of the semi-conductors; a brief description of which follows in the subsequent sections.

A solar cell constitutes the basic unit of a PV generator, which, in turn, is the main component of a solar generator. A PV generator is the total system consisting of all PV modules which are connected in series or parallel or a combination of both series and parallel with each other.

RSC Energy Series No. 2
Fundamentals of Photovoltaic Modules and Their Applications
By G. N. Tiwari and Swapnil Dubey

Published by the Royal Society of Chemistry, www.rsc.org

Solids can be divided into three categories, on the basis of electricity conduction through them. They are: conductors, semi-conductors and insulators. The gap between the valence band and the conduction band (forbidden energy band) in the case of insulators ($h\upsilon < E_g$, h is the Planck constant and υ is the frequency) is very large. Thus it is not possible for the electrons in the valence band to reach the conduction band; hence there is no conduction of current. In the case of semi-conductors ($h\upsilon > E_g$), the gap is moderate and the electrons in the valance band may acquire energy sufficient enough for them to cross the forbidden (Figure 3.1) region. While, in the case of conductors ($E_g \approx 0$), no forbidden gap exists and electrons can easily move to the conduction band.

The semi-conductor can again be divided into two categories: intrinsic and extrinsic. Intrinsic (pure) semi-conductors have a Fermi-level in the middle of the conduction and valence band. In this case the densities of free electrons in the conduction band and free holes in the valence band are equal $n = p = n_i$ and each is proportional to exp $(-E_g/2kT)$.

Example 3.1

Determine the band gap in a silicon crystal at 40 °C.

Solution

The variation of band gap with temperature is given by the relation:

$$E_g(T) = E_g(0) - \frac{aT^2}{T + b}$$

where, a and b for different materials are as follows:

Material	$E_g(0)$	a	b
Silicon (Si)	1.16 eV	7×10^{-4} eV K^{-1}	1100 K
Gallium arsenide (GaAs)	1.52 eV	5.8×10^{-4} eV K^{-1}	300 K

Substituting the appropriate values in the above equation, we get

$$E_g(T) = 1.16 - \frac{7 \times 10^{-4} \times (313)^2}{313 + 1100} = 1.11 \text{ eV}$$

Solar cells are classified into three generations, which indicate the order in which each became prominent. At present there is concurrent research into all three generations while the first-generation technologies are most highly represented in commercial production, accounting for 89.6% of 2007 production.[1]

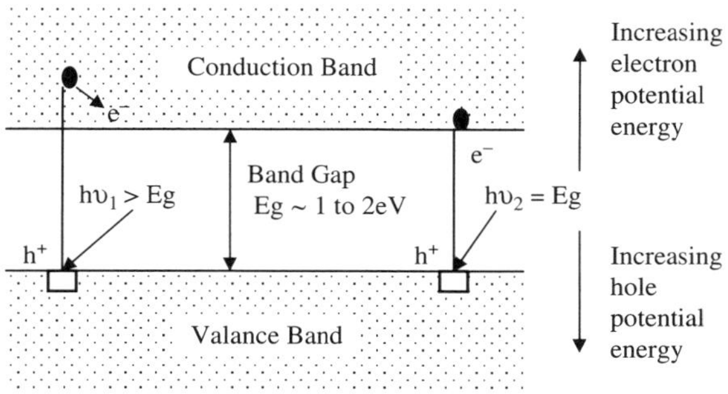

Figure 3.1 Semi-conductor band structure of intrinsic material. Photon absorption $h\upsilon < E_g$, no photoelectric absorption. $h\upsilon_1 - E_g$, excess energy dissipated as heat. $h\upsilon_2 = E_g$, photon energy equals band gap.

3.1.1 First Generation

First-generation cells consist of large-area, high-quality and single junction devices. First-generation technologies involve high energy and labour inputs which prevent any significant progress in reducing production costs. Single junction silicon devices are approaching the theoretical limiting efficiency of 33%[2] and achieve cost parity with fossil fuel energy generation after a pay back period of 5–7 years.

3.1.2 Second Generation

Second-generation materials have been developed to address energy requirements and production costs of solar cells. Alternative manufacturing techniques such as vapour deposition and electroplating are advantageous as they reduce high-temperature processing significantly. It is commonly accepted that as manufacturing techniques evolve, production costs will be dominated by constituent material requirements,[2] whether this be a silicon substrate or a glass cover. Second-generation technologies are expected to gain market share in 2008.[1]

The most successful second-generation materials have been cadmium telluride (CdTe), copper indium gallium selenide, amorphous silicon and micromorphous silicon.[1] These materials are applied in a thin film to a supporting substrate such as glass or ceramics, reducing material mass and therefore costs. These technologies do hold promise of higher conversion efficiencies and offer significantly cheaper production costs.

3.1.3 Third Generation

Third-generation technologies aim to enhance poor electrical performance of second-generation thin-film technologies while maintaining very low production costs. Current research is targeting conversion efficiencies of 30–60% while retaining low cost materials and manufacturing techniques.[2] There are a few approaches to achieving these high efficiencies:[3]

1. Multijunction photovoltaic cell
2. Modifying incident spectrum (concentration)
3. Use of excess thermal generation to enhance voltages or carrier collection

3.2 Doping

In order to increase the conductivity of intrinsic semi-conductors, controlled quantities of specific impurity ions are added to the intrinsic semi-conductor to produce doped (extrinsic) semi-conductors. Impurity ions of valency less than the semi-conductor enter the semi-conductor lattice and become electron acceptor sites that trap free electrons. These traps have an energy level within the band gap, but near the valence band. The absence of free electrons produces positively charged states called holes that also move through the material as free carriers. Such a material is called a *p*-type material, having holes as majority carriers and electrons as minority carriers. If impurity ions of a valency greater than that of the semi-conductor are added then an *n*-type material results, which has electrons as majority carriers and holes as minority carriers. Both *p*- and *n*-type extrinsic semi-conductors have higher electrical conductivity than the intrinsic basic material.

3.3 Fermi Level

The Fermi level is the apparent energy level within the forbidden band gap from which majority carriers (electrons in *n*-type and holes in *p*-type) are excited to become charge carriers. The probability for the majority carrier excitation varies as $\exp[-e\varphi/(kT)]$, where e is the charge of the electron and hole and φ is the electric potential difference between the Fermi level and the valence or conduction band, T is the temperature (K) and k is the Boltzmann constant, 1.38×10^{-23} J/K.

For an *n*-type material:

$$E_{\mathrm{F}} = E_{\mathrm{c}} + \mathrm{kT} \ln \frac{N_0}{N_{\mathrm{c}}} \tag{3.1}$$

where E_{F} is the Fermi-energy level, E_{c} the conduction band energy; k the Boltzmann constant; N_0 the donor concentration and N_{c} the effective density of states in conduction band, and is constant at fixed temperature T.

For p type material,

$$E_{\mathrm{F}} = E_V - \mathrm{kT} \ln \frac{N_{\mathrm{A}}}{N_{\mathrm{V}}} \tag{3.2}$$

where E_{V} is the valence band energy, N_{A} is the acceptor ion concentration and N_{V} is the effective density of states in the valence band.

Example 3.2

Calculate the shift in Fermi energy level in a silicon crystal doped with a V group impurity of concentration 10^{15} cm^3, given that the effective density of states in the conduction band is 2.82×10^{19} cm^3 and the band gap is 1.1 eV; room temperature is 27 °C.

Solution

From eqn 3.1, we have

$$E_F = E_C + kT \ln(N_D/N_C)$$

If the valence band is taken as the reference level, then $E_C = 1.1$ eV. Substitution of the values gives

$$\begin{aligned} E_F &= 1.1 + (1.38 \times 10^{-23}/1.6 \times 10^{-19}) \times 300 \ln(10^{15}/2.82 \times 10^{19}) \\ &= 1.1 - 0.1152 = 0.9848\,\text{eV} \end{aligned}$$

The shift is $0.9848 - 0.55 = 0.4348$ eV.

3.4 *p-n* Junction

The basic requirement for photovoltaic energy conversion is an electronic asymmetry in the semi-conductor structure known as a junction. When *n*- and *p*-type semi-conductors are brought in contact, then electrons from the *n*-region near the junction would flow to the *p*-type semi-conductor, leaving behind a layer which is positively charged. Similarly holes will flow in the opposite direction leaving behind a negatively charged layer. A steady state is finally reached, resulting in a junction, which contains practically no mobile charges, hence the name depletion region.

The *p-n* junction (Figures 3.2 and 3.3) may be connected to a battery in two ways: (i) in forward bias (Figure 3.4a), the positive conventional circuit current

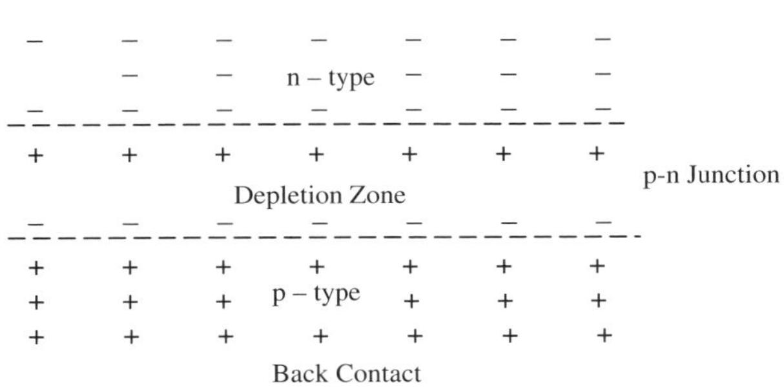

Figure 3.2 *p-n* junction energy levels in a *p-n* junction.

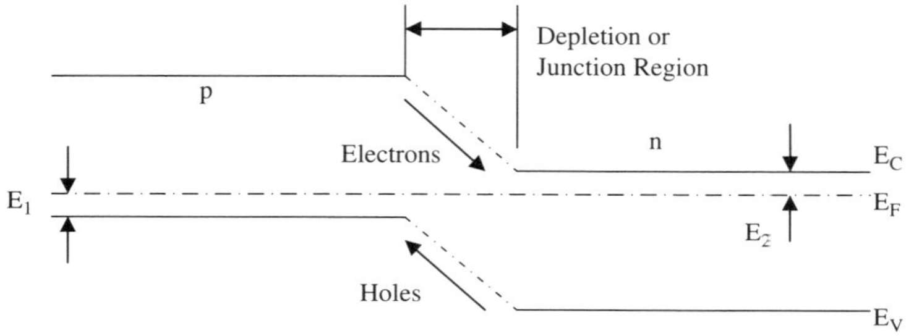

Figure 3.3 Energy levels in a *p-n* junction.

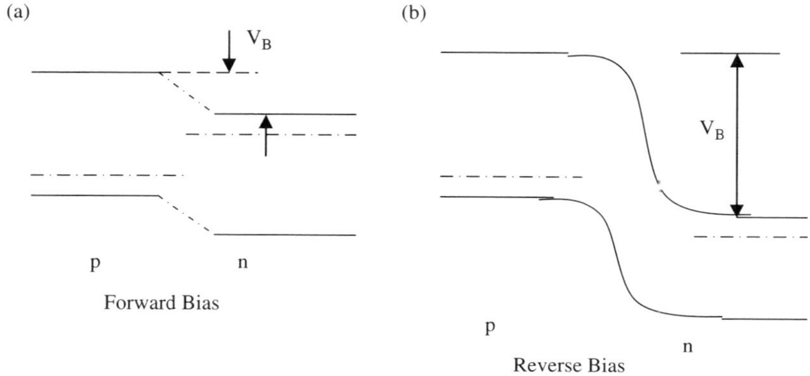

Figure 3.4 Energy levels in a *p-n* junction with (a) forward bias and (b) reverse bias.

passes from the p to the n material across a reduced-band potential difference V_B, (ii) in reverse bias (Figure 3.4b), the conventional positive current has an increased-band potential difference V_B to overcome. Thermally or otherwise generated electrons and holes recombine after a typical relaxation time τ, having moved a typical diffusion length L through the lattice. In intrinsic material the relaxation time can be long, $\tau \sim 1$s, but for commercial doped materials relaxation times are much shorter, $\tau \sim 10^{-2}$ to 10^{-8} s.

3.4.1 Forward Bias

Forward-bias occurs when the p-type semi-conductor material is connected to the positive terminal of a battery and the n-type semi-conductor material is connected to the negative terminal. With a battery connected this way, the holes in the p-type region and the electrons in the n-type region are pushed towards the junction. This reduces the width of the depletion zone. The positive charge applied to the p-type material repels the holes, while the negative charge applied to the n-type material repels the electrons. As electrons and holes are pushed towards the junction, the distance between them decreases. This lowers the barrier in the potential. With increasing forward-bias voltage, the depletion zone eventually becomes thin enough that the zone's electric field can't counteract the charge carrier motion across the *p-n* junction, consequently reducing electrical resistance. The electrons which cross the *p-n* junction into the *p*-type material (or holes which cross into the *n*-type material) will diffuse in the near-neutral region. Therefore, the amount of minority diffusion in the near-neutral zones determines the amount of current that may flow through the diode.

3.4.2 Reverse Bias

Connecting the *p*-type region to the negative terminal of the battery and the *n*-type region to the positive terminal produces the reverse-bias effect. Because the *p*-type material is now connected to the negative terminal of the power supply, the 'holes' in the *p*-type material are pulled away from the junction, causing the width of the depletion zone to increase. Similarly, because the *n*-type region is connected to the positive terminal, the electrons will also be pulled away from the junction. Therefore the depletion region widens, and does so increasingly with increasing reverse-bias voltage. This increases the voltage barrier, causing a high resistance to the flow of charge carriers thus allowing minimal electric current to cross the *p-n* junction. The strength of the depletion zone electric field increases as the reverse-bias voltage increases. Once the electric field intensity increases beyond a critical level, the *p-n* junction depletion zone breaks down and current begins to flow, usually by either the Zener or the avalanche breakdown processes. Both of these breakdown processes are non-destructive and are reversible, so long as the amount of current flowing does not reach levels that cause the semi-conductor material to overheat and cause thermal damage.

Electrons and holes may be generated thermally or by light, and become carriers in the material (Figure 3.5). Minority carriers in the depletion region are pulled across electrostatically down their respective potential gradients. The

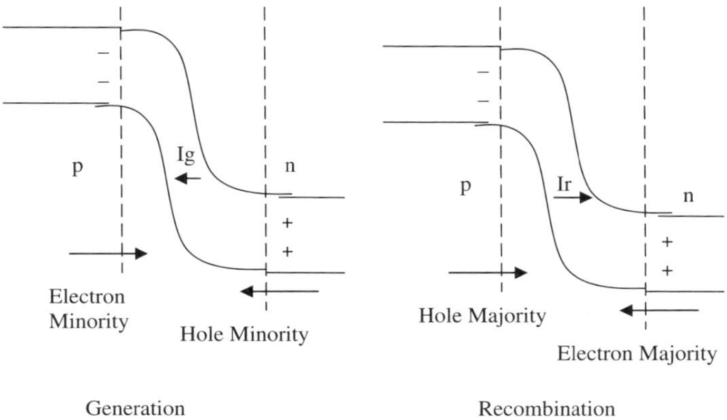

Figure 3.5 Generation and recombination currents at *p-n* junction.

minority carriers that cross the region become majority carriers in the adjacent layer. The passage of these carriers causes the generation current, I_g, which is mainly controlled by temperature in a given junction without illumination.

In an isolated junction, there can be no overall imbalance of current across the depletion region. Thus, a reverse recombination current I_r of equal magnitude occurs from the bulk material, which restores the normal internal electric field. The band potential V_B is slightly reduced by I_r. The recombination current I_r can be varied by external bias as explained earlier (Figure 3.6).

3.5 *p-n* Junction Characteristics

The *p-n* junction characteristics have been given in Figure 3.7.

With no external bias ($V = 0$).

$$I_r = I_g \tag{3.3}$$

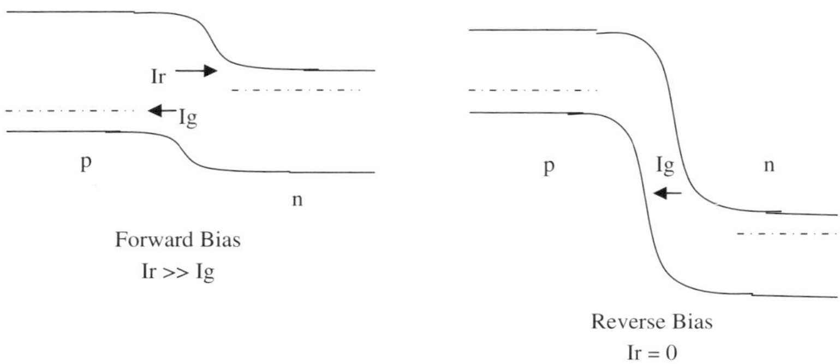

Figure 3.6 Generation and recombination currents with external bias.

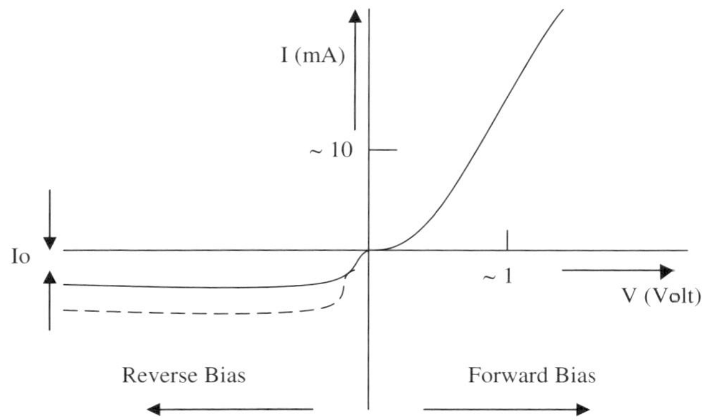

Figure 3.7 *p-n* junction dark characteristics.

with a forward bias of voltage V, the recombination current becomes an increased forward current.

$$I_r = I_g \exp(eV/(kT)) \tag{3.4}$$

The total current (with no illumination) is

$$I_D = I_r - I_g = I_g[\exp(eV/kT) - 1] \tag{3.5}$$

The above equation is the Shockley equation and can be written as

$$I_D = I_0[\exp(eV/kT) - 1] \tag{3.6}$$

where $I_0(=I_g)$ is the saturation current under reverse bias, before avalanche breakdown occurs. It is also known as leakage or diffusion current. For good solar cells $I_0 \sim 10^{-8}\,A\,m^{-2}$. Its value increases with temperature (Figure 3.7, dotted curve).

Example 3.3

Determine the value of saturation current for silicon at 40 °C.

Solution

The dependence of saturation current on temperature is given by the relation:

$$I_0 = AT^3 \exp(E_g/kT)$$

Here, A is the non-ideality factor and its value is taken as 1,

$$E_g = 1.11\,eV = 1.11 \times 1.6 \times 10^{-19}\,J$$

Substituting the known values in the above equation, we get

$$I_0 = (40 + 273)^3 \exp\left(-\frac{1.11 \times 1.6 \times 10^{-19}}{1.38 \times 10^{-23} \times 313}\right) = 4.26 \times 10^{-11}\,A\,m^{-2}$$

Example 3.4

Determine the value of dark current in the limiting case $V \to 0$.

Solution

From eqn 3.6:

as $V \to 0$, exp $(eV/kT) \to 1$ and hence dark current $I_D \to 0$.

3.6 Photovoltaic Effect

When the solar cell (*p-n* junction) is illuminated, electron-holes pairs are generated and acted upon by the internal electric fields, resulting in a photo current (I_L). The generated photocurrent flows in a direction opposite to the forward dark current. Even in the absence of external applied voltage, this photocurrent continues to flow, and is measured as the short circuit current (I_{sc}). This current depends linearly on the light intensity, because absorption of more light results in additional electrons flowing in the internal electric field force.

The overall cell current I is determined by subtracting the light induced current I_L from the diode dark current I_D.

$$I = I_D - I_L \tag{3.7}$$

$$\text{Then, } I = I_0\left[\exp\left(\frac{eV}{kT}\right) - 1\right] - I_L \tag{3.8}$$

This phenomenon is called the photovoltaic effect.

Example 3.5a

Determine the value of the overall cell current in the limiting case $V \rightarrow 0$.

Solution

From eqn (3.8)
as $V \rightarrow 0$, exp (eV/kT) $\rightarrow 1$ and hence, $I \rightarrow -I_L$.

Example 3.5b

Find out the voltage for zero overall cell current.

Solution

Substituting $I = 0$ in eqn (3.8), we get

$$I_0\left[\exp\left(\frac{eV}{kT}\right) - 1\right] - I_L = 0$$

$$\exp\left(\frac{eV}{kT}\right) = \frac{I_L}{I_0} + 1$$

$$V = \frac{kT}{e}\ln\left[\frac{I_L}{I_0} + 1\right]$$

3.7 Photovoltaic Material

Solar cells are made of various materials and with different structures in order to reduce the cost and achieve maximum efficiency. There are various types of solar cell material, single crystal, polycrystalline and amorphous silicon, compound thin-film material and other semi-conductor absorbing layers, which give highly efficient cells for specialized applications.

Crystalline silicon cells are most popular, though they are expensive. The amorphous silicon thin-film solar cells are less expensive. The amorphous silicon layer is used with both hydrogen and fluorine incorporated in the structure. These a-Si: F: H alloys have been produced by the glow discharge decomposition of SiF_4 in the presence of hydrogen. The efficiency of an a-Si module is about 6–8%.

A variety of compound semi-conductors can also be used to manufacture thin-film solar cells. These compound materials are $CuInSe_2$, CdS, CdTe, Cu_2S and InP. The $CuInSe_2$ solar cell stability appears to be excellent. The combinations of different band gap materials in tandem configurations lead to photovoltaic generators of much higher efficiencies.

3.7.1 Silicon

The most prevalent bulk material for solar cells is crystalline silicon (c-Si), also known as 'solar grade silicon'. Bulk silicon is separated into multiple categories according to crystallinity and crystal size in the resulting ingot, ribbon or wafer.

1. Monocrystalline silicon (c-Si): often made using the Czochralski process. Single-crystal wafer cells tend to be expensive and, because they are cut from cylindrical ingots, do not completely cover a square solar-cell module without a substantial waste of refined silicon. Hence most c-Si panels have uncovered gaps at the corners of four cells.
2. Poly- or multicrystalline silicon (poly-Si or mc-Si): made from cast square ingots – large blocks of molten silicon carefully cooled and solidified. These cells are less expensive to produce than single-crystal cells but are less efficient. Polycrystalline silicon wafers are made by wire-sawing block-cast silicon ingots into very thin (180 to 350 micrometre) slices or wafers. The wafers are usually lightly *p*-type doped. To make a solar cell from the wafer, a surface diffusion of *n*-type dopants is performed on the front side of the wafer. This forms a *p-n* junction a few hundred nanometres below the surface.
3. Ribbon silicon: formed by drawing flat thin-films from molten silicon and having a multicrystalline structure. These cells have lower efficiencies than poly-Si, but save on production costs due to a great reduction in silicon waste, as this approach does not require sawing from ingots.

Anti-reflection coatings, which increase the amount of light coupled into the solar cell, are typically applied next. Over the past decade, silicon nitride has gradually replaced titanium dioxide as the anti-reflection coating of choice because of its excellent surface passivation qualities. It is typically applied in a layer several hundred nanometres thick using plasma-enhanced chemical vapour deposition (PE-CVD). Some solar cells have textured front surfaces that, like anti-reflection coatings, serve to increase the amount of light coupled into the cell. Such surfaces can usually only be formed on single-crystal silicon, though in recent years methods of forming them on multicrystalline silicon have been developed.

Silicon thin-films are mainly deposited by chemical vapour deposition (typically plasma-enhanced (PE-CVD)) from silane gas and hydrogen gas. Depending on the deposition's parameters, this can yield:

1. amorphous silicon (a-Si or a-Si:H)
2. protocrystalline silicon or
3. nanocrystalline silicon (nc-Si or nc-Si:H).

These types of silicon present dangling and twisted bonds, which results in deep defects (energy levels in the band gap) as well as deformation of the valence and conduction bands. The solar cells made from these materials tend to have lower energy conversion efficiency than bulk silicon, but are also less expensive to produce. The quantum efficiency of thin-film solar cells is also lower due to the reduced number of collected charge carriers per incident photon.

Amorphous silicon has a higher band gap (1.7 eV) than crystalline silicon (c-Si) (1.1 eV), which means it absorbs the visible part of the solar spectrum more strongly than the infrared portion of the spectrum. As nc-Si has about the same band gap as c-Si, the two materials can be combined in thin layers, creating a layered cell called a tandem cell. The top cell in a-Si absorbs the visible light and leaves the infrared part of the spectrum for the bottom cell in nanocrystalline Si.

Recently, solutions to overcome the limitations of thin-film crystalline silicon have been developed. Light trapping schemes, where the incoming light is obliquely coupled into the silicon and the light traverses the film several times, enhance the absorption of sunlight in the films. Thermal processing techniques enhance the crystallinity of the silicon and pacify electronic defects. The result is a new technology – thin-film Crystalline Silicon on Glass (CSG).[4] CSG solar devices represent a balance between the low cost of thin films and the high efficiency of bulk silicon.

A silicon thin-film technology is being developed for building integrated photovoltaics (BIPV) in the form of semi-transparent solar cells which can be applied as window glazing. These cells function as window tinting while generating electricity. Despite the numerous attempts at making better solar cells by using new and exotic materials, the reality is that the photovoltaics market is still dominated by silicon wafer-based solar cells (first-generation solar cells). The aim of the research is to achieve the lowest \$/watt solar cell design that is suitable for commercial production.

3.7.2 Cadmium Telluride (CdTe)

Cadmium telluride is an efficient light-absorbing material for thin-film solar cells. Compared to other thin-film materials, CdTe is easier to deposit and more suitable for large-scale production. Despite much discussion of the toxicity of CdTe-based solar cells, this is the only technology (apart from amorphous silicon) that can be delivered on a large scale, as shown by First Solar and Antec Solar. Other companies such as Primestar Solar, AVA Technologies as well as Arendi SRL have also started CdTe divisions respectively. There is a 40-megawatt plant in Ohio (USA) and a 10-megawatt plant in Germany. First Solar is scaling up to a 100-megawatt plant in Germany and has started building another 100-megawatt plant in Malaysia (2007).

The perception of the toxicity of CdTe is based on the toxicity of elemental cadmium, a heavy metal that is a cumulative poison. Scientific work, particularly by researchers of the National Renewable Energy Laboratories (NREL) in the USA, has shown that the release of cadmium to the atmosphere is lower with CdTe-based solar cells than with silicon photovoltaics and other thin-film solar cell technologies.[5]

3.7.3 Copper-Indium Selenide ($CuInSe_2$)

The materials based on $CuInSe_2$ that are of interest for photovoltaic applications include several elements from Groups I, III and VI in the periodic table. These semi-conductors are especially attractive for thin-film solar cell application because of their high optical absorption coefficients and versatile optical and electrical characteristics.

3.7.4 Gallium Arsenide (GaAs) Multijunction

High-efficiency cells have been developed for special applications such as satellites and space exploration. These multijunction cells consist of multiple thin films produced using molecular beam epitaxy. A triple-junction cell, for example, may consist of the semi-conductors GaAs, Ge and $GaInP_2$.[6] Each type of semi-conductor will have a characteristic band-gap energy which causes it to absorb light most efficiently at a certain colour or, more precisely, to absorb electromagnetic radiation over a portion of the spectrum. The semi-conductors are carefully chosen to absorb nearly the entire solar spectrum, thus generating electricity from as much of the solar energy as possible.

GaAs multijunction devices are the most efficient solar cells to date, reaching a record high of 40.7% efficiency under solar concentration and laboratory conditions.[7] These devices use 20 to 30 different semi-conductors layered in series.

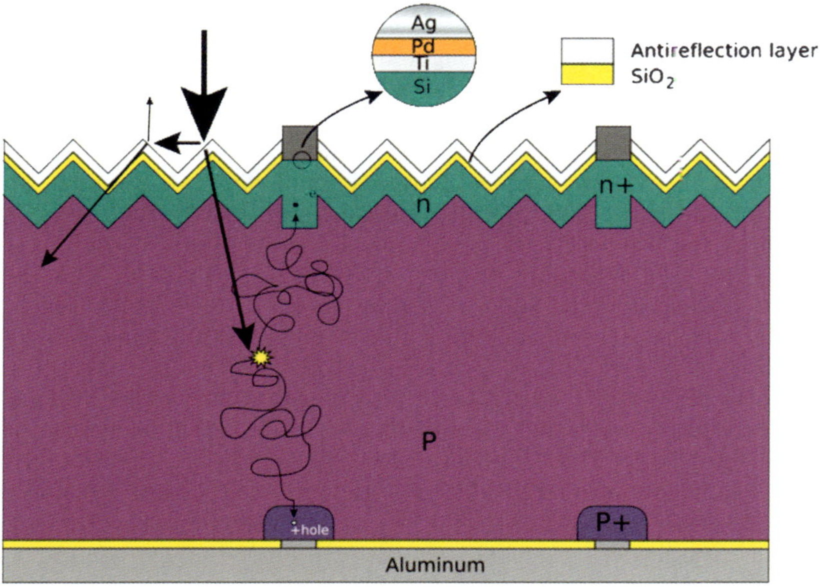

Figure 3.8 The structure of a silicon solar cell and working mechanism.[16]

3.7.5 Single Crystal Solar Cell

Single-crystalline solar cells made from high-purity material (solar grade) show excellent efficiencies and long-term stability but they are generally considered to be too expensive for large-scale mass production.

Figure 3.8 shows a diagram of a silicon solar cell structure and mechanism. The electric current generated in the semi-conductor is extracted by contact to the front and rear of the cell. The cell is covered with a thin layer of dielectric material, the anti-reflecting coating or ARC (to minimize the reflection from the top surface).

The total series resistance of the cell can be expressed as:

$$R_s = R_{cp} + R_{bp} + R_{cn} + R_{bn} \tag{3.9}$$

where R_{cp} is the metal contact to p-type semi-conductor resistance. R_{bp} is the bulk p-type resistance (bulk of p-type region is where most electron/hole pairs are generated by the absorption of light and where minority carriers (electrons) are transported by diffusion and partially lost by recombination), R_{cn} is the contact to n-type semi-conductor resistance and R_{bn} is the bulk n-type resistance.

The idealized junction current is given as

$$I = I_0\left[\exp\frac{e(V + IR_s)}{kT} - 1\right] \tag{3.10}$$

In addition, a shunt path may exist for current flow across the junction due to surface effect or poor junction region. This alternate path for current constitutes a shunt resistance R_p across the junction. Then

$$I = I_L - I_0\left[\exp\left(\frac{e(V - IR_S)}{AkT}\right) - 1\right] - \left(\frac{V - IR_S}{R_p}\right) \tag{3.11}$$

where A is an empirical non-ideality factor and is usually 1.

3.7.6 Light-absorbing Dyes

Typically a ruthenium metal organic dye (Ru-centred) is used as a monolayer of light-absorbing material. The dye-sensitized solar cell (DSSC) depends on a mesoporous layer of nanoparticulate titanium dioxide to greatly amplify the surface area (200–300 $m^2 g^{-1}$ TiO_2, as compared to approximately 10 $m^2 g^{-1}$ of flat single crystal). The photogenerated electrons from the light-absorbing dye are passed on to the n-type TiO_2, and the holes are passed to an electrolyte on the other side of the dye. The circuit is completed by a redox couple in the electrolyte, which can be liquid or solid.[8]

This type of cell allows a more flexible use of materials, and is typically manufactured by screen printing, with the potential for lower processing costs than those used for bulk solar cells. However, the dyes in these cells also suffer from degradation under heat and UV light, and the cell casing is difficult to seal due to the solvents used in assembly. In spite of the above, this is a popular emerging technology with some commercial impact forecast within this decade.

3.7.7 Organic/Polymer Solar Cells

Organic solar cells and polymer solar cells are built from thin films (typically 100 nm) of organic semi-conductors such as polymers and small-molecule compounds like polyphenylene vinylene, copper phthalocyanine (a blue or green organic pigment) and carbon fullerenes.[9] Energy conversion efficiencies achieved to date using conductive polymers are low compared to inorganic materials, with the highest reported efficiency of 6.5%[7] for a tandem cell architecture. However, these cells could be beneficial for some applications where mechanical flexibility and disposability are important.

3.7.8 Nanocrystalline Solar Cells

These structures make use of some of the same thin-film light absorbing materials but are overlain as an extremely thin absorber on a supporting matrix of conductive polymer or mesoporous metal oxide having a very high surface area to increase internal reflections (and hence increase the probability of light absorption). Using nanocrystals allows one to design architectures on the length scale of nanometres, the typical exciton diffusion length. In particular, single-nanocrystal (channel) devices, an array of single *p-n* junctions between the electrodes and separated by a period of about a diffusion length, represent a new architecture for solar cells and potentially high efficiency.

3.7.9 Low-cost Solar Cells

Dye-sensitized solar cells (DSSC) are considered the lowest-cost solar cells. These cells are extremely promising because they are made of low-cost materials and do not need elaborate apparatus to manufacture, so they can be made in a DIY way allowing more players to produce them than any other type of solar cell. In bulk they should be significantly less expensive than older solid-state cell designs. They can be engineered into flexible sheets. Although their conversion efficiency is less than the best thin-film cells, their price/performance ratio should be high enough to allow them to compete with fossil fuel electrical generation.

Example 3.5c

What is the condition for zero idealized junction current ($I = 0$).

Solution

Substituting $I = 0$ in eqn 3.10, we get

$$\exp\left(\frac{\mathrm{eV}}{\mathrm{kT}}\right) = 1 \Rightarrow \mathrm{V} = 0$$

3.8 Basic Parameters of Solar Cells

There are certain parameters to be mentioned in the *I-V* characteristics of a solar cell.

3.8.1 Overall Current (*I*)

Overall current is determined by subtracting the light-induced current from the diode dark current and can be expressed as:

Overall current (I) = Diode dark current (I_D) – light-induced current (I_L)

$$I = I_0\left[\exp\left(\frac{eV}{kT}\right) - 1\right] - I_L \quad (3.12)$$

where I_0 is the saturation current, which is also known as the leakage or diffusion current ($I_0 \approx 10^{-8}\,\mathrm{A\,m^{-2}}$ for good solar cells); e is the charge on an electron and hole and k is Boltzmann's constant.

Both I_L and I_0 depend on the structure of solar cells.

3.8.2 Short Circuit Current (I_{sc})

Short circuit current is the light-generated current or photo current, I_L. It is the current in the circuit when the load is zero in the circuit. It can be achieved by connecting the positive and negative terminals by copper wire.

3.8.3 Open Circuit Voltage (V_{oc})

Open circuit voltage is obtained by setting $I=0$ in the expression for overall current *i.e.* $I=0$ when $V=V_{oc}$.

$$V_{oc} = \frac{kT}{e}\ln\left(\frac{I_L}{I_0}+1\right) \tag{3.13}$$

The open circuit voltage is the voltage for maximum load in the circuit.

3.8.4 *I–V* Characteristics

The current equation for a solar cell is given by,[10] $I = I_0\left[\exp\frac{e(V-IR_s)}{kT} - 1\right]$ and shown in Figure 3.9. For a good solar cell, the series resistance, R_s, should be very small and the shunt (parallel) resistance, R_p, should be very large. For commercial solar cells, R_p is much greater than the forward resistance of a diode so that it can be neglected and only R_s is of interest. The following are a few of the characteristics parameters that have been discussed.

The optimum load resistance R_L (P_{max}) $= R_{pmax}$ is connected, if the PV generator is able to deliver maximum power.

$$P_{max} = V_{P\,max}\, I_{P\,max} \tag{3.14}$$

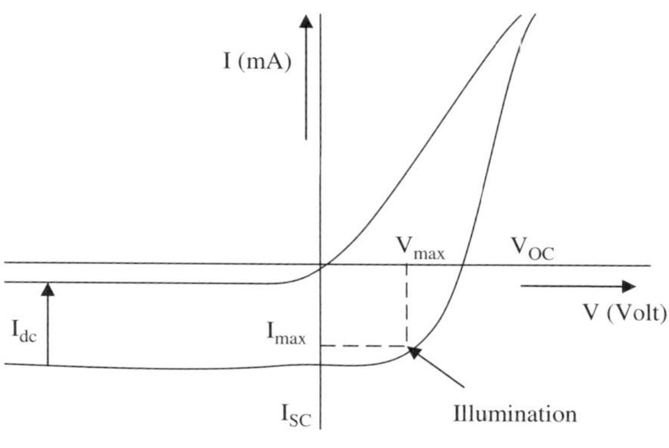

Figure 3.9 *I-V* characteristics of a solar cell.

$$\text{and,}\ R_{\text{P max}} = \frac{V_{\text{P max}}}{I_{\text{P max}}} \tag{3.15}$$

The efficiency is defined as

$$\eta = P/\Phi \tag{3.16}$$

where $P = V \times I$ is the power delivered by the PV generator.
$\Phi = I_T \times A$ is the solar radiation falling on the PV generator.
I_T is the solar intensity and A is the surface area irradiated.

3.8.5 Fill Factor (FF)

The fill factor, also known as the curve factor (Figure 3.10), is a measure of sharpness of the knee in an I-V curve. It indicates how well a junction was made in the cell and how low the series resistance has been made. It can be lowered by the presence of series resistance and tends to be higher whenever the open circuit voltage is high. The maximum value of the fill factor is one, which is not possible. Its maximum value in Si is 0.88.

$$FF = \frac{P_{\max}}{V_{\text{oc}} \times I_{\text{sc}}} = \frac{I_{\max} \times V_{\max}}{V_{\text{oc}} \times I_{\text{sc}}} \tag{3.17}$$

3.8.6 Maximum Power ($P_{\max}$)

No power is generated under short or open circuit. The power output is defined as

$$P_{\text{out}} = V_{\text{out}} \times I_{\text{out}} \tag{3.18}$$

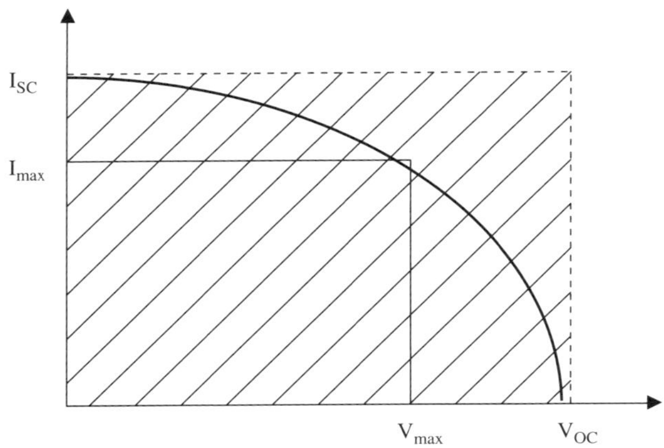

Figure 3.10 Characteristic curve for determining the fill factor.

The maximum power P_{max} provided by the device is achieved at a point on the characteristics, where the product IV is maximum. Thus

$$P_{max} = I_{max} \times V_{max} \tag{3.19a}$$

The maximum possible output can also be given as

$$P_{max} = V_{oc} \times I_{sc} \times \mathrm{FF} \tag{3.19b}$$

where FF is the fill factor given by eqn (3.17).

3.8.7 Solar Cell Efficiency (η_{ec})

The solar cell power conversion efficiency can be given as

$$\begin{aligned}\eta_{ec} = \frac{P_{max}}{P_{in}} &= \frac{I_{max} \times V_{max}}{\textit{Incident solar radiation} \times \textit{Area of solar cell}}\\ &= \frac{V_{OC} \times I_{SC} \times FF}{I(t) \times A_C}\end{aligned} \tag{3.20}$$

where I_{max} and V_{max} are the current and voltage for maximum power, corresponding to solar intensity (I(t)).

Example 3.6

Calculate the fill factor for a solar cell which has the following parameters:

$$V_{oc} = 0.2\,V,\ I_{sc} = -5.5\,\mathrm{mA},\ V_{max} = 0.125\,\mathrm{V},\ I_{max} = -3\,\mathrm{mA}$$

Solution

Substituting the appropriate values in eqn 3.17, we get

$$\text{Fill factor} = \frac{V_{max} I_{max}}{V_{oc} I_{sc}} = \frac{0.125 \times 3}{0.2 \times 5.5} = 0.34$$

Example 3.7

Calculate the maximum power and cell efficiency of the cell at an intensity of 200 W m^{-2}, given $V_{oc} = 0.24$ V, $I_{sc} = -9$ mA, $V_{max} = 0.14$ V and $I_{max} = -6$ mA, $A_C = 4\,\mathrm{cm}^2$.

Solution

From eqn (3.19a), we have

$$P_{max} = V_{max} \times I_{max} = 0.14 \times (-6) = -0.84\,\text{mW}$$

and from eqn (7.6), we have

$$\begin{aligned}\text{Cell efficiency} &= \text{output/input} = (0.14 \times 6 \times 10^{-3})/(200 \times 4 \times 10^{-4}) \\ &= 0.0105 \\ &= 1.05\%.\end{aligned}$$

Example 3.8

Calculate the power output from a solar cell under standard test conditions ($I(\text{t}) = 1000\,\text{W m}^{-2}$ and $T_c = 25\,°\text{C}$), when $\eta = 16\%$, FF = 0.782, aperture area = $4.02 \times 10^{-4}\,\text{m}^2$.

Solution

$$\text{Power output} = 0.16 \times 1000 \times 4.02 \times 10^{-4} \times 0.782 = 0.05\,\text{W}.$$

3.8.8 Limits to Cell Efficiency

Photovoltaic cells are limited in efficiency by many losses; some of these are avoidable but others are intrinsic to the system and may be described as follows:

(i) The electric current leaves the top surface by a web of metal contacts arranged to reduce series resistance losses in the surface. These contacts have a finite area and thus cover part of the active surface and block the incident solar radiation.

(ii) Without special precautions, the reflectance from semi-conductors is high (about 40% of the incident solar radiation). However, this may be reduced to 3% or less by the use of a thin-film surface.

(iii) Photons of quantum energy $h\nu < E_g$ cannot contribute to photovoltaic current generation. For silicon, the inactive wavelengths include 23% of the insolation.

(iv) The excess energy of active photons ($h\nu - E_g$) appears as heat. This loss is about 33% of the insolation.

(v) Quantum efficiency – the fraction of incident absorbed active photons producing electron-hole pairs is usually very high. The design of the cell should be such that at least 95% absorption takes place.

(vi) Collection efficiency is defined as the proportion of radiation-generated electron-hole pairs that produce current in the external circuit.

For 10% overall efficiency cells, the collection efficiency factor is usually about 0.7. Increasing this to about 0.9 would produce more than 20% overall efficiency cells.

(vii) Each absorbed photon produces electron-hole pairs with an electric potential difference of E_g/e (1.1 V in Si). However, only a part (V_B) of this potential is available for the e.m.f. of an external circuit. The voltage factor F_v is equal to eV_B/E_g. The missing e.m.f. occurs because in the open circuit the Fermi level across the junction equates at the dopant n and p levels and not at the displaced conduction to valence band levels. Increased dopant concentration increases F_v. The loss due to the voltage factor is about 20% of the insolation.

(viii) The solar cell I-V characteristics are strongly influenced by the p-n diode characteristics. Thus, as the solar cell (Figure 3.11) output is raised towards V_{oc} the diode becomes increasingly forward biased, so increasing the internal recombination current I_r across the junction. This necessary behaviour is treated as a fundamental loss in the system. The loss due to the curve factor is about 4% of the insolation.

(ix) In practice, the cell characteristics do not follow eqn (3.12) and are better represented as

$$I = I_0 \exp\left[\frac{eV}{AkT} - 1\right] - I_L$$

The factor A results from increased recombination in the junction and tends to change V_{oc} and I_0, so, in general, optimum output would occur if $A = 1$.

Within the cell, recombination is lessened if:

(a) diffusion paths are long (> 50 to 100 μm in Si); this requires long minority carrier lifetimes;
(b) the junction is near the top surface (within 0.15 μm);
(c) the material has few defects other than the dopant.

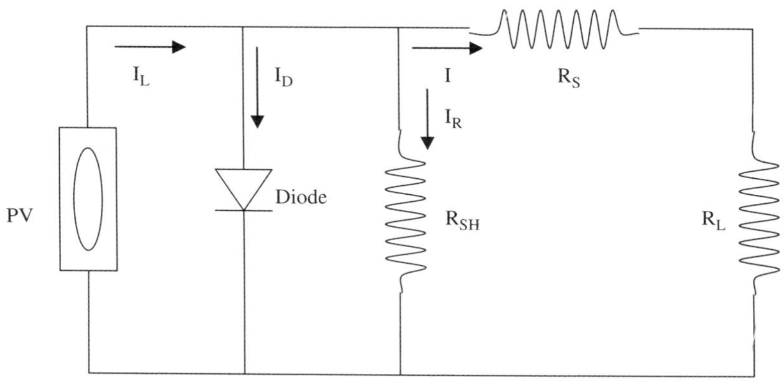

Figure 3.11 Equivalent circuit of solar cell.

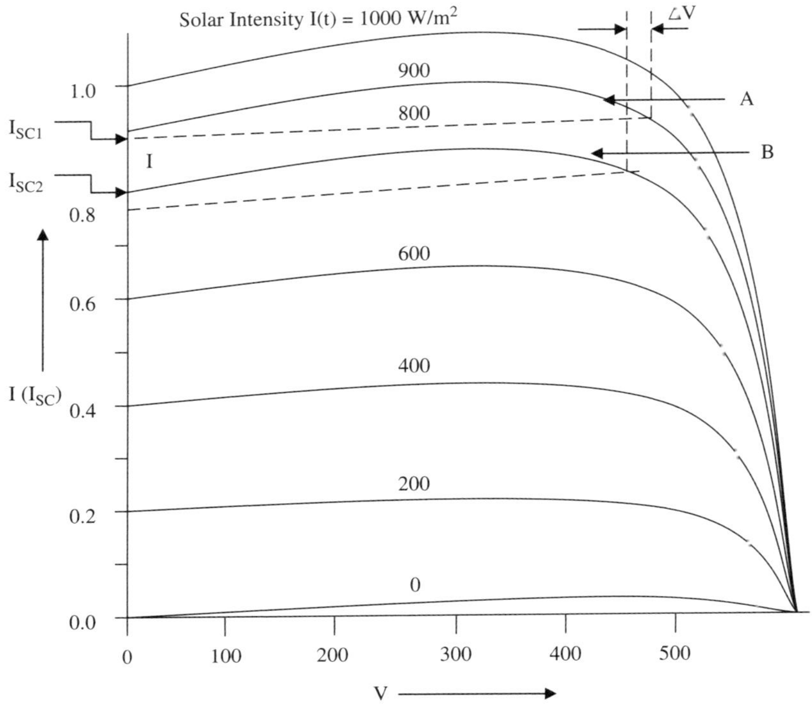

Figure 3.12 Characteristic curve by varying the solar intensity.

3.8.9 Determination of R_s

For determination of R_s, I–V curves at the same temperature but for two different solar intensities I_{T1} and T_{T2} are plotted (Figure 3.12). A point A is selected cn the higher intensity curve corresponding to a voltage slightly greater than V_{Pmax}.

$$I = I_{sc1} - I(\text{A})$$

$$\text{or,}\ \ I(\text{A}) = I_{sc1} - I \tag{3.21}$$

Next a point B is selected on the lower intensity curve.

$$I(\text{B}) = I_{sc2} - I$$

The voltage difference corresponding to the voltages at A and B is

$$\Delta V = V(\text{B}) - V(\text{A})$$

$$R_{s1} = \frac{\Delta V}{I_{sc1} - I_{sc2}} \tag{3.22}$$

This process can be repeated to obtain other values of R_s and the mean of these values gives R_s.

3.8.10 Determination of R_p

R_p can be determined from the slope of an I–V curve at the short circuit point.

$$\left.\frac{dI}{dV}\right|_{V=0} = -\frac{1}{R_P}$$

3.8.11 Thin-film Solar Cell

Thin-film solar cells are efficient for large-scale photovoltaic energy conversion. This not only reduces the semi-conductor material required but is also beneficial for production of a large area module.

Semi-conductor material for thin-film solar cells should have a high absorption coefficient ($\alpha > 10^4\,cm^{-1}$). Two groups of material meet this requirement.

(i) Compound semi-conductor with direct band gap and polycrystalline structure.
(ii) Amorphous semi-conductor.

3.8.12 Amorphous Si Solar Cells (a-SiH)

Hydrogenated amorphous silicon film represents an extremely suitable material for the solar cell mainly due to its optical properties. Only a thin film of about 0.7 μm thickness absorbs a large fraction of the incident solar radiation due to a high absorption coefficient. The optical band gap of pure a-SiH is well matched with the solar spectrum.

3.8.13 Tandem Solar Cells

A tandem system can be realized as a stack of cells with decreasing band gap in the direction of the light path.

3.8.14 Concentrating Solar Cells

The most advanced solar cells, for concentrator applications, are based on the crystalline silicon and AlGaAs/GaAs single junction cells. The most successful Si concentrator cells are p^+-n-n^+ or n^+-p^+ configuration.

3.9 Effect of Cell Temperature on Cell Efficiency

The temperature of operation of a PV module can be determined by an energy balance. The solar energy absorbed by a module is converted partly into thermal energy and partly into electrical energy. The electrical energy is removed from the cell through the external circuit. The thermal energy is dissipated by a combination of heat-transfer mechanisms; the upward losses and back losses.[10] Back losses, in this case, are more important, as the heat transfer from the module should be maximized so that the cell operates at the lowest possible temperature.

An energy balance on a unit area of module, cooled by losses to the surroundings can be written as

$$\tau\alpha I_{\mathrm{T}} = \eta_{\mathrm{c}} I_{\mathrm{T}} + U_{\mathrm{L}}(T_{\mathrm{c}} - T_{\mathrm{a}}) \tag{3.23}$$

where τ is the transmittance of any cover that may be over the cells, α is the fraction of the radiation incident on the surface of the cells that is absorbed and η_{c} is the efficiency, of the module, of conversion of incident radiation into electrical energy. The efficiency will vary from zero to a maximum, depending on how close to the maximum power point the module is operating. The loss coefficient, U_{L}, will include losses by convection and radiation from top and bottom and by conduction through any mounting framework that may be present, to the ambient temperature T_{a}.

The nominal operating cell temperature (NOCT) is defined as that cell or module temperature which is reached when the cells are mounted in their normal way at a solar radiation level of 800 W m^{-2}, a wind speed of 1 m s^{-1}, an ambient temperature of 20 °C and no load operation (*i.e.* with $\eta_{\mathrm{c}} = 0$).

$$\alpha\tau / U_L = (T_{\mathrm{C,NOCT}} - T_{\mathrm{a}}) / I_{\mathrm{T,NOCT}} \tag{3.24}$$

Knowing T_{a}, $I_{\mathrm{T,NOCT}}$ and $T_{\mathrm{C,NOCT}}$, $\tau\alpha/U_{\mathrm{L}}$ can be calculated. Then treating $\tau\alpha/U_{\mathrm{L}}$ as a constant, the temperature at any other condition can be found from the relation:

$$T_{\mathrm{c}} = T_{\mathrm{a}} + (I_{\mathrm{T}}\tau\alpha / U_{\mathrm{L}})(1 - \eta_{\mathrm{c}} / \tau\alpha) \tag{3.25}$$

The electrical efficiency (η_{el}), as a function of temperature, is given by:[11]

$$\eta_{\mathrm{el}} = \eta_0[1 - \beta_0(T_{\mathrm{c}} - 298)] \tag{2.26}$$

where $\eta_{\mathrm{el}} = \eta_{\mathrm{ec}}$, η_0 is the efficiency of the PV module at a temperature of 298 K, β_0 is the silicon efficiency temperature coefficient (0.0045 K^{-1} or 0.0064 K^{-1}) and T_{c} is the cell temperature (K).

3.10 Current Research on Materials and Devices

There are currently many research groups active in the field of photovoltaics in universities and research institutions around the world. This research can be divided into three areas: making current technology solar cells cheaper and/or more efficient to effectively compete with other energy sources; developing new technologies based on new solar cell architectural designs; and developing new materials to serve as light absorbers and charge carriers.

3.10.1 Silicon Processing

One way of reducing the cost is to develop cheaper methods of obtaining silicon that is sufficiently pure. Silicon is a very common element, but is normally bound in silica, or silica sand. Processing silica (SiO_2) to produce silicon is a very high-energy process – at current efficiencies, it takes over two years for a conventional solar cell to generate as much energy as was used to make the silicon it contains.[12] More energy-efficient methods of synthesis are beneficial not only to the solar industry, but also to industries surrounding silicon technology as a whole.

The current industrial production of silicon is *via* the reaction between carbon (charcoal) and silica at a temperature around 1700 °C. In this process, known as carbothermic reduction, each tonne of silicon (metallurgical grade, about 98% pure) is produced with the emission of about 1.5 tonnes of carbon dioxide.

Solid silica can be directly converted (reduced) to pure silicon by electrolysis in a molten salt bath at a fairly mild temperature (800 to 900 °C).[4,12] While this new process is in principle the same as the FFC Cambridge Process, which was first discovered in late 1996, the interesting laboratory finding is that such electrolytic silicon is in the form of porous silicon which turns readily into a fine powder (with a particle size of a few micrometres), and may therefore offer new opportunities for the development of solar cell technologies.

Another approach to reduce the amount of silicon used, and thus the cost, is by micromachining wafers into very thin, virtually transparent layers that could be used as transparent architectural coverings. The technique involves taking a silicon wafer, typically 1 to 2 mm thick, and making a multitude of parallel, transverse slices across the wafer, creating a large number of slivers that have a thickness of 50 micrometres and a width equal to the thickness of the original wafer. These slices are rotated 90°, so that the surfaces corresponding to the faces of the original wafer become the edges of the slivers. As a result of this rotation, the electrical doping and contacts that were on the face of the wafer are located the edges of the sliver, rather than the at the front and rear as is the case with conventional wafer cells. This has the interesting effect of making the cell sensitive from both the front and rear of the cell (a property known as bifaciality).[13] Using this technique, one silicon wafer is enough to build a 140 watt panel, compared to about 60 wafers needed for conventional modules of the same power output.

3.10.2 Thin-film Processing

Thin-film solar cells use less than 1% of the raw material (silicon or other light absorbers) compared to wafer-based solar cells, leading to a significant price drop per kWh. One particularly promising technology is crystalline silicon thin films on glass substrates. This technology makes use of the advantages of crystalline silicon as a solar cell material, with the cost savings of using a thin-film approach. Another interesting aspect of thin-film solar cells is the possibility to deposit the cells on all kind of materials, including flexible substrates, which opens a new dimension for new applications.

3.10.3 Polymer Processing

The invention of conductive polymers may lead to the development of much cheaper cells that are based on inexpensive plastics. However, all organic solar cells made to date suffer from degradation upon exposure to UV light, and hence have lifetimes which are far too short to be viable. The conjugated double-bond systems in the polymers, which carry the charge, are always susceptible to breaking up when radiated with shorter wavelengths. Additionally, most conductive polymers, being highly unsaturated and reactive, are highly sensitive to atmospheric moisture and oxidation, making commercial applications difficult.

3.10.4 Nanoparticle Processing

Experimental non-silicon solar panels can be made of quantum heterostructures, *e.g.* carbon nanotubes or quantum dots, embedded in conductive polymers or mesoporous metal oxides. In addition, thin films of many of these materials on conventional silicon solar cells can increase the optical coupling efficiency into the silicon cell, thus boosting the overall efficiency. By varying the size of the quantum dots, the cells can be tuned to absorb different wavelengths. Researchers at the University of California, San Diego, have come up with a way of making solar photovoltaic cells more efficient by making them fuzzy with indium phosphide nanowires. It sounds similar to a project announced by a consortium of German universities, working in concert with Harvard University Science department.[14]

3.10.5 Transparent Conductors

Many new solar cells use transparent thin films that are also conductors of electrical charge. The dominant conductive thin films used in research now are transparent conductive oxides (TCO), and include fluorine-doped tin oxide (SnO_2:F, or FTO), doped zinc oxide (*e.g.* ZnO:Al) and indium tin oxide (ITO). These conductive films are also used in the LCD industry for flat panel displays. The dual function of a TCO allows light to pass through a substrate window to the active light-absorbing material beneath, and also serves as an ohmic contact to transport photo-generated charge carriers away from that light-absorbing material. The present TCO materials are effective for research, but perhaps are not yet optimized for large-scale photovoltaic production. They require very special deposition conditions at high vacuum, they can sometimes suffer from poor mechanical strength and most have poor transmittance in the infrared portion of the spectrum (*e.g.* ITO thin films can also be used as infrared filters in aircraft windows). These factors make large-scale manufacturing more costly.

A relatively new area has emerged using carbon nanotube networks as a transparent conductor for organic solar cells. Nanotube networks are flexible and can be deposited on surfaces in a variety of ways. With some treatment, nanotube films can be highly transparent in the infrared, possibly enabling efficient low band gap solar cells. Nanotube networks are *p*-type conductors, whereas traditional transparent conductors are exclusively *n*-type. The availability of a *p*-type transparent conductor could lead to new cell designs that simplify manufacturing and improve efficiency.

3.10.6 Silicon Wafer-based Solar Cells

Despite the numerous attempts at making better solar cells by using new and exotic materials, the reality is that the photovoltaics market is still dominated by silicon wafer-based solar cells (first-generation solar cells). This means that most solar cell manufacturers are equipped to produce these types of solar cells. Therefore, a large body of research is currently being done all over the world to create silicon wafer-based solar cells that can achieve higher conversion efficiency without an exorbitant increase in production cost. The aim of the research is to achieve the lowest cost per watt solar cell design that is suitable for commercial production.

IBM has a semi-conductor wafer reclamation process that uses a specialized pattern removal technique to repurpose scrap semi-conductor wafers to a form used to manufacture silicon-based solar panels. The new process was recently awarded the '2007 Most Valuable Pollution Prevention Award' from The National Pollution Prevention Roundtable (NPPR).

Infrared solar cells

Researchers have devised an inexpensive way to produce plastic sheets containing billions of nano-antennas that collect heat energy generated by the Sun and other sources. The technology, developed at the US DOE's Idaho National Laboratory, is the first step toward a solar energy collector that could be mass produced on flexible materials. While methods to convert the energy into useable electricity still need to be developed, the sheets could one day be manufactured as lightweight 'skins' that power everything from hybrid cars to computers and iPods with higher efficiency than traditional solar cells. The nano-antennas also have the potential to act as cooling devices that draw waste heat from buildings or electronics without using electricity. The nano-antennas target mid-infrared rays, which the Earth continuously radiates as heat after absorbing energy from the Sun during the day; also double-sided nano-antenna sheets can harvest energy from different parts of the Sun's spectrum. In contrast, traditional solar cells can only use visible light, rendering them idle after dark.[15]

Problems

3.1 Calculate the fill factor if a solar cell of area $4\,cm^2$ is irradiated with an intensity of $100\,W\,m^{-2}$, given $V_{OC}=0.24\,V$, $I_{SC}=-10\,mA$, $V_{max}=0.14\,V$, $I_{max}=-6.5\,mA$. Also calculate R_{op}. Hint: use eqn (3.16) and use $R_{op}=V_{max}/I_{max}$.

3.2 What will be the solar cell current if dark and light induced current are equal. Hint: use eqn (3.12).

3.3 Calculate the fill factor for a given solar cell for a solar intensity of 300 $W\ m^{-2}$. Hint: use Figure 3.10 and eqn (3.16).

3.4 Draw the curve between efficiency of a solar cell and solar intensity for Figure 3.10. Hint: use eqn (3.20).

3.5 Calculate R_{op} for the solar cell given in Example 3.6. Hint: $R_{op}=V_{max}/I_{max}$.

3.6 Determine the band gap in gallium arsenide. Hint: see Example 3.1 and its table.

3.7 How does dark current vary with potential 'V'? Hint: see Figure 3.7.

3.8 Plot the variation of Fermi energy level for n-type and p-type materials with concentration of doping materials. Hint: use eqns (3.1) and (3.2), respectively.

3.9 Find out the temperature for zero band gap for silicon and gallium arsenide. Hint: put $E_g\ (T)=0$ in the equation of Example 3.1

3.10 What should be the acceptor ion concentration for the same shift in Fermi level, for a given p-type material at different temperatures? Hint: use eqn (3.2) and vary T between 273 and 300 K.

3.11 What will be the acceptor ion concentration at $-273\ °C$?

3.12 What will be the acceptor ion concentration for extrinsic p-type material ($E_F=E_V$)?

3.13 Calculate the dark current for a solar cell for reverse and forward bias mode. Hint: use eqn (3.12) for different V in reverse and forward bias mode, for a given room temperature.

3.14 Calculate the fill factor for a given solar cell for a solar intensity of $300\,W\,m^{-2}$. Hint: use Figure 3.11b1 and eqns (3.13) and (3.17).

References

1. W. P. Hirshman, G. Hering and M. Schmela, *Cell and Module Production 2007: Photon International*, 2008, 152.
2. M. A. Green, *Physica E Low-dimensional Systems and Nanostructures*, 2002, **14**(1–2), 65–70.

3. Third Generation Photovoltaics, http://www.pv.unsw.edu.au/Research/3gp.asp, accessed 12 September 2008.
4. X. Jin, P. Gao, D. Wang, X. Hu and G. Z. Chen, *Angew. Chem. Int. Ed. Engl.*, 2004, **43**(6), 733.
5. V. M. Fthenakis, *Renew. Sustain. Energ. Rev.*, 2004, **8**, 303–334.
6. J. AbuShama, S. Johnston, T. Moriarty, G. Teeter, K. Ramanathan and R. Noufi, *Progress in Photovoltaics: Research and Applications*, 2004, **12**, 39.
7. J. Y. Kim, K. Lee and N. E. Coates, *Science J.*, 2007, **317**(5835), 222.
8. S. A. McDonald, G. Konstantatos, S. Zhang, P. W. Cyr, E. J. Klem, L. Levina and E. H. Sargent, *Nat. Mater.*, 2005, **4**(2), 138.
9. A. Mayer, *Mater. Today*, 2007, **10**(11), 28.
10. G. N. Tiwari, *Solar Energy Fundamentals, Design, Modelling and Applications*, Narosa Publishing House, New Delhi, India, 2004.
11. E. Radziemska, *Progr. Energ. Combust. Sci.*, 2003, **29**(5), 407–424.
12. T. Nohira, K. Yasuda and Y. Ito, *Nat. Mater.*, 2003, **2**(6), 397–401.
13. Sliver Technology Research at the Australian National University, http://solar.anu.edu.au/level_1/research/sliver.php, accessed 12 March 2008.
14. Chia Pet meets the solar cell, http://blog.makezine.com/archive/2008/05/chia_pet_meets_the_solar.html?CMP=OTC-0D6B48984890, accessed 3 September 2008.
15. The energy of innovation: Idaho National Laboratory, https://inlportal.inl.gov/portal/server.pt?open=512&objID=255&mode=2, accessed 14 September 2008.
16. Silicon solar cell structure and mechanism, http://en.wikipedia.org/wiki/Image:Silicon_Solar_cell_structure_and_mechanism.svg, accessed 10 September 2008.

CHAPTER 4

PV Array Analysis

4.1 Introduction

A photovoltaic array is a linked collection of photovoltaic modules, which are in turn made of multiple interconnected solar cells. The cells convert solar energy into direct current electricity *via* the photovoltaic effect. The power that one module can produce is seldom enough to meet the requirements of a home or a business, so the modules are linked together to form an array. Most PV arrays use an inverter to convert the DC power produced by the modules into alternating current that can plug into the existing infrastructure to power lights, motors and other loads. The modules in a PV array are usually first connected in series to obtain the desired voltage; the individual strings are then connected in parallel to allow the system to produce more current. Solar arrays are typically measured by the electrical power they produce, in watts, kilowatts or even megawatts.

The electrical output of the module depends on the size and number of cells, their electrical interconnection and, of course, the environmental conditions to which the module is exposed. Solar electric panels come in all shapes and sizes, and may be made from different materials. However, the most commonly used module is a 'glass-plate-sandwich' that has 36 PV cells connected in series to produce enough voltage to charge a 12-volt battery. The purpose of the structure is to provide a rigid package and protect the inter-cell connections from the environment. Plus (+) and minus (–) connectors are located on the back of the module for interconnection. The modules may have an individual metal frame or be protected by a rubber gasket and intended for installation in a larger mounting system designed to hold several modules.

There are four factors that determine any solar electric panel's output – efficiency of the photovoltaic cells, the load resistance, solar irradiance and cell temperature. The solar cell efficiency is set by the manufacturing process – today's commercially available modules are from 9% to 17% efficient at converting the solar energy to electrical energy. The load resistance determines where on the current and voltage (*I–V*) curve the module will operate. The

RSC Energy Series No. 2
Fundamentals of Photovoltaic Modules and Their Applications
By G. N. Tiwari and Swapnil Dubey

Published by the Royal Society of Chemistry, www.rsc.org

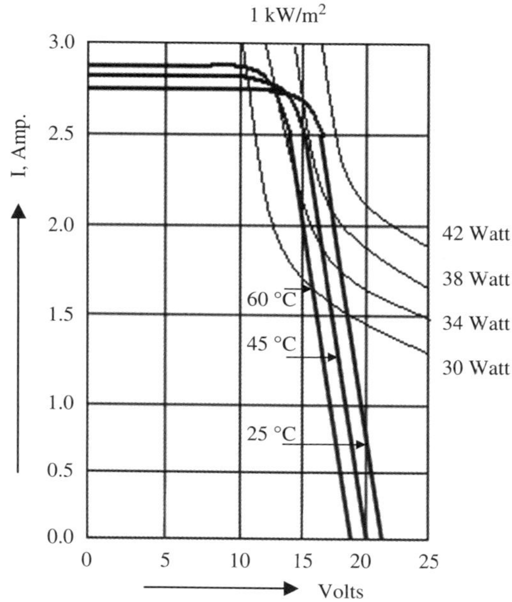

Figure 4.1 Output of any PV generator.

obvious preferred operating point is where maximum power (power is calculated by multiplying the current by the voltage; see Chapter 3) is generated – called the peak power point. Study the *I–V* curve shown in Figure 4.1. This curve (Figure 4.1) represents the output of any PV generator – from a cell to the largest array.

For a given solar-cell area, the current generated is directly proportional to the solar irradiance $I(t)$ and is almost independent of temperature (T). Thus, as the Sun's brightness increases the output voltage and power decrease as temperature increases. The voltage of crystalline cells decreases about 0.5% per degree centigrade temperature increase. Therefore, arrays should be mounted in the sunniest place (no shading) and kept as cool as possible by ensuring air can move over and behind the array.

4.2 Photovoltaic (PV) Module and Array

A photovoltaic module is a packaged interconnected assembly of photovoltaic cells, also known as solar cells. An installation of photovoltaic modules or panels is known as a photovoltaic array or a solar panel. Photovoltaic cells typically require protection from the environment. For cost and practicality reasons a number of cells are connected electrically and packaged in a photovoltaic module, while a collection of these modules that are mechanically fastened together, wired and designed to be a field-installable unit, sometimes with a glass covering and a frame and backing made of metal, plastic or fibreglass, are known as a photovoltaic panel or simply a solar panel. A photovoltaic installation typically includes an array of photovoltaic modules or panels, an inverter, batteries (for off grid) and interconnection wiring.

Most solar PV panels have 30 to 36 cells connected in series. Each cell produces about 0.5 V in sunlight, so a panel produces 15 V to 18 V. These panels are designed to charge 12-V batteries. A 30-cell panel (15 V) can be used to charge the battery without a controller, but it may fail to charge the battery completely. A 36-cell panel (18 V) will do better, but needs a controller to prevent overcharging. The current depends on the size of each cell, and the solar radiation intensity. Most cells produce a current of 2 A to 3 A in bright sunlight. The current is the same in every cell because the cells are connected in series.

Panels are rated in peak watts (Wp), namely the power produced in an optimally matched load with incident solar radiation 1000 W m^{-2}. A typical panel rating is 40 Wp. In a tropical climate a 40 Wp may produce an average of 150 Wh of electricity per day, but as the weather changes the energy varies, typically between 100 Wh and 200 Wh per day.

If two 40-Wp panels, each giving 2.5 A at 16 V in bright sunlight, are connected in parallel they give 5 A at 16 V. If they are connected in series they give 2.5 A at 32 V. In both cases the power is the same: 80 W.

Since the intensity of sunlight is rarely at the peak value, the power output from a panel is usually much less than the peak rating. At low solar radiation intensities the voltage remains almost the same, but the current is low.

Panels should normally be mounted facing the point where the celestial equator crosses the meridian, but should be tilted at least 5° to allow rain to drain off. Since the power output of solar cells is reduced by high temperatures there should be at least 100 mm clearance for ventilation under the panels. There must be no shading of the panels by obstructions, and the panels should be kept clean. Even partial shading of one or more panels can create a resistance in the circuit and reduce the performance of the system.

4.2.1 Theory and Construction

The majority of modules use wafer-based crystalline silicon cells or a thin-film cell based on cadmium telluride or silicon crystalline silicon, which is commonly used in the wafer form in photovoltaic (PV) modules. It is derived from silicon, a relatively multi-faceted element.

In order to use the cells in practical applications, they must be:

- connected electrically to one another and to the rest of the system;
- protected from mechanical damage during manufacture, transport and installation and use (in particular against hail impact, wind and snow loads). This is especially important for wafer-based silicon cells which are brittle;
- protected from moisture, which corrodes metal contacts and interconnects, (and for thin-film cells the transparent conductive oxide layer) thus decreasing performance and lifetime;

- electrically insulated including under rainy conditions; and
- mountable on a substructure.

Most modules are rigid, but there are some flexible modules available, based on thin-film cells. Electrical connections are made in series to achieve a desired output voltage and/or in parallel to provide a desired amount of current source capability. Diodes are included to avoid overheating of cells in case of partial shading. Since cell heating reduces the operating efficiency it is desirable to minimize the heating. Very few modules incorporate any design features to decrease temperature; however, installers try to provide good ventilation behind the module. New designs of module include concentrator modules in which the light is concentrated by an array of lenses or mirrors onto an array of small cells. This allows the use of cells with a very high cost per unit area (such as gallium arsenide) in a cost-competitive way. Depending on construction the photovoltaic array can cover a range of frequencies of light and can produce electricity from them, but cannot cover the entire solar spectrum. Hence much of incident sunlight energy is wasted when used for solar panels, although they can give far higher efficiencies if illuminated with monochromatic light. Another design concept is to split the light into different wavelength ranges and direct the beams onto different cells tuned to the appropriate wavelength ranges. This is projected to raise efficiency to 50%.[1] Sunlight conversion rates (module efficiencies) can vary from 5–18% in commercial production.

A group of researchers at MIT has recently developed a process to improve the efficiency of luminescent solar concentrator (LSC) technology, which redirects light along a translucent material to PV modules located along its edge. The researchers have suggested that efficiency may be improved by a factor of ten over the old design in as little as three years. Three of the researchers involved have now started their own company, called Covalent Solar, to manufacture and sell their innovation in PV modules.[2]

4.2.1.1 Rigid Thin-film Modules

In rigid thin-film modules, the cell and the module are manufactured in the same production line. The cell is created directly on a glass substrate or superstrate, and the electrical connections are created *in situ*, a so-called 'monolithic integration'. The substrate or superstrate is laminated with an encapsulant to a front or back sheet, usually another sheet of glass. The main cell technologies in this category are CdTe, amorphous silicon, micromorphous silicon (alone or tandem) or CIGS (or variant). Amorphous silicon has a sunlight conversion rate of 5–9%.

4.2.1.2 Flexible Thin-film Modules

Flexible thin-film cells and modules are created on the same production line by depositing the photoactive layer and other necessary layers on a flexible substrate. If the substrate is an insulator (*e.g.* polyester or polyimide film) then monolithic integration can be used. If it is a conductor then another technique for electrical connection must be used. The cells are assembled into modules by laminating them to a transparent colourless fluoropolymer on the front side (typically ETFE or FEP) and a polymer suitable for bonding to the final substrate on the other side. The only commercially available (in MW quantities) flexible module uses amorphous silicon triple junction (from Unisolar). So-called Inverted Metamorphic (IMM) multi-junction solar cells made on compound-semi-conductor technology is just becoming commercialized in July 2008.

4.2.2 Single Crystal Solar Cells Module

After testing solar cells under test conditions and sorting to match current and voltage, about 36 solar cells are interconnected and encapsulated to form a module (Figure 4.2). A module consists of the following components: (i) front cover low iron tempered glass, (ii) encapsulate, transparent, insulating, thermoplastic polymer, the most widely used one is EVA (ethylene vinyl acetate), (iii) the solar cell and metal interconnected and (iv) back cover usually a foil of tedlar or Mylar.

Cells are usually mounted in modules and multiple modules are used in arrays. Individual modules may have cells connected in series and parallel combinations

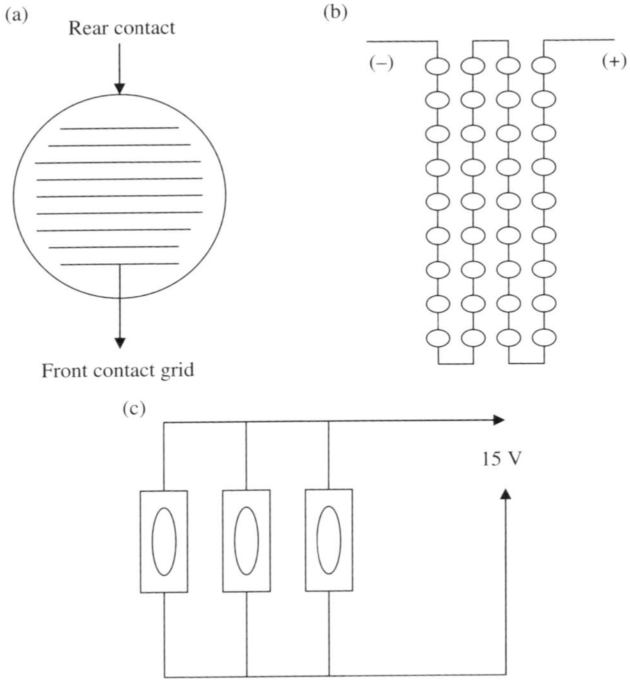

Figure 4.2 Typical arrangements of commercial Si solar cells; (a) cell, (b) module of 36 cells array.

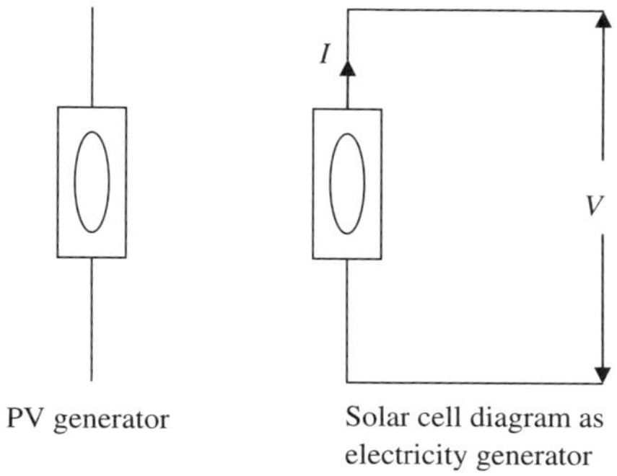

Figure 4.3 Technical signs for various units of PV generator.

to obtain the desired voltage. Arrays of modules may also be arranged in series and parallel depending upon the requirement of current and voltage.

Photovoltaic generators, Figure 4.3, may be used to drive machines such as electric pumps, refrigerators and other devices. PV arrays mounted on the rooftops offer the possibility of large-scale power generation in decentralized medium-size grid-connected units. The PV system supplies the electricity need of the building, feeds the surplus electricity need of the building, feeds the surplus electricity to the grid, to earn revenue, and draws electricity from the grid at low insolation.

4.2.3 Packing Factor (β_c) of a PV Module

The packing factor is defined as the ratio of the total solar cell area to the total module area and can be expressed as:

$$\beta_c = \frac{area\ of\ solar\ cells}{area\ of\ PV\ module} \quad (4.1)$$

It is clear that β_c is less than unity (pseudo solar cell) and it has a maximum value of one when all the area is covered by a solar cell (rectangular solar cell).

4.2.4 Efficiency of a PV/T Module

The electrical efficiency of a PV module can be expressed as:

$$\eta_{em} = (\eta_{ec} \times \beta_c) \times 100 \quad (4.2a)$$

It can also be expressed as (Chapter 3):

$$\eta_{em} = \left(\frac{FF \times I_{sc} \times V_{oc}}{A_m \times I_p}\right) \times 100 \quad (4.2b)$$

where A_m = area of PV module and I_p = incident solar intensity on PV module. The maximum value of the fill factor (FF) in Si is 0.88.

The equivalent thermal efficiency of the PV module may be expressed as:

$$\eta e_{th} = \left(\frac{\eta_e}{0.38}\right) \times 100 \tag{4.3}$$

The electrical load efficiency may be expressed as:

$$\eta_{load} = \left(\frac{I_L \times V_L}{A_m \times I_p}\right) \times 100 \tag{4.4}$$

The overall thermal efficiency of the hybrid PV/T system may be written as:

$$\eta_{ov,th} = \eta_{th} + \frac{\eta_e}{0.38} \tag{4.5}$$

where η_{th} is thermal efficiency.

The overall exergy efficiency of the hybrid PV/T system may be written as:

$$\eta_{ov,ex} = \eta_{ex} + \eta_e \tag{4.6}$$

where η_{ex} is the exergy efficiency $= \eta_{th}\left(1 - \frac{T_{\sin k}}{T_{source}}\right)$ and T is the temperature in Kelvin.

Example 4.1

Calculate the packing factor of a PV module (36 solar cells) of area 0.605 m^2, each pseudo-solar cell having an area of 0.015 m^2.

Solution

From eqn (4.1), we get

$$\beta_c = \frac{0.54}{0.605} \times 100 = 89.2\%$$

Example 4.2

Calculate the efficiency of a PV module at an intensity of 400 W m^{-2}, given:

FF = 0.8, $I_{SC} = 3.2\,A$, $V_{oc} = 16\,V$, $I_L = 1\,A$, $V_L = 14\,V$, area of module $= 1$ m^2.

Solution

From eqn (4.2b), we have

$$\eta_{em} = \frac{0.8 \times 3.2 \times 16}{400 \times 1} \times 100 = 10.24\%$$

Example 4.3

Using Example 4.1, calculate the load efficiency of a PV module.

Solution

From eqn (4.4), we have

$$\eta_{em} = \frac{1 \times 14}{400 \times 1} \times 100 = 3.5\%$$

4.2.5 Applications

In urban and suburban areas, photovoltaic arrays are commonly used on rooftops to measure power use; often the building will have a pre-existing connection to the power grid, in which case the energy produced by the PV array will be sold back to the utility in some sort of net metering agreement. In more rural areas, ground-mounted PV systems are more common. The systems may also be equipped with a battery backup system to compensate for a potentially unreliable power grid. In agricultural settings, the array may be used to directly power DC pumps, without the need for an inverter. In remote settings, such as mountainous areas, islands or other places where a power grid is unavailable, solar arrays can be used as the sole source of electricity, usually by charging a storage battery. Satellites use solar arrays for their power. In particular the International Space Station uses multiple solar arrays to power all the equipment on board. Solar photovoltaic panels are frequently applied in satellite power. However, costs of production have been reduced in recent years for more widespread use through production and technological advances. For example, single crystal silicon solar cells have largely been replaced by less expensive multicrystalline silicon solar cells, and thin-film silicon solar cells have also been developed recently at lower costs of production. Although they are reduced in energy conversion efficiency from single crystalline Si wafers, they are also much easier to produce at comparably lower costs. Together with a storage battery, photovoltaics have become commonplace for certain low-power applications, such as signal buoys or devices in remote areas or simply where connection to the electricity mains would be impractical.

4.2.5.1 PV in Buildings

Building integrated photovoltaics (BIPV) are increasingly incorporated into new domestic and industrial buildings as a principal or ancillary source of electrical power, and are one of the fastest growing segments of the photovoltaic industry. Typically, an array is incorporated into the roof or walls of a building, and roof tiles with integrated PV cells can now be purchased. Arrays can also be retrofitted into existing buildings; in this case they are usually fitted on top of the existing roof structure. Alternatively, an array can be located separately from the building but connected by cable to supply power for the building. Where a building is at a considerable distance from the public electricity supply (or grid) – in remote or mountainous areas – PV may be the preferred possibility for generating electricity, or PV may be used together with wind, diesel generators and/or hydroelectric power. In such off-grid circumstances batteries are usually used to store the electric power.

4.2.5.2 PV in Transport

PV has traditionally been used for auxiliary power in space. PV is rarely used to provide motive power in transport applications, but is being used increasingly to provide auxiliary power in boats and cars. Recent advances in solar-cell technology, however, have shown the cell's ability to administer significant hydrogen production, making it one of the top prospects for alternative energy for automobiles.

4.2.5.3 PV in Stand-alone Devices

PV has been used for many years to power calculators and novelty devices. Improvements in integrated circuits and low-power LCD displays make it possible to power a calculator for several years between battery changes, making solar calculators less common. In contrast, solar-powered remote fixed devices have seen increasing use recently, due to the increasing cost of labour for connection of mains electricity or a regular maintenance programme. In particular, it is used in parking meters, emergency telephones and temporary traffic signs.

4.2.5.4 PV in Agriculture

PV systems are used effectively worldwide to pump water for livestock, plants or humans. Water pumping appears to be most suitable for solar PV applications as water demand increases during dry days when plenty of sunshine is available. A Solar Photovoltaic (SPV) water pumping system is expected to deliver a minimum of 15,000 litres per day for 200 Wp and 170,000 litres per day for 2,250 Wp panel from a suction of 7 m and/or total head of 10 m on a clear sunny day. PV is also used to power remote electric fences on farms.

4.2.5.5 Medical Refrigeration

For life-saving vaccines, the World Health Organization (WHO) has laid down ground rules to maintain the cold chain from the point of their manufacture to their application. WHO has specified technical details for PV-based refrigeration. This has resulted in the success of WHO-sponsored immunization programmes in those countries/remote areas where electricity is not available.

4.2.5.6 PV in Street Lights

Solar PV street lights can be used as yard lighting, peripheral lighting for industries, street lights in layout, compound lights, *etc.* The photovoltaic modules charge the batteries during the day time. At dusk an automotive sensor switches on a powerful high-efficiency light and at dawn the lamp is switched off automatically. A photograph of a solar PV street light is shown in Figure 4.4.

4.2.6 PV Performance

At high noon on a cloudless day at the equator, the power of the Sun is about $1\,kW\,m^{-2}$, on the Earth's surface, to a plane that is perpendicular to the Sun's

Figure 4.4 Photograph of a solar PV street light.

rays. As such, PV arrays can track the Sun through each day to greatly enhance energy collection. However, tracking devices add cost, and require maintenance, so it is more common for PV arrays to have fixed mounts that tilt the array and face due south in the northern hemisphere (in the southern hemisphere, they should point due north). The tilt angle, from horizontal, can be varied for season, but if fixed it should be set to give optimal array output during the peak electrical demand portion of a typical year. For large systems, the energy gained by using tracking systems outweighs the added complexity (trackers can increase efficiency by 30% or more). PV arrays that approach or exceed one megawatt often use solar trackers. Accounting for clouds, and the fact that most of the world is not on the equator, and that the Sun sets in the evening, the correct measure of solar power is insolation – the average number of kilowatt-hours per square metre per day. A typical '150-watt' solar panel is about a square metre in size. Such a panel may be expected to produce 1 kWh every day, on average, after taking into account the weather and the latitude.

Manufacturers of photovoltaic panels typically provide electrical parameters at only one operating condition. Photovoltaic panels operate over a large range of conditions so the manufacturer's information is not sufficient to determine their overall performance. Designers need a reliable tool to predict energy production from a photovoltaic panel under all conditions in order to make a sound decision on whether or not to incorporate this technology.[3] For grid-connected photovoltaic systems, an optimum PV/inverter sizing ratio is important for maximizing the PV performance.[4,5] The sizing ratio (S_R) is defined as the ratio of the PV array capacity at standard test conditions (STC) to the rated inverter input DC power given as

$$S_R = \frac{P_{PV,\ rated}}{P_{Inverter,\ rated}} \tag{4.7}$$

The optimal PV/inverter sizing depends on local climate, PV surface orientation and inclination, inverter performance and PV/inverter cost ratio.[6–8] In solar photovoltaic arrays simple series-parallel, total-cross-tied arrays (TCT) and bridge-linked (BL) solar-cell interconnection configurations are used. The cross-tied type of solar-cell interconnection networks (BL and TCT) are better types of networks in controlling the effects of electrical mismatches Furthermore, the bridge-linked type of configuration tends to be optimal in minimizing power dissipation due to both mismatched and shadowed cells.[9]

A photovoltaic (PV) system should be installed to maximize the solar contribution to a particular load. Optimum PV inclination and orientation depends on local climate, load consumption, temporal profile and latitude.[10–12] Incident insolation and PV output were maximum for a surface with inclination 30° facing due south and minimum for a vertical surface with orientation 90° east or west from south. The monthly optimum collection angle maximizing incident insolation varied from 10° to 70°.[13] Generally, a surface with tilt angle equal to the latitude of a location receives maximum insolation. However, some locations experience a weather pattern where winter is typically cloudier than summer or the average morning and afternoon insolation is not symmetric. The maximum available energy may then be received by a surface whose azimuth angle is either east or west of due south (in the northern hemisphere). The optimum tilt angle is thus site dependent and calculation of this angle requires solar radiation data for that particular site for the whole year. Normally, during summer, the incident insolation is maximized for a surface with an inclination 10–15° less than the latitude and, during winter, 10–15° more than the latitude.[14]

4.2.7 Solar Photovoltaic Panels on Spacecraft

Spacecraft operating in the inner solar system usually rely on the use of photovoltaic solar panels to derive electricity from sunlight. In the outer solar system, where the sunlight is too weak to produce sufficient power, radioisotope thermal generators (RTGs) are used as a power source.[15] The first spacecraft to use solar panels was the Vanguard 1 satellite, launched by the USA in 1958.

Solar panels need to have a lot of surface area that can be pointed towards the Sun as the spacecraft moves. More exposed surface area means more electricity can be converted from light energy from the Sun. Since spacecraft have to be small, this limits the amount of power that can be produced.

Spacecraft are built so that the solar panels can be pivoted as the spacecraft moves. Thus, they can always stay in the direct path of the light rays no matter how the spacecraft is pointed. Spacecraft are usually designed with solar panels that can always be pointed at the Sun, even as the rest of the body of the spacecraft moves around, much as a tank turret can be aimed independently of where the tank is going. A tracking mechanism is often incorporated into the solar arrays to keep the array pointed towards the Sun.[15]

Sometimes, satellite operators purposefully orient the solar panels to 'off point', or out of direct alignment from the Sun. This happens if the batteries are completely charged and the amount of electricity needed is lower than the amount of electricity made; off-pointing is also sometimes used on the International Space Station for orbital drag reduction.

Gallium arsenide-based solar cells are typically favoured over silicon in industry, due to the fact that they have a higher efficiency. The most efficient solar cells currently in production are multijunction cells. These use a combination of several layers of both gallium arsenide and silicon to capture the largest spectrum of light possible. Leading-edge multijunction cells are capable of nearly 29% efficiency under ideal conditions.

Solar power, other than for propulsion, has been practical for spacecraft operating no farther from the Sun than the orbit of Mars. For example, Magellan, Mars Global Surveyor and Mars Observer used solar power as does the Earth-orbiting Hubble Space Telescope. The Rosetta space probe, launched March 2, 2004, will use solar panels as far as the orbit of Jupiter (5.25 AU); previously the furthest use was the Stardust spacecraft at 2 AU. Solar power for propulsion was also used on the European lunar mission SMART-1 with Hall Effect Thrusters.

The upcoming Juno mission will be the first mission to Jupiter to use solar panels instead of the traditional RTGs (Radioisotope Thermoelectric Generators) that were used by previous outer solar system missions.[16] In 2005 Rigid-Panel Stretched Lens Arrays were producing 7 kW per wing. Solar arrays producing 300 W kg^{-1} and 300 W m^{-2} from the Sun's 1366 W m^{-2} energy near the Earth are available. Entech Inc. hopes to develop 100 kW panels by 2010 and 1 MW panels by 2015.

4.3 Series and Parallel Combinations

PV modules are connected in series or parallel to increase the current and voltage ratings. When modules are connected in series, it is desirable to have each module's maximum power production occurring at the same current and voltages of each module add up. When modules are connected in parallel, it is desirable to have each module's maximum power production occurring at the same voltage and currents of each module add up. Thus, while interconnecting the modules; the installer should have this information available for each module. A solar panel is a group of several modules connected in series–parallel combination in a frame that can be mounted on a structure.

Series and parallel connection of modules in a panel is shown in Figure 4.5. In parallel connection, blocking diodes are connected in series with each series string of modules, so that if any string should fail, the power output of the remaining series string will not be absorbed by the failed string. Also bypass diodes are installed across each module, so that if one module should fail, the power output of the remaining modules in a string will bypass the failed

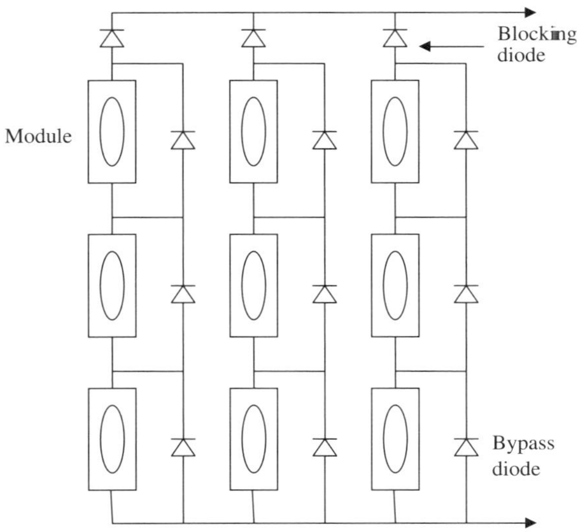

Figure 4.5 Series and parallel connection of modules in a panel.

module. Some modern PV modules come with such internally embedded bypass diodes. A large number of interconnected solar panels is known as a solar PV array.

Example 4.4

Calculate the daily load for domestic use and how many 40-Wp PV panels are required in the array.

Solution

Four 40-W lamps used 4 hours per day: 640 Wh
One 15-W television used 4 hours per day: 60 Wh
Two 35-W fans used 6 hours per day: 420 Wh
One 60-W refrigerator used all day, compressor on 50% of the time: 720 Wh
Total daily load = 1840 Wh.
Assuming each panel produces 150 Wh per day, then
= 1840 Wh/150 Wh = 12.3.
Therefore, a 12-V system needs 13 panels connected in parallel.

4.4 Balance of PV Array

The balance of PV system (BOS) components include mounting materials for the module, wire and all wiring components which includes distribution panel, junction box and miscellaneous connectors, lighting protectors, grounding connections, battery fuses, battery cables and battery containers. In some cases connected loads are also considered to be part of the BOS, for example, when the system is installed to operate a specific load. Certain BOS components are regulated by codes or standards. For example, array mounts must meet the wind-loading requirements of applicable building codes and battery compartments are covered under the National Electrical Code (NEC). All the BOS components should be appropriate for environmental considerations.

4.5 Partial Shading of Solar Cell and Module

PV modules are very sensitive to shading. Partial shadowing has been identified as a main cause for reducing the energy yield of grid-connected photovoltaic systems. Shading of a single cell within a PV-module, which itself is part of a string containing a number of modules connected in series, leads to a reverse-bias operation of the cell, which may result in hot-spots and potential breakdown of the shaded cell. In order to avoid this threat, bypass diodes are inserted into the modules, which take over the string current in case of a partially shaded module.[17]

Unlike a solar thermal panel, which can tolerate some shading, many brands of PV modules cannot even be shaded by the branch of a leafless tree. When even a small portion of a cell, module or array is shaded, while the remainder is in sunlight, the output falls dramatically due to internal 'short-circuiting' (the electrons reversing course through the shaded portion of the *p-n* junction).

Shading obstructions can be defined as soft or hard sources. If a tree branch, roof vent, chimney or other item is shading from a distance, the shadow is diffuse or dispersed. These soft sources significantly reduce the amount of light reaching the cell(s) of a module. Hard sources are defined as those that stop light from reaching the cell(s), such as a blanket, tree branch, bird dropping or the like, sitting directly on top of the glass. If even one full cell is hard shaded, the voltage of that module will drop to half of its unshaded value in order to protect itself. If enough cells are hard shaded, the module will not convert any energy and will in fact become a tiny drain of energy on the entire system.

Partial-shading even one cell of a 36-cell module will reduce its power output. Because all cells are connected in a series string, the weakest cell will bring the others down to its reduced power level. Therefore, whether 1/2 of one cell is shaded, or 1/2 a row of cells is shaded as shown above, the power decrease will be the same and proportional to the percentage of area shaded, in this case 50%.

When a full cell is shaded, it can act as a consumer of energy produced by the remainder of the cells, and trigger the module to protect itself. The module will route the power around that series string. If even one full cell in a series string is shaded, as seen on the right, it will likely cause the module to reduce its power level to $\frac{1}{2}$ of its full available value. If a row of cells at the bottom of a module is fully shaded, as seen in Figure 4.6, the power output may drop to zero. The best way to avoid a drop in output power is to avoid shading whenever possible.

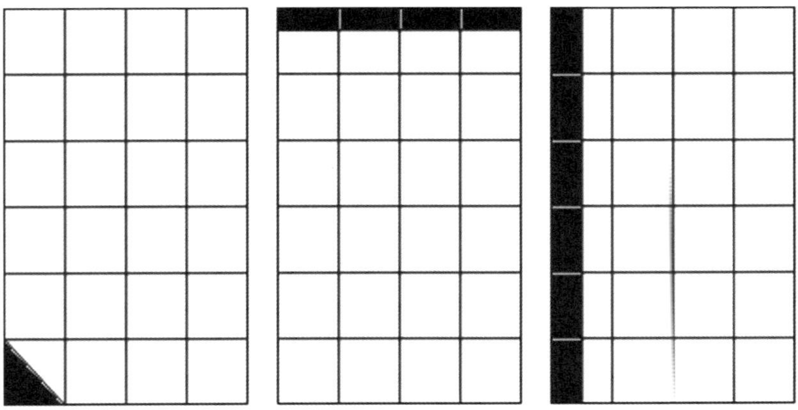

Figure 4.6 Examples of partial cell shading of a module that will reduce a solar electric panel's power by 50%.

Alonso-García *et al.*[18] have simulated the shading effects in arrays with different string configurations and concluded that the increase of shading rate over one cell produces higher deformations in the *I–V* characteristics; cells with higher shunt conductances (lower shunt resistance) cause smaller deformation in the resulting *I–V* characteristics; the increase in the number of shaded cells in the same string does not affect mpp (maximum power point), nevertheless when cells are placed in different strings power losses are considerably increased and bypass diodes should be included to investigate the influence of the mismatching effects in the power–voltage characteristics of a PV array.[19] The effect on current and voltage by increasing the number of shaded cells is shown in Figure 4.7.

However, since it is impossible to prevent occasional shading, the use of bypass diodes around series-connected modules is recommended. Almost all panels of the solar panels that are offered come with these diodes integrated right into the module itself. Bypass diodes are not required if all the modules are in parallel, *i.e.* a 12-volt array using 12-volt modules, and many designers do not use them on 24-volt arrays. However, for array voltages higher than 24 volts, bypass diodes should be used around each module to provide an alternative current path in case of shading. Many module manufacturers will provide modules with the bypass diodes integrated into the module junction box. Using bypass diodes may postpone failure, but it does not prevent the loss of energy production from the shading. It is important to check for potential shading before installing the PV array. Consider the seasonal changes in foliage and Sun angle. After installation, the area must be maintained to prevent weeds or tree branches from shading the array.

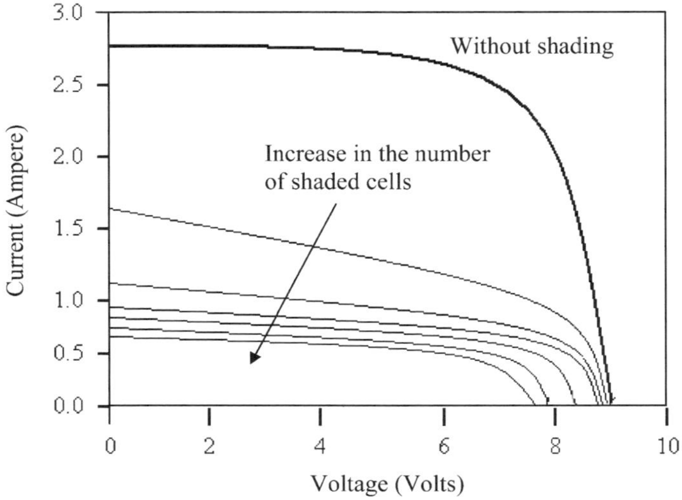

Figure 4.7 Effect of increasing the number of shaded cells.

4.6 Maximum Power Point Tracker (MPPT)

A maximum power point tracker (or MPPT) is a high-efficiency DC-to-DC converter, which functions as an optimal electrical load for a photovoltaic (PV) cell, most commonly for a solar panel or array, and converts the power to a voltage or current level which is more suitable to whatever load the system is designed to drive. PV cells have a single operating point where the values of the current (I) and Voltage (V) of the cell result in a maximum power output. These values correspond to a particular resistance, which is equal to V/I as specified by Ohm's Law. A PV cell has an exponential relationship between current and voltage, and the maximum power point (MPP) occurs at the knee of the curve, where the resistance is equal to the negative of the differential resistance ($V/I = -\mathrm{d}V/\mathrm{d}I$). Maximum power point trackers utilize some type of control circuit or logic to search for this point and thus to allow the converter circuit to extract the maximum power available from a cell.

MPPT is not a mechanical tracking system that 'physically moves' the modules to make them point more directly at the Sun. MPPT is a fully electronic system that varies the electrical operating point of the modules so that the modules are able to deliver maximum available power. Additional power harvested from the modules is then made available as increased battery charge current. MPPT can be used in conjunction with a mechanical tracking system, but the two systems are completely different.

Batteryless grid-tied PV inverters utilize MPPTs to extract the maximum power from a PV array, convert this to alternating current (AC) and sell excess energy back to the operators of the power grid. MPPT charge controllers are desirable for off-grid power systems to make the best use of all the energy generated by the panels.

The benefits of MPPT regulators are greatest during cold weather, on cloudy or hazy days or when the battery is deeply discharged. Solar MPPTs can also be used to drive motors directly from solar panels. The benefits are huge, especially if the motor load is continuously changing. This is due to the fact that the AC impedance across the motor is related to the motor's speed. The MPPT will switch the power to match the varying resistance.

4.7 International Status of PV Power Generation

World solar photovoltaic (PV) market installations reached a record high of 2.8 gigawatts peak (GWp) in 2007. The three leading countries (Germany, Japan and the USA) represent nearly 89% of the total worldwide PV installed capacity. On 1 August, 2007, word was published of construction of a production facility in China, which is projected to be one of the largest wafer factories in the world, with a peak capacity of around 1500 MW. Germany was the fastest-growing major PV market in the world during 2006 and 2007. In 2007, over 1.3 GWp of PV was installed. The German PV industry generates over 10,000 jobs in production, distribution and installation. Some of the largest photovoltaic plants in the world are in Germany, which has a 10-MW

Table 4.1 The world's largest PV power plants.[20]

DC Peak Power	*Location*	*Description*	*GWh year^{-1}*
154 MW	Mildura/Swan Hill, Australia	Heliostat Concentrator Photovoltaic technology	270
62 MW	Moura, Portugal	BP, Yingli Green Energy	88
40 MW	Muldentalkreis, Germany	550,000 thin-film modules (First Solar)	40
23 MW	Murcia, Spain	Hoya de Los Vincentes	41.6
21 MW	Calavéron, Spain	Solar park Calaveron	40
20 MW	Trujillo, Spain	Planta Solar La Magascona SunPower trackers 120,000 Atersa modules	–
20 MW	Beneixama, Spain	Tenesol, Aleo and Solon solar modules with Q-Cells cells	30
18 MW	Olivenza, Spain	SunPower T20 tracking system	32
14 MW	Nellis AFB, Nevada	SunPower T20 tracking system	30
13.8 MW	Salamanca, Spain	Planta Solar de Salamanca	–
12.7 MW	Murcia, Spain	Lobosillo Solar Park	–
12 MW	Arnstein, Germany	1464 SOLON mover	14
11 MW	Serpa, Portugal	52,000 solar modules	–
10 MW	Pocking, Germany	57,912 solar modules	11.5
9.5 MW	Milagro, Spain	Monte Alto photovoltaic power plant	14

photovoltaic system in Pocking, and a 12-MW plant in Arnstein, with a 40-MW power station planned for Muldentalkreis; Portugal, which has an 11-MW plant in Serpa and a 62-MW power station planned for Moura. A 20-MW power plant is also planned for Beneixama, Spain. The photovoltaic power station proposed for Australia will use heliostat concentrator technology and will not come into service until 2010. It is expected to have a capacity of 154 MW when it is completed in 2013. The details of the world's largest PV power plants are given in Table 4.1. Many of these plants are integrated with agriculture and some use innovative tracking systems that follow the Sun's daily path across the sky to generate more electricity than conventional fixed-mounted systems. There are no fuel costs or emissions during operation of the power stations.

In India, a total of 32 grid-interactive solar PV power plants have been installed with financial assistance from the Federal Government. These plants, with aggregate capacity of 2.1 MW, are estimated to generate about 2.52 million units of electricity in a year. In addition, solar PV systems with an aggregate capacity of 12 MW were installed for applications such as lighting, water pumping, communications, *etc*. These systems are capable of generating 18 million kWh of electricity per year. In 2003 alone, India added 2.5 MW of solar PVs for rural electrification as well as employment and income generation. The Ministry of New and Renewable Energy (MNES) has been implementing installation of solar PV water-pumping systems for irrigation and drinking-water applications through subsidy since 1993–1994. Typically, a 1800

Wp PV array capacity solar PV water-pumping system, which cost about Rs. 3.65 lakh, is being used for irrigation purposes. The Ministry is providing a subsidy of Rs. 30 per watt of PV array capacity used, subject to a maximum of Rs. 50,000 per system. The majority of the pumps fitted with a 200 watt to 3000 watt motor are powered with 1800 Wp PV arrays, which can deliver about 140,000 litres of water/day from a total head of 10 metres. By 30th September, 2006, a total of 7068 solar PV water pumping systems have been installed.

Problems

4.1 What is the effect of partial or complete shadowing of a cell in a PV module?
4.2 What is the importance of MPPT in an SPV system? Explain various strategies used for operation of an MPPT.
4.3 Calculate the load and no-load efficiency of a PV module at an intensity of 400 W m^{-2}, given: FF = 0.8, I_{SC} = 3.2 A, V_{oc} = 16 V, I_L = 1 A, V_L = 14 V, area of module = 1 m^2. Hint: use eqns (4.2b) and (4.4).
4.4 Describe the classification of solar cells based on the type of active material used.
4.5 Define the sizing ratio (S_R) of the PV array capacity.
4.6 Describe the theory and construction of PV modules and their applications.
4.7 Calculate the daily load for domestic use and how many 75-Wp PV panels are required in the array. Hint: see Example 4.4.
4.8 Describe the national and international status of PV power generation.

References

1. B. Pierce, Very high efficient solar cells, http://www.arpa.mil/sto/smallunitops/vhesc.html, accessed 25 July 2008.
2. J. Hance, Breakthrough in solar energy, http://news.mongabay.com/2008/0710-hance_solar.html, accessed 18 August 2008.
3. W. De Soto, S. A. Klein and W. A. Beckman, *Sol. Energ.*, 2006, **80**, 78–88.
4. J. D. Mondol, Y. G. Yohanis and B. Norton, *Sol. Energ.*, 2006, **80**, 1517–1539.
5. B. Decker, U. Jahn, U. Rindelhardt and W. Vaaben, in *11th European Photovoltaic Solar Energy Conference*, Montreux, Switzerland, 1992, pp. 1497–1500.
6. M. H. Macagnan and E. Lorenzo, in *11th European Photovoltaic Solar Energy Conference*, Montreux, Switzerland, 1992, pp. 1167–1170.
7. M. Jantsch, H. Schmidt and J. Schmid, in *11th Photovoltaic Solar Energy Conference*, Montreux, Switzerland, 1992, pp. 1589–1593.
8. A. Louche, G. Notton, P. Poggi and G. Peri, in *12th European Photovoltaic Solar Energy Conference*, Amsterdam, The Netherlands, 1994, pp. 1638–1641.

9. N. K. Gautam and N. D. Kaushik, *Energy*, 2002, **27**, 347–361.
10. P. Tsalides and A. Thanailakis, *Sol. Cell.*, 1985, **14**, 83–94.
11. J. Kern and I. Harris, *Sol. Energ.*, 1975, **17**, 97–102.
12. S. Bari, *Energ. Convers. Manag.*, 2000, **41**, 855–60.
13. J. D. Mondol, Y. G. Yohanis and B. Norton, *Renew. Energ.*, 2007, **32**, 118–140.
14. J. A. Duffie and W. A. Beckman, *Solar Engineering of Thermal Processes*, Wiley, 1991.
15. NASA JPL Publication: Basics of Space Flight, Chapter 11, http://www2.jpl.nasa.gov/basics/bsf11-3.html, accessed 5 May 2008.
16. NASA JPL Publication: Basics of Space Flight, Chapter 11, http://www2.jpl.nasa.gov/basics/bsf11-4.html#propulsion, accessed 5 May 2008.
17. A. Woyte, J. Nijsa and R. Belmansa, *Sol. Energ.*, 2003, **74**(3), 217–233.
18. M. C. Alonso-García, J. M. Ruiz and W. Herrmann, *Renew. Energ.*, 2006, **31**, 1986–1993.
19. E. Karatepe, M. Boztepe and M. Colak, *Sol. Energ.*, 2007, **81**, 977–992.
20. Greenpeace Energy, http://www.pvresources.com/en/top50pv.php, accessed 8 May 2008.

CHAPTER 5

Role of Batteries and Their Uses

5.1 Introduction

To many people a battery is a very useful but rather mysterious device. It delivers electric power for a multitude of purposes, but is silent, has no moving parts and gives no visual evidence of its operation. The advantages of batteries are:

i) They provide a portable source of electric power. This power is available in considerable quantity for use on moving equipment or where no power lines are accessible. They are unaffected by cords or cables.
ii) They are capable of delivering very large quantities of power for short periods and being recharged at low rates over extended times. Thus heavy surges on power are available when required, without heavy demands on a power system or equipment.
iii) They provide the most reliable known source of emergency power, instantaneously when normal power fails. They can thus enable light or power to continue when the need is greatest.
iv) They provide a source of pure direct current for laboratory and other specific purposes, either as a separate and independent supply or by acting as filter in a normal supply system.

These and other distinctive attributes of a battery make it the optimum selection for an almost infinite number of applications.

In many types of stand-alone photovoltaic (PV) systems for continuous power supply, batteries are required to even out irregularities in the solar irradiation. Today, nickel-cadmium (NiCd) and lead-acid (PbA) batteries are commonly used in PV systems. Some emerging battery technologies may also be suitable for storage of renewable energy, such as different types of redox flow batteries and high-temperature sodium-sulfur batteries. Identification of the important parameters in PV applications can be used to direct research and product improvements, and comparison of different battery technologies can be used to guide battery choice for specific user conditions.

RSC Energy Series No. 2
Fundamentals of Photovoltaic Modules and Their Applications
By G. N. Tiwari and Swapnil Dubey

Published by the Royal Society of Chemistry, www.rsc.org

The energy produced during the day, which was not consumed by loads, is saved in batteries. Saved energy can be used at night or during days with bad weather conditions. Batteries in photovoltaic systems are often charged/discharged, therefore they must meet stronger requirements. Most often used classic lead-acid (PbA) batteries are produced especially for PV systems, where deep discharge is required. Other battery types, such as nickel-cadmium (NiCd) or nickel metal hydride (NiMH), are rarely used, except in portable devices. Hermetical batteries often consist of an electrolyte in gel form. Such batteries do not require maintenance. Typical solar system batteries' lifetimes span from three to five years, depending heavily on charging/discharging cycles, temperature and other parameters. The more often the battery is charged/discharged, the shorter the lifetime.

Lifetime depends on charge/discharge cycle rates numbers. The deeper the battery is discharged, the shorter the lifetime. The most important battery parameter is battery capacity, which is measured in ampere-hours (Ah). Battery capacity depends on discharging current; the higher the discharging current the lower the capacity, and *vice versa*. Batteries can be charged in many different ways, for example with constant current, with constant voltage *etc*., which depends on the battery type used. The charging characteristics are recommended and prescribed by different standards. The prices of solar batteries are higher than the prices of classic car batteries, but their advantages are longer lifetime and lower discharging rates. Consequently, the maintenance costs of the photovoltaic system are lower.

The battery's capacity for holding energy is rated in amp-hours: 1 amp delivered for 1 hour = 1 amp-hour.

Battery capacity is listed in amp-hours at a given voltage, *e.g*. 220 amp-hours at 6 volts. Manufacturers typically rate storage batteries at a 20-hour rate:

A 220-amp-hour battery will deliver 11 amps for 20 hours.

This rating is designed as a means to compare different batteries to the same standard. Batteries are electrochemical devices sensitive to climate, charge/discharge cycle history, temperature and age. The performance of a battery depends on climate, location and usage patterns. For every 1.0 amp-hour removed from a battery, about 1.25 amp-hours will need to be pumped back in to return the battery to the same state of charge. This figure also varies with temperature, battery type and age.

Batteries used in PV applications are fundamentally required to operate differently from those used in normal stationary or motive power applications. Unlike other conventional uses of storage batteries, the batteries meant for PV applications are characterized by a small or fractional change in state-of-charge (SOC) level on daily charge/discharge cycles, while exhibiting a sharp decline in SOC during certain periods in the year, depending on climatic conditions and season. In addition, typical stand-alone and remote PV installations require roughness and environmental flexibility and to be capable of unattended operation, easy installation and reliability. These conditions require that the sub-system including the battery should also meet the same criteria as set for the PV module. The batteries specially developed for such applications, usually called solar or photovoltaic batteries, are therefore designed to have the following characteristics:[1]

a) high cycle life;
b) good reliability under cyclic discharge conditions;
c) high capacity appreciation at slow rate of discharge;
d) low equalizing and boost charging requirement;
e) low self-discharge;
f) high watt-hour efficiency and ampere-hour efficiency at different SOC levels;
g) wide operating temperature range;
h) highly cost effective;
i) long life, robust design and low maintenance requirement;
j) manufacturing under stringent quality controls.

The technical performance and energy requirements for production and transportation of a stand-alone photovoltaic (PV)-battery system at different operating conditions are presented by Rydh and Sandén.[2] The energy requirement for battery production and transport is dominant for systems based on NiCd, NiMH and PbA batteries. Production and transport of batteries contribute 24–70% to the energy requirements, and the PV array contributes 26–68%. The contribution from other system components is less than 10%. For a PV-battery system with a service life of 30 years, this corresponds to energy pay back times between 2.5 and 13 years. The energy pay back time is 1.8–3.3 years for the PV array and 0.72–10 years for the battery.[3] The overall battery efficiency, including direct energy losses during operation and also energy requirements for production and transport of the charger, is 0.41–0.80 for battery and inverter, respectively.[3]

5.2 Fundamental Principles

A lead-acid storage battery is fundamentally a very simple thing. A laboratory model of a battery cell can be made by anyone in just a few minutes. Simply take two strips of metallic lead and hang them in and on opposite sides of a small glass jar and fill the jar with dilute sulfuric acid. Connect a source of direct current to these strips or plates and allow them to charge. In a short time the surface of one strip will become increasingly dark brown in colour while the other will retain its original lead colour. The brown plate has become covered with a layer of lead peroxide and is the positive plate of the cell. The unchanged plate is negative. When the DC charging source is removed, a sensitive voltmeter will indicate a voltage of approximately 2 volts between the terminals of the cells. If an electrical load is connected to the terminals, a current will flow from positive to negative and the cell will deliver power to the circuit. The thickness of this surface film, and therefore the cell's capacity, can be somewhat increased by alternate cycles of charge and discharge.

Of course, such a cell has no practical value because the available surface area of the two lead strips is not large enough to accumulate sufficient active material, these being the brown lead peroxide of the positive and metallic sponge lead of the negative. The primary problem in the development of batteries has been to increase the effective area of the plate surface to achieve greater and greater capacity for industrial use.

The most common method, however, of attaining large areas of active materials is to use very finely powdered lead oxides made up into pastes. These are in the form of a sponge with the electrolyte filling all the pores and thus coming into contact with the active material over an area many times the size of the evident surface of the pastes.

The active materials alone have no rigid mechanical form or strength and, particularly the positive, are very poor conductors of electricity. It is necessary, therefore, to mount them in some sort of lead alloy frame or grid to achieve and retain a physical shape and to conduct the current to all parts of the material. This lead grid usually takes the form of either a lattice-work into which paste is pressed, or a series of spines or core rods, each surrounded by a perforated rubber, plastic or glass fabric tube with the active material in the annular space between. The lattice type is commonly known as a lat-plate or pasted-plate type. This construction is nearly always used for the negative plates and can be used for positives also. The spine-and-tube construction is known as a tubular plate and is used only for positives.

5.2.1 Electro-chemical Action

In a lead-acid type cell, two different kinds of lead are acted upon electrochemically by a solution of dilute sulfuric acid (H_2SO_4). When the battery is fully charged, the active material of the positive plate is lead peroxide (dioxide) (PbO_2); the negative plate is sponge lead (Pb). As the cell is discharged, the electrolyte (H_2SO_4) divides into H_2 and SO_4. The H_2 combines with some of the oxygen formed at the positive plate to produce water (H_2O), which reduces the amount of the acid in the electrolyte. The SO_4 combines with lead (Pb) of both plates, forming lead sulfate ($PbSO_4$).

When the cell is discharged this action is reversed, and the lead sulfate ($PbSO_4$) on the positive and negative plates is converted to lead peroxide (PbO_2) and sponge lead (Pb), respectively. The strength of the electrolyte increases as the SO_4 from the plates combines with hydrogen from the water to form H_2SO_4.

$$PbO_2 + Pb + 2H_2SO_4 \underset{\text{Charge}}{\overset{\text{Discharge}}{\rightleftharpoons}} 2PbSO_4 + 2H_2O$$

In a fully charged battery, all of the active material of the positive plates is lead peroxide, and that of the negative plates is pure sponge lead. All the acid is in the electrolyte and the specific gravity is at its maximum. As the battery discharges, some of the acid separates from the electrolyte, which is in the pores of the plate, forming a chemical combination with the active material, changing it to lead sulfate and producing water. As the discharge continues, additional acid is withdrawn from the electrolyte and further sulfate and water is formed. As this process continues, it can be readily understood that the specific gravity of the electrolyte will gradually decrease because the proportion of acid is decreasing and that of water is increasing.

When the battery is placed on charge, the reverse action takes place. The acid in the sulfated active material of the plates is driven out and back into the electrolyte. This return of the acid to the electrolyte reduces the sulfate in the plates and increases the specific gravity of the electrolyte. The specific gravity will continue to rise until all the acid is driven out of the plate and back into the electrolyte. There will then be no sulfate in the plates.

After all the acid is returned to the electrolyte, additional charging will not raise the gravity higher. All of the acid in the cells is in the electrolyte and the battery is said to be fully charged. The material of the positives is again lead peroxide, the negatives are sponge lead and the specific gravity is at a maximum. On discharge the plates absorb acid and on charge they return the acid absorbed back to the electrolyte. As the cells approach full charge they cannot absorb all of the energy from the charging current and the excess acts to break up water from the electrolyte into its two components, hydrogen and oxygen, which are liberated from the cells as gases. This is the primary reason for the required addition of water to battery cells.

5.3 Physical Construction

The positive and negative elements are invariably in the form of a comparatively thin plate with grid structure usually of lead-antimony alloy. The addition of antimony to the lead gives it greater physical strength and rigidity and offers greater resistance to formation or corrosion by the electrolyte action with the acid. These plates are arranged parallel to each other, alternately positives and negatives. All the positives are joined and thus connected together by an alloy strap, and likewise the negatives. This strap, through its post, leads to the external circuit.

The length, width, thickness and numbers of plates in a cell are determined by the capacity required for the desired application. It is common practice to have a negative plate at each end of the element, thus making one more negative than positive plates in the cell. Thus a 15-plate cell has 7 positive and 8 negative plates. As mentioned, this is merely common practice; there is no technical reason for it. The two outside negative plates are frequently thinner as the outer surface gets very little use. The positive and negative plates must not come into contact with each other and are prevented from doing so by a separator. Separators are usually in sheet form and are commonly made of rubber, glass or plastic. They must be microporous in structure to permit the electrolyte to permeate them. The element consisting of the positive and negative plates and separator is placed in a jar or multicell container, which holds the electrolyte, this being, as mentioned above, dilute sulfuric acid, and a cover is placed over the element and sealed to the top of the jar to exclude dirt or foreign material and reduce the evaporation of water from the electrolyte. The cover has a vent plug which has small holes for the escape of gases and which can be removed for the purpose of adding water and taking hydrometer readings. The above assembly constitutes a cell. One or more cells together for a given application constitute a battery.

5.3.1 Voltage

The voltage of a cell is a fundamental characteristic of the elements that constitute it. Almost any two dissimilar metals or elements in a conducting electrolyte will produce some voltage. The vast majority of such combinations, however, have no practical or commercial value. The lead-acid cell has the highest voltage (per cell) of any commercial type. It is generally referred to as has having a nominal voltage of 2 volts.

Thus, a 3-cell battery is usually referred to as a 6-volt battery or as a 120-volt battery *etc*. The voltage on an open circuit (with no current flowing in either direction, and after sufficient time for the voltage to stabilize) is a direct function of the specific gravity and is presented very closely by the formula

$$\text{volts} = \text{specific gravity} + 0.84$$

Thus, the open circuit of a cell with a specific gravity of 1.210 will be 2.50 volts; one with a gravity of 1.280 will be 2.12 volts.

As soon as a cell starts to discharge, there is a decrease in voltage due to the effective internal resistance of the cell. This voltage drop increases with increase in discharge current, thus lowering the output voltage of the cell by that amount. Also at a continuous given rate of discharge, the voltage gradually becomes lower as the discharge progresses until, as the cell nears exhaustion, the voltage drops very rapidly to and below a value where it is no longer effective for the final voltage. It varies with the rate of discharge being lower with higher ampere rates. A representative value of 1.75 volts is, however, commonly used for a large proportion of typical battery applications. When a discharged battery is placed on charge, its voltage immediately rises, the extent of this rise increasing with the charging rate. With commonly used rates, the voltage will rise within a matter of minutes to 2.10 or 2.15 volts and then increase gradually until the charge is perhaps three-quarters complete. Near that point the voltage rises more sharply, and then levels off at a maximum when the battery is fully charged. The voltage at this point is about 2.6 volts per cell at the normally used finish-rate of charge.

5.3.2 Specific Gravity

The value of specific gravity of a battery when fully charged is a matter of design and is affected by many factors. In the first place the gravity must be high enough for the electrolyte to contain a sufficient amount of actual sulfuric acid to fulfil the chemical requirement of the cell. On the other hand, if the gravity is too high the acid content may be strong enough to have a direct chemical effect on certain parts of the cell. Between two extremes there are other factors, such as capacity, temperature and battery life, *etc.*, which dictate the particular gravity best suited to a given purpose. The full-charge gravities most commonly used (usually expressed as a range of plus or minus 10 points) and certain representative applications are as follows:

1.290 – Heavily worked or cycled batteries such as electric industrial trucks.
1.260 – Automotive services.
1.245 – Partially cycled batteries such as railway car lighting and large engine starting batteries, *etc.*
1.215 – Batteries in stationary standby or emergency service.

The electrolyte of a lead-acid cell takes a direct part in the chemical reaction, decreasing in gravity as the battery discharges and increasing to its original value as the battery is recharged. Thus, its value at any particular time is an approximate indication of the state of charge of the battery. This is determined by comparing the gravity as read with the full-charge value and the published specific gravity drop, which is the decrease from full charge to nominal discharge. The change in specific gravity is directly proportional to the charge or discharge (in ampere-hours).

5.3.3 Specific Gravity Corrections

The specific gravity varies with changes in temperature. This is not due to any characteristic of the battery but merely to the fact that the electrolyte expands as the battery temperature is lowered and the gravity rises. This change is equal to one point (0.001) in gravity for every 1.7 °C change in temperature.

Similarly, the gravity will vary as the electrolyte level falls and rises with the use and addition of water. As the water is consumed by gassing and evaporation, the level falls and the remaining electrolyte contains a greater proportion of acid, thus the specific gravity is higher (after water is added and becomes mixed it will return to its previous value). In a certain type of cell, for example, the gravity may rise 15 points with each 1/2" drop in level. In order to accurately compare specific gravity readings taken at different times and different temperatures and electrolyte levels, such readings are corrected to the normal reference temperature of 42.7 °C and the normal level. Such corrected specific gravity readings indicate what the gravity would be if the temperature and level were at the above normal values. To make this correction on the above type of cell:

- add one point of gravity for each 1.7 °C above 42.7 °C or subtract one point of gravity for each 1.7 °C below 42.7 °C.
- Subtract 15 points of gravity for each 1/2" below the normal level or add 15 points for each 1/2" above the normal level. Example: specific gravity of a cell reads 1.235 at 49.5 °C and 1/2" low level.

49.5–42.7 = 6.8 °C/1.7 = 4 points to be added.
1/2″ low level = 15 points to be subtracted.
Net result: subtract 11 points: corrected gravity is 1.224.

5.3.4 Capacity

The capacity of a storage battery is its ability to deliver energy and it is usually expressed in ampere-hours, which is simply the product of the discharge in amperes over a numbers of hours. However, a simple figure of say 200 ampere-hours has very little significance unless it is qualified by the many factors which influence a battery's capacity and also by the customary usage of the application in which it is applied. The principal factors which influence capacity are:

Discharge rate: The higher the discharge rate in amperes, the fewer total ampere-hours a battery will deliver under otherwise similar conditions. This relationship will vary somewhat with different types of plate and cell construction. Figure 5.1 shows a nominal relationship of a typical commercial cell. During discharge, the only portion of the electrolyte which is useful is

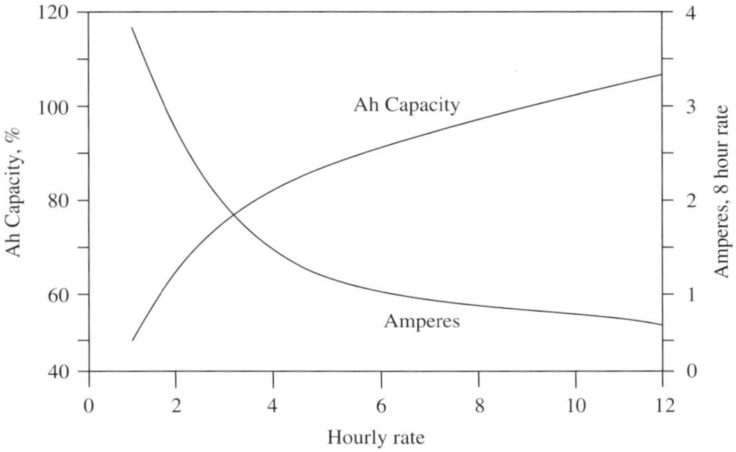

Figure 5.1 Capacity-rate curve based on 8-hour rate.

that in the pores of the plate in actual contact with the active material. As the acid in this portion becomes depleted or exhausted, the electrolyte must diffuse or circulate in order to bring more acid to the active material where it is needed. The higher the rate of discharge, the more rapid this circulation must be to maintain normal cell voltage. As the rate increases, however, this circulation or diffusion does not increase in the same proportion, with the result that the electrolyte in the pores of plates is less dense and the cell voltage decreases more rapidly, thus limiting the total capacity.

Another result of higher current rates is the increase in voltage drop within the cell. All the cells have a certain internal ohmic resistance. The higher the current, the greater the voltage drop or the loss in this resistance within the cell, thus reducing its external or useful voltage which supplies the load. The rate most commonly used as a standard is the 8-hour rate which can be expressed, for example, either as 100 Ah at the 8-hour rate or 12.5 amperes for 8 hours. Cranking and reserve capacity and motive power (industrial truck) types are rated on a 6-hour basis. Any correct rating is quite proper to use as long as it is properly specified and understood. Manufacturers usually list several hourly ratings, nearly always including the 8-hour, for the convenience of users in making comparisons and conducting tests.

Specific Gravity: This likewise affects cell capacity as electrolytes of different gravities have different amounts of actual acid per unit of volume. Thus, an electrolyte of higher gravity has more actual acid in contact with the active material and available for chemical reactions than an electrolyte of lower gravity. With given total acid requirements, the need is met more readily by high gravity and with less rapid diffusion or circulation. Also the higher gravity electrolyte has a lower electrical resistance, which better maintains the terminal voltage of the cell. The degree to which specific gravity affects cell capacity will vary considerably with different types of designs but a rule of thumb frequently applied is that a difference of 25 points in gravity will change the capacity 8–10%. For example, if a certain cell has a capacity of 100 ampere-hours with full charge gravity of 1.275 its capacity will be 90–92 ampere-hours if the full charge gravity is reduced to 1.250.

Temperature: Many chemical reactions are accelerated at high temperatures. Also the resistance and viscosity of the electrolyte are reduced at higher temperatures, thus reducing the voltage drop or loss within the cell and maintaining its terminal voltage at higher value. These combine to increase the battery's capacity at higher temperatures and reduce it at lower temperatures.

Final Voltage: This term is used to designate the minimum useful and accepted voltage at various rates of discharge, and is the value at which the maximum number of ampere-hours can be obtained before the cell voltage begins its rapid decline as the point of exhaustion is approached. It is just over the knee of the discharge curve and is lower with higher rates of discharge. The final voltage selected or listed for a particular cell depends largely on its application.

5.4 Discharge Characteristics

In general, a battery may be discharged without harm at any rate of current it will deliver but the discharge should not be continued beyond the point where the cell approaches exhaustion or where the voltage falls below a useful value. Discharging at a constant current value, the initial voltage depends on the rate of discharge and the normal characteristics of the cell. As the discharge continues the cell voltage will slowly decrease during perhaps the first 70 to 80% of the total time period. Then it will fall rapidly passing over the knee of the curve to the final voltage as full time and capacity are reached. This knee is more pronounced at low rates of discharge. The total ampere-hours available varies with the rate of discharge, being at higher rates. This lower ampere-hour value does not, however, represent any specific loss of energy – it simply means that the cell voltage falls to its minimum useful value in a shorter period of time. To illustrate this, assume a cell rated at 100 ampere-hours at the 8-hour rate, which means that it will deliver 12.5 amperes for 8 hours. The 2-hour capacity is about 66 ampere-hours or 33 amperes for 2 hours. If it discharged at this latter rate, the voltage would fall to its established minimum or final voltage in 2 hours, but if the discharge rate is then decreased, the voltage will recover or rise and further capacity (ampere-hours) can be obtained before the voltage again falls to the same minimum value. In fact, if the current is reduced to 5.5 amperes for the remaining 6 hours, the total 100 ampere-hours (or nearly that amount) can be still be obtained over the 8-hour period to the same final voltage. Figure 5.2 shows the approximate effect of discharging a cell at successively lower rates, carrying each one to the same final voltage. This result is not obtained when the higher rates are at the end or latter part of the discharge period as there is then

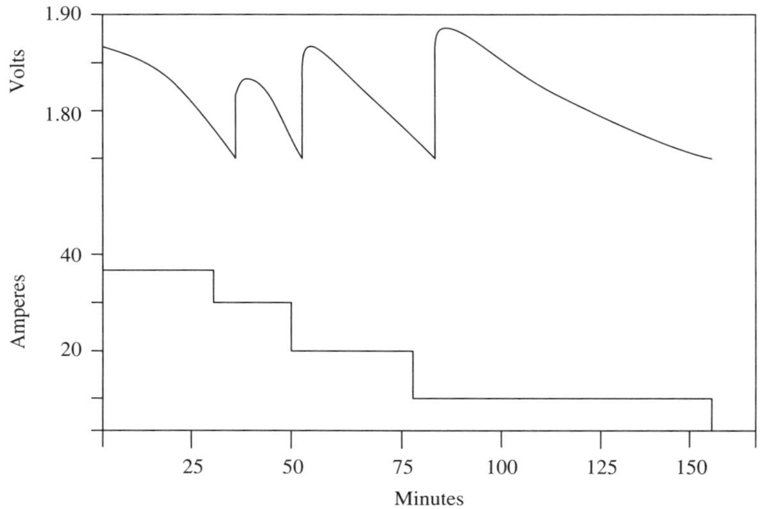

Figure 5.2 Effect of discharge rates on cell voltage.

no opportunity for sufficient diffusion of the electrolyte to maintain the cell voltage.

During discharge there is normally a rise in battery temperature, depending on the rate of discharge and the type of battery assembly from the standpoint of heat dissipation. The higher the ampere discharge rate, the greater the temperature rise effect. The actual chemical reactions on discharge absorb a small amount of heat, which would tend to cool the battery slightly, but the heat due to the internal resistance (I^2R) of the cell is greater so that the net result is an increase in temperature.

As mentioned above, a battery should not be discharged beyond the point where the cells approach exhaustion. This is referred to as over-discharging and can have harmful results; if the battery is not promptly recharged during all discharge, a certain amount of lead sulfate is formed, this being a perfectly normal and necessary part of the chemical reaction. This lead sulfate occupies more space than the sponge lead of the negative plate, so that, during discharge, the plate material expands slightly. If the discharge is carried too far, the material may expand to the point where portions of it separate and lose proper contact with the grid, and therefore with the electrical circuit. It cannot receive the charge remaining as sulfate, instead returning to its normal full-charged state as sponge lead. This can also occur to some extent if the battery is normally discharged but allowed to remain in that condition for a long period before being recharged. In this case some of the normal sulfate may become crystalline in nature and difficult to return to its original state. Any material subtracts from the capacity of the cell, tends to wash from the surface of the plate and falls to the bottom of the cell as sediment.

5.5 Charging Characteristics

A battery may be charged at any rate in amperes that will not produce excessive gassing. Another index is that any rate that does not result in a cell voltage of more than 2.4 volts is safe, while the current is above the normal or finished rate of charge. The current may be continued at the finish rate whenever charging is required, regardless of the cell voltage. The manufacturer usually determines and publishes such a normal or finish rate in amperes for each type and size of cell made. This rate is a current value, which can safely be used at any time that charging is required and which can be continued to the completion of the charge without causing excessive gassing or high temperature. This finish rate is usually between 4 and 10 amperes per 100 ampere-hours of the battery's capacity (8 hours) depending on the type of the cell assembly. Where a number of high-capacity cells are assembled as compact mass the available surface for heat dissipation is much less than for separate individual cells, and comparatively lower finish rates must be used in order to avoid high temperature. A battery which is partially or completely discharged can safely absorb much higher currents than the finish rate, up to possibly 10 times that value, but as it approaches full charge, the current must be reduced, either gradually or in one or more steps, to the finish rate or less at the end of charge. In practical applications, it is seldom necessary to use currents of more than four or five times the finish rate to charge in the time available. When the charge is complete, it should be stopped or reduced to a very low value. In any type of service, a battery should receive the correct amount of charge, sufficient to fully charge it and/or maintain it in that condition, but no more. In other words, undercharge or overcharge should be avoided to whatever extent is practical under the conditions in use. An insufficient amount of charge, even to a small degree but continued, will cause gradual sulfation of the negative plates with eventual loss of capacity and reduction of battery life.

An excessive amount of charge will tend to form up (corrode) the grid of the positive plates into lead peroxide, thus weakening them physically and increasing their electrical resistance. If the overcharging is at comparatively high rates, the gassing will be excessive and this tends to wash the positive active material from the plates. All of these results reduce the capacity and shorten the life of the battery. With the time operation, there are reasonably simple checks to determine whether or not the amount of charge is correct. If the proper amount of charge is being given, the specific gravity will reach its approximate full charge value at the end of a recharge or remain at that value in floating or similar service. Also, the amount of water required by the cells will be a normal minimum.

It is difficult to specify in general terms the normal water requirements, as they vary with batteries of different full-charge gravities and with the type of service from the standpoint of the amount of cycling (charge and discharge) which the battery receives.

5.6 Selection of PV Battery

In most cases the choice of battery is based on lowest price. Because of this, an inadequate and improper battery is selected, which reduces the system's reliability and durability. Many approaches can be followed for the selection of a PV battery. Cycle life, performance at extreme temperature, effect of rate of discharge, self-discharge rate, battery voltage and maximum current drain capacity in ampere-hours, watt-hours per weight, maintenance requirements, watt-hours per unit volume and cost per watt-hour are a few critical parameters which can be optimally combined to select the right battery for any particular PV installation. Conventionally, a lead-acid automotive battery has been used in most PV installations. Recently, industrial lead-acid battery types with pasted, plante or tubular plates, having grids with low or high antimony content or of pure lead or calcium alloys, are frequently used. Further, vented, gelled and recombination types make the selection even wider. In addition, according to promoters the nickel-cadmium battery has better performance characteristics over the lead-acid battery. Other alkaline battery systems also compete with lead-acid batteries for PV applications for their longer service life and completely maintenance-free operation.[1]

5.6.1 Batteries Commonly Used for PV Applications

The most commonly used storage battery for PV applications is the lead-acid type. Alkaline batteries are also suitable for PV applications, however at present only nickel-cadmium has acceptable performance characteristics and life-cycle costs for these applications.[4] Automotive, traction, stationary and maintenance-free gelled electrolyte batteries have found their use in different PV applications. Automotive batteries (also known as SLI; Starting, Lighting and Ignition batteries) have traditionally been used for daily shallow depth-of-discharge (DOD) PV applications, *e.g.* street lighting, although they have only a 2–4 years life span and a poor cycling ability. A stationary battery is frequently used for applications involving telecommunications, navigational aids, emergency lights, uninterrupted power supply systems, *etc*. These are capable of occasional deep discharge. Rechargeable traction or motive power batteries are used in electric vehicles, which can also be powered by a photovoltaic array. Maintenance-free batteries are increasingly required in automotive, traction or stationary applications. Gelled electrolyte or sealed maintenance-free batteries are suitable for PV applications, which require completely unattended operations. Research and development on sealed lead-acid batteries for PV power application has recently led to the development of a tubular-type battery featuring acid immobilization using silica gel, antimony-free Pb grids and thicker plates compared to conventional ones.[5] Batteries with lead plates strengthened by calcium or small amounts of antimony are relatively cheap and exhibit good properties for remote applications. Self-discharge accelerated by antimony is reduced by using pure lead grids. As per the experience of the Electricity Generating Authority of Thailand (EGAT) and BP Solar, Australia, a battery with low antimony content is the best choice for PV applications.[6,7]

5.6.2 Battery Installation, Operation and Maintenance

In order to investigate the 'Battery charge control and management in PV systems', the Commission of the European Community (CEC) initiated concerted efforts in 1987 in this direction. The objective of this work was to identify battery operating problems based on experiences with 16 PV power plants.[8,9] The main problems found in these studied plants were due to poor operation and maintenance procedures; an inadequate battery charging system; improper sizing of the battery; and inadequate information on the condition of the batteries. In several PV plants, batteries were found to be damaged due to deep discharge, ageing and structural failure of the cell casing. A few cases of excessive overcharging and the large number of operating cycles in five years of their operation were observed. In addition to these, cases of explosions caused by a build-up of hydrogen in the cells were also observed. The investigation revealed that in most of the studied plants, the operation and maintenance procedure was not documented and routine tests of voltage, temperature, specific gravity and periodic visual inspections were not carried out. Some of the observed problems could have been detected and avoided, if proper operation and maintenance procedures were adopted. Several other studies have also reported the significance of adequate management of battery storage in PV installations. One such study emphasizes the need to install peripheral components for the acquisition and processing of battery specific parameters in addition to adopting active measures for battery management.[10] The battery in a PV installation is subjected to two distinct cycles, namely:

a) a daily cycle characterized by varying profile and amplitude depending upon the PV energy supplied and electrical energy given to the load;
b) a seasonal cycle depending on the variation in average insolation during the year.

These cycles cause several stresses and ageing mechanisms in the battery. The most commonly observed problem areas are the following:[1]

i) Overcharging the battery causes corrosion of positive grid plates and excessive gassing resulting in loosening of the active material. Due to this, loosened material deposits as sediment at the bottom of the cell. Overcharging may also cause temperature to rise to a permanently destructive level.
ii) Consistent undercharging of the battery leads to a gradual running down of the cell, which is indicated by the reduced specific gravity readings and the tendency of plates to become light coloured. Excessive undercharging also causes sedimentation of white lead sulfate powder. The strain on the plates caused by the lead sulfate, which occupies more space than the original active material on the plates, results in their buckling.
iii) Presence of non-conducting materials, which form a layer between the battery terminal and the connector, may offer an increased resistance to the passage of large currents through the load. Corroded terminals, however, may not ordinarily interfere with the charging of the battery or with the discharging at low discharge currents.
iv) Short circuits may be caused by a breakdown of separators and excessive sedimentation, due to a phenomenon called 'treeing', in which tree-like structures of lead are formed from the negative to positive plates. Treeing may be due to the presence of certain materials in the grid, *e.g.* cadmium. It may also be due to 'mossing', in which the sediment brought to the surface of the electrolyte by the gas settles on top of the plates leading to the formation of bridges over the separator tops.
v) When a battery is either operated at partial SOC for several days without equalization or it remains unused for any length of time in fully or partially discharged conditions, the deposition of large lead sulfate crystals instead of normal tiny ones on the plates takes place. The phenomenon called sulfation also occurs when there are temperature variations in the battery. These large crystals tend to increase the

internal resistance of the cell, which results in low discharge and high charge voltages.

vi) When the battery reaches full charge, the rise in plate potential beyond a certain cut-off voltage leads to the decomposition of water to hydrogen and oxygen gas (water loss). The quantity of gas formed depends on the amount of excess charging current which is not absorbed by the battery.

It is recommended that a battery (conventional flooded type) meant for PV applications is installed in a separate room in order to avoid accidents due to the formation of hazardous gases. Adequate ventilation and moderate temperature must be provided in accordance with the supplier's instructions. Batteries are most commonly designed for floor placement over wooden or plastic planks. In some cases, installation is also done on steel step stands with acid-resistant paint on them. A sealed maintenance-free battery can be housed in a usual working area with normal ventilation. It can be installed on slotted iron racks, although battery suppliers recommend specific installation guidelines, including battery-room designs based on the type and construction of the supplied battery. The standard guidelines for installation and maintenance of lead-acid batteries for PV applications and of nickel-cadmium batteries for generating stations and substations are available from the IEEE in the form of the American National Standards. These standards describe in detail the safety precautions, installation procedures, installation and design criteria and maintenance requirements.

A photograph of a battery bank and solar inverter is shown in Figure 5.3. The system is installed at Solar Energy Park (SEP), IIT Delhi. The total capacity of the system is 2320 Wp. This power is used for lighting the SEP, water pumping using a 0.35kW DC motor and also for street lights.

5.6.3 Battery Protection and Regulating Circuits

Proper battery operation in a PV system requires voltage-regulating protection circuitry to prevent overcharging and excessive discharging. Permanent damage can be done to a battery if it is charged too fast and for too long. Similarly, forcing higher charging currents into a battery when it is fully charged will cause the battery to gas. Excessive discharging will cause the plates to disintegrate and should be avoided. The use of voltage-regulating circuits to maintain the battery voltage within an acceptable range or window is therefore necessary. A few elementary regulator currents are discussed below.[4]

Shunt Regulator

The regulator which is connected in parallel to the PV generator dissipates excess energy through a resistor and power components. There is no voltage drop in the charging unit and the power consumption by the regulator is negligible during the non-regulation period. Any failure in the regulator does not interrupt the battery charging.

Figure 5.3 Photograph of a battery bank and solar inverter at Solar Energy Park, IIT Delhi.

Series Regulator with Semi-conductor

The series regulator uses a transistor in series with the PV generator. The regulator behaves like a variable resistance, whose value is a function of the state-of-charge (SOC) of the battery. The dissipated power at the transistor terminals is low compared to PV peak power. However, during the non-regulation period, the regulator introduces a voltage drop and thereby current consumption in the circuit.

Series Regulator by Electromechanical Cut-off

This regulator stops the battery charging by an electromechanical cut-off when it reaches the maximum acceptable voltage level. It is reset for charging automatically when the threshold voltage is reached. There is no power dissipated in the regulator.

Automatic Circuit Breaking

This regulator is used in cases of weak sunlight, over-consumption, *etc.*, when it becomes necessary to cut off the load to limit the depth of battery discharge. Below a certain threshold voltage level, the load is cut off and is reset automatically when the battery reaches a sufficient charge level.

5.6.4 Battery Simulation and Sizing

Stand-alone PV systems with battery storage are usually designed to ensure array energy output exceeding the load demand year round. The system is also expected to maintain a continuous supply of energy during cloudy days and for night-time loads. The battery size is dependent upon the load energy requirement and weather patterns on the site, the latter necessitating increased storage and PV capacity during the heavily overcast sky and low insolation period of the year. Consequently, during peak sunshine days, the battery will remain near a fully charged state with the array generating excess energy. In order to prevent the battery from overcharging it needs to be either disconnected or dissipated. A major concern in designing any PV power system, therefore, is to obtain optimum capacities of the PV array and the battery storage for the supply of energy at the chosen reliability. In order to match the battery behaviour properly with the array, as well as with the load, a modelling exercise is performed. This modelling exercise gives parameters characterizing the battery's state, *e.g.* current accepted and lost, internal e.m.f., voltage or terminal voltages, state of charge, internal resistance, *etc.* The application of the simulation technique in battery sizing results in an optimum battery capacity required to satisfy the given load with an expected reliability. Several researcher groups have developed battery models describing the relation between battery voltage, current and SOC. The University of Utrecht, The Netherlands, has carried out studies on models applicable for both technical design and economic analysis of the PV battery system.[11] The model describing a relation between the voltage, current and SOC of a battery is needed for its design. The discharging current is useful for designing the control system, a model for relating the capacity of the battery. The ageing model describing the lifetime of a battery is useful for an economic analysis.

5.7 Battery Lifetime in a PV System[12]

In PV systems, the average currents are relatively low compared with the battery capacity (discharges are often between the 100- and 300-h rates). The daily cycling is often very shallow. However, in order to get the required data in a reasonable time, cycle lives are usually measured at relatively high rates (10-h rate or higher current) and at high depth of discharge (DOD), often 80%. The 80% DOD in published cycle life data usually refers to a percentage of either the capacity at the standard discharge rate (often the 10-h rate), or to the actual rate at which the cycling test was carried out. In PV use, the low discharge rates mean that available capacity can be much higher than the nominal capacity, especially for tubular plate vented batteries, which have a large reserve of free acid.[12]

In order to translate manufacturers' cycle life data into meaningful numbers for estimating PV cycle life, some assumptions have been made. The total number of Ah discharged over the whole cycle life is a constant, independent of the DOD.[13] Where cycle life data at different DOD is reported, it is often claimed that the product of cycle life and DOD is higher at low DOD. By using a constant value which is taken from high DOD data (*e.g.* 80%), we should obtain a conservative value. The total number of Ah discharged over the whole cycle life is a constant, independent of the discharge rate. As far as we know, there is no experimental evidence for this at very low discharge rates. However, it is reasonable if we assume that the cycle life is dependent on the volume (*i.e.* density) changes in the battery plates. Note that the second assumption is not equivalent to the statement that the same number of cycles at a certain DOD is obtained at any discharge rate. To give a concrete example, a tubular plate battery may have a nominal capacity of 500 Ah at the 10-h discharge rate, and it may give 1000 cycles of 80% DOD at this discharge rate. At the 120-h discharge rate, the available capacity may be 725 Ah. However, the battery will not give 1000 cycles at 80% of 725 Ah per cycle. Using assumptions, at the 120-h rate it would either give 1000 cycles at 80% of 500 Ah (which is about 55% DOD referenced to the higher 120-h capacity of 725 Ah) or it would give approximately 690 cycles at 80% DOD based on a capacity of 725 Ah.

These assumptions have been represented by the following equation:[12]

$$L_c = N_c \times C_s \times D_{od} \times X_c \tag{5.1}$$

where N_c is the number of reported cycles at depth of discharge D_{od} relative to the standard (nameplate) capacity rating C_s (often the 10-h capacity). L_c is the total number of Ah for all cycles over the whole cycle life. X_c is an arbitrary correction factor used to de-rate the manufacturers' data (which refers to continuous and regular cycling) for PV use (where the cycling is not continuous or regular).

The predicted cycle life (in years) is then simply given by:[12]

$$Y_c = \frac{L_c}{365 \times D_c} \tag{5.2}$$

where D_c is the average daily cycling Ah that the battery experiences. If the electrical load is only switched on during the night (*e.g.* for lighting), then D_c is equal to the average load Ah in a 24-h period, since the charging occurs during the daytime, and all the discharging is at night. If the load is only switched on during the day, then D_c will be very small and the predicted L_c will be very large. If the load is continuous, D_c will be reduced from the total daily load Ah by a factor $(24–H_c)/24$, where H_c is the number of hours during the day when the PV array is charging the battery (*i.e.* when the array current is larger than the load). An approximation for this for any day is:[12]

$$H_c = H_{dl}\left(\frac{Ah_l}{Ah_a}\right) \tag{5.3}$$

where H_{dl} is the number of hours of daylight, Ah_l, is the total daily Ah load and Ah_a is the total Ah of battery charging that the PV array could give.

If there are no detailed data for the day length and available charging from the PV array, then a reasonable guess for H_c is between 6 and 10 h for most systems. Stationary batteries are often used in float service, where the voltage is held constant by a mains charger, the small float current keeping the battery completely charged for any emergency situation. Cycling is effectively zero in this application. In the limit of very shallow PV cycling, we might expect the conditions to be rather similar. The lifetime of a battery in float service is quoted by manufacturers as the expected service life at a particular float voltage and, very importantly, at a particular ambient temperature. The battery lifetime limitation in this case is corrosion of the positive grid material, which is dependent on both the applied voltage and the temperature. It is generally accepted in the battery industry that an increase in temperature of 10 °C will lead to a halving of the expected life in float service. The lifetime based on this temperature-dependent corrosion process is given as:[12]

$$Y_t = L_s \times X_t \left[2^{\frac{T_{av} - T_s}{10}} \right] \quad (5.4)$$

where X_t is the predicted lifetime in years, L_s is the stated lifetime for float service, at the standard ambient temperature T_s, and T_{av} is the average temperature of the battery environment. X_t is another arbitrary correction factor to compensate for the fact that the battery voltage is not truly constant. If T_{av} is less than T_a, a higher battery lifetime than the quoted standard lifetime L_s would be predicted. The value of T_{av} depends on the type of battery installation. For unheated buildings or for battery boxes mounted in the shade of a PV array, the average temperature can normally be approximated by the average outside air temperature considering that the battery itself generates negligible heat. At the low rates common in PV systems this is certainly justifiable for vented batteries. In the case of valve-regulated batteries, the heat produced on overcharge can be significant if the battery enclosure cannot reject heat to the surroundings easily. If the batteries are mounted in an equipment shelter, where the electrical load itself is a heat source, then the battery temperature can be expected to be higher than the outside air temperature. In the case when a battery is mounted in the same room as a PV-powered vaccine refrigerator, the refrigerator itself will cause some heating of the room and again the average battery temperature is a few degrees higher than the outside air temperature.

5.8 Charging State of PV-powered Storage Batteries[14]

Stand-alone photovoltaic (PV) applications, such as domestic and street lighting systems, usually include a storage battery which is subjected to a daily charge/discharge cycle. During such a cycle, the battery charges during the day and loses a percentage of its charge to the load at night. Knowledge of the battery state-of-charge (SOC) during charging is important, since it leads to design information about the desired size of the PV array and battery capacity to satisfy a given load.

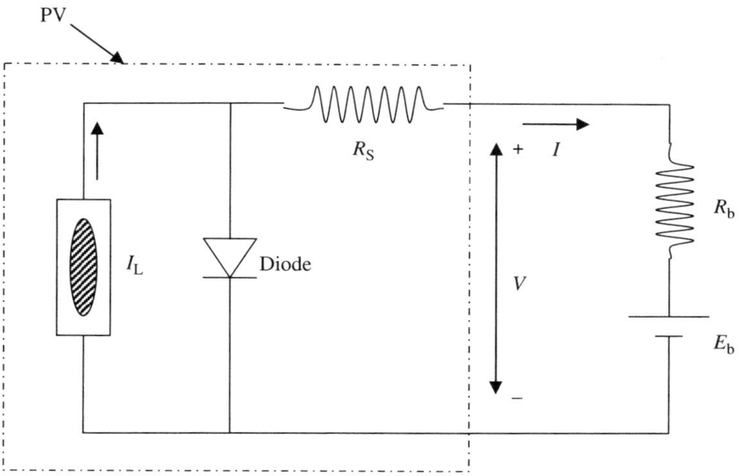

Figure 5.4 Equivalent circuit for charging a battery from a PV array.

Consider a simple stand-alone PV system configuration consisting of a storage battery as a daytime load to be charged from the PV array during the day. The equivalent circuit of such a system is shown in Figure 5.4, where the PV cell/module/array is represented by the single-exponential lumped-constant parameters model for which the solar cell I–V characteristic is described as

$$I = I_L - I_o(\exp B(V - IR_s) - 1) \tag{5.5}$$

where I is the output current, I_L is the light-generated current, I_o is the diode reverse saturation current, V is the terminal voltage, R_s is the lumped-effective series resistance and $B = q/nkT$, where q is the electronic charge, n is the diode ideality factor, k is the Boltzmann constant and T is the absolute temperature. When charged, the storage battery, represented by its open-circuit voltage E_b in series with its internal resistance R_b, has an I–V characteristic described by:

$$V = E_b + IR_b \tag{5.6}$$

which can be written as:

$$I = -E_b/R_b + V/R_b \tag{5.7}$$

In order to obtain the system operating point, eqns (5.5) and (5.7) have to be mathematically solved. A much simpler approach would be to solve the problem graphically by determining the intersection of the I–V curve representing the PV array with the straight line representing the battery load (eqn (5.7)). This is shown in Figure 5.5 for a certain radiation level and battery SOC. The operating point defines the charging current I_k flowing into the battery and the charging voltage V_k at the instant k depicted. Under actual operating conditions, the I–V

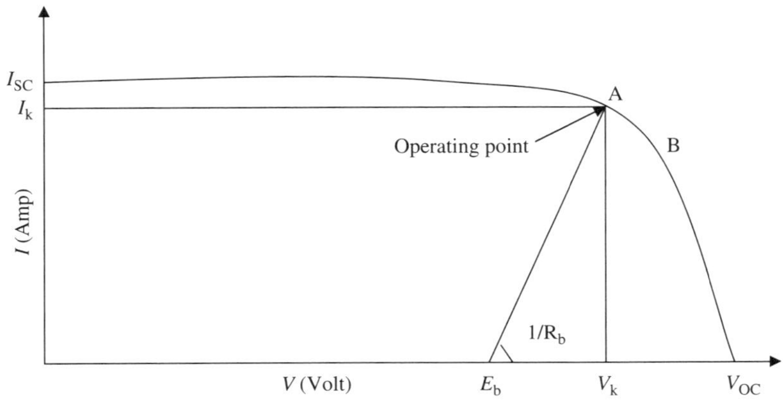

Figure 5.5 Intersection point of the PV array curve (B) with the battery load line (A).

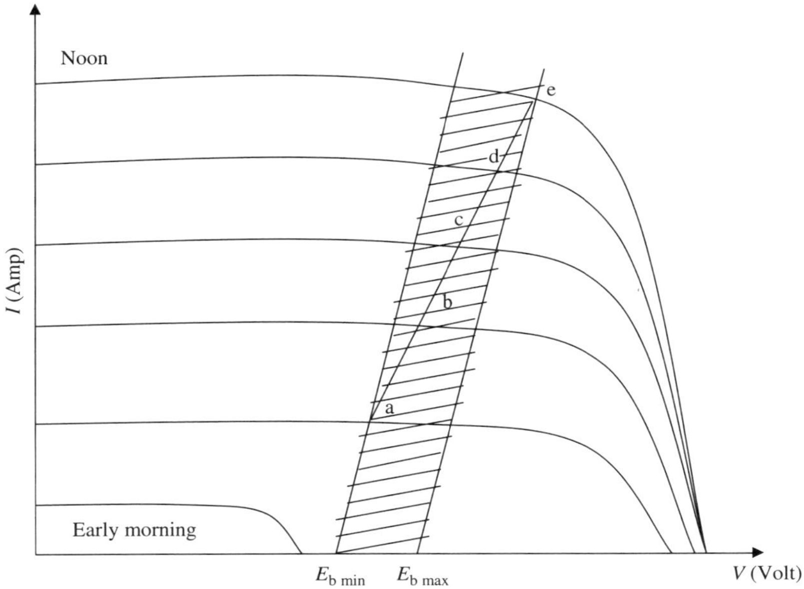

Figure 5.6 System operation region during charging.

characteristic of the PV array changes in response to the variations in solar radiation and cell temperature, resulting in a family of curves similar to curve B of Figure 5.5. Also, as the battery is charged, its open-circuit voltage increases, resulting in a family of load lines parallel to line A of Figure 5.5. This situation is depicted in Figure 5.6, where the shaded area represents the charging region in which the battery load line is allowed to exist, namely between the limits $E_{b\ min}$

and $E_{b\ max}$. The position of the load line depends on the charging rate which dictates the battery SOC and hence its open circuit voltage.

5.9 General Terms

5.9.1 Efficiency

The efficiency of a battery also defined as energy input *versus* energy output is widely dependent upon the circumstances of use. A small amount of energy is required to maintain it, even without any use, so that the greater the amount of proper use, the higher the efficiency. Normally the relation between a normal discharge and the necessary recharge is the basis on which efficiency is considered. This may be expressed in two ways: as the ampere-hour efficiency or as the watt-hour efficiency. In terms of ampere-hours, it is usually considered that the recharge should equal 110% of the discharge giving an efficiency of about 91%. However, the average voltage on charge is considerably higher than on discharge, in an approximate proportion of 17–18%, giving a voltage efficiency of 85%. Combining these two (91×0.85) results in a watt-hour (or total energy) efficiency of 77–78%, which can be considered as a representative figure.

5.9.2 Local Action

This is the term used to refer to the internal losses of a battery standing on an open circuit or when on float charge, and without considering any losses incidental to any discharge. As the term implies, this is due to the local chemical action between component parts of the plates and is almost entirely in the negative plates. For example, the negative material – pure lead and the antimony of the grid and any other constituents of the alloy react with the electrolyte as a "cell". It is practical to use a pure lead grid and eliminate every trace of impurity in the cell; there would be virtually no local action or loss. The degree of local action may be expressed either as the percentage loss in capacity per month on an open circuit, or by the amount of current required on float or trickle to overcome it and keep the battery fully charged. In either case, this varies with temperature, being greater at high temperatures and less with low.

5.9.3 Gassing

A battery cell cannot absorb all the energy from the charging towards the end of the charge, and the excess energy dissociates water by electrolysis into its component gases, hydrogen and oxygen. The oxygen is liberated at the positive plate and the hydrogen at the negative. When a battery is completely charged, all of the energy, except the small resistance loss, is consumed in the electrolysis.

During a recharge, gassing is first noticed when the cell voltage reaches 2.30–2.35 volts per cell and increases as the charge progresses. At full charge, when most of the energy goes into gas, the amount of hydrogen liberated is about one cubic foot per cell for each 63 ampere-hours input. In as much as a

4%-content of hydrogen in the air may be hazardous, the above may be used to relate the maximum amount from a given battery to the size of the room in which it is located.

5.9.4 Mossing

This is the term used to describe the possible deposition of a sponge-like layer of lead on the negative plates or strap. This material was originally shed from the plates (mostly the positive) in very fine particles and circulated throughout the cell by gassing, falling on both the positive and negative plates. When in contact with either plate it is changed to the active material of the plate. That on the positive is loose and non-cohesive in nature and simply washes off again from the gassing of the cell. Such material on the negative plate, however, is cohesive in nature and thus adheres to and builds up on the top edge and possibly along the side edges of the plate. It can accumulate to such an extent that it bridges over or around the separators, touching an adjacent positive plate and causing a partial short circuit. The accumulation of any appreciable amount of moss is usually an indication of overcharging in ampere-hours and/or high charging currents in amperes.

5.9.5 Sediment

There is a tendency for some of the active material on the surface of the plate to separate from the main body of material and fall or settle to the bottom of the container. This is counteracted in various ways. The material may contain a binding agent or it may be held in place by the various types of tubular construction, or on flat plates by perforated glass, rubber or plastic sheets or mats known as retainers.

Despite these means, a small amount of such material may fall from the plates. Most of it is usually from the positives and a certain space in the bottom of the container, below the plates, is usually reserved for this sediment. With a proper floating type of operation, this sediment is entirely negligible and may amount to hardly more than a layer of dust after years of operation. In active cycle service, an appreciable quantity may accumulate after years of use but the size of the sediment space is designed to accommodate all that will fall during the battery's life. Thus it should never be necessary to remove or clean the sediments from a battery.

5.9.6 Temperature

The operating temperature of a battery should preferably be in the normal range of 33 °C to 44 °C. A higher temperature gives some additional capacity at the time, but will reduce the total battery life. A very, very high temperature (70 °C) can damage some of the battery components and cause early failure. A low temperature reduces capacity but will prolong battery life under floating operation or in storage. A very low temperature may freeze the electrolyte, but only if the battery is discharged (low in specific gravity) at the time. At the temperatures shown in the following table, the electrolyte will not freeze unless the specific gravity is lower than indicated:

	Specific gravity (approx)	
Temperature	*At same temperature*	*Corrected to 42.7°C*
+20	1.100	1.080
+10	1.150	1.130
0	1.185	1.160
–10	1.210	1.180
–20	1.235	1.200
–30	1.250	1.215
–40	1.265	1.225

5.9.7 Internal Resistance

The actual ohmic value of the internal resistance of a cell is sometimes requested, as this varies with (1) the state of charge, (2) specific gravity, (3) cell size in amperes, (4) temperature, (5) physical construction and (6) its condition or the degree to which it is worn out. While it can be estimated for a given set of conditions, resistance value has little importance in practical value in the application or operation of a battery. The voltage and current characteristics on discharge and charge always are used to solve any practical problems.

5.9.8 Testing

The actual testing of battery capacity can be done only by conducting a discharge under controlled and recorded conditions. Manufacturers regularly do this in their laboratories in line research and production checking. The test discharges are conducted in the following manner.

The battery is first properly and completely charged. Temperature and specific gravity must be at their normal or standard values or corrections applied to allow for any difference. A discharge rate is selected, depending on the time and load equipment available. Usually a rate between the 31-hour and 8-hour discharge rates is chosen. The discharge rate in amperes is held constant at the chosen value and the total battery voltage read falls to the pre-selected final voltage value. The capacity is expressed as the percentage of time at which final voltage was reached.

5.9.9 Dry-charged Batteries

When it is desired to keep or store new batteries for a considerable time before they are required, they are frequently manufactured or prepared in a dry-charged condition. This consists essentially of charging and drying the plates, before assembly in an atmosphere devoid of air or oxygen. All elements of the assembled cell are completely dry and the cell is partially or completely sealed to keep out any moisture. Such batteries must be stored in a cool, dry location until ready for use, and under these conditions the plates will retain most of their charge for as long as perhaps two years. The battery may be used shortly thereafter although it may not have full capacity depending upon the length and condition of storage. In any case it is preferable first to give it a thorough equalizing charge after filling. Once it has been properly prepared, its capacity, characteristics and life are the same as a new wet-battery.

5.9.10 Maintenance

The routine maintenance of storage batteries varies widely with the type of battery and its use.

- Proper charging is the most important factor in battery service and life and the proper method for each application should be carefully followed. A battery in frequent cycle service need not necessarily be completely recharged each time but should be given a proper equalizing charge weekly. A battery in floating or standby service or in storage should be kept fully charged or as nearly as conditions will permit.
- Water should be added at necessary intervals to keep the electrolyte level between normal upper and lower limits. The plates must not be allowed to become dry.
- Batteries must be kept clean and dry to the extent that no corrosion, dust or moisture offer a conducting path to partially short-circuit the cell or contact ground.

Lead batteries do not require any routine overhaul or solution changes during their entire life except as a result of accidental or similar damage.

5.9.11 Lead-Calcium Cell

The lead calcium cell is recognized as an improvement over antimony for batteries in certain types of float applications. They are best utilized where the discharge requirements are light. The advantage of calcium batteries is their very low local action or self-discharge as compared to the antimony battery. There are very low amounts of calcium in the grids as compared to the large amount of antimony required. This is the reason for the low local action. This reduces the current required to maintain full charge on float and therefore reduces the water consumption. This results in a reduction of power required and substantially increases the watering interval for the calcium battery. Calcium and antimony batteries will both provide reliable and satisfactory life service but the specific application must be established to select the proper battery. Such batteries in standby floating service draw less current due to the lower local action and thus require less frequent water addition. Because of these lower losses, it is assumed that they will have longer life in years.

On the other hand the nearly pure lead grid of the positive plate is more susceptible to "*formation*" (corrosion) from charging and all unnecessary charging must be carefully avoided. This eliminates them "*from*" any regular cycle type of operation as the regular recharges would soon form the grid to the point where it would have high resistance and eventually crack and crumble. They also develop a higher voltage near the end of charge, which means that in order to fully charge them either a higher charger voltage or a longer time is required.

Problems

5.1 Explain the characteristics of photovoltaic batteries.
5.2 Explain electrochemical action in a battery, using a chemical formula.
5.3 How do the specific gravity, discharge rate and temperature affect the capacity of a battery? Explain with a diagram.
5.4 Explain discharge characteristics with a diagram.
5.5 Calculate the battery lifetime in a PV system of a tubular plate battery having capacity of 500 Ah at the 10-h discharge rate, which gives 1000 cycles of 80% DOD. Hint: use eqns (5.1) and (5.2).
5.6 Explain the charging state of PV-powered storage batteries with a diagram.
5.7 What are gassing and mossing in a battery?

References

1. A. Chaurey and S. Deambi, *Renew. Energ.*, 1992, **2**(3), 227–235.
2. C. J. Rydh and B. A. Sandén, *Energ. Convers. Manag.*, 2005, **46**, 1957–1979.
3. C. J. Rydh and B. A. Sandén, *Energ. Convers. Manag.*, 2005, **46**, 1980–2000.
4. F. Lasnier and T. G. Ang, *Photovoltaic Engineering Handbook*, Adam Hilger, Bristol New York, 1990, pp. 101–137.
5. M. Iwate, *Battery and Fuel Cells*, 1991, **3**, 114.
6. S. J. Lancashire, in *Proceedings of the 20th IEEE PV Specialists Conference*, Las Vegas, Nevada, USA, 1988, pp. 1157–1163.
7. C. Jivacate, *in Proceedings of the Third Asian Battery Conference*, Bangkok, 1989, pp. 14–16.
8. S. McCarthy, in *9th European Commission Photovoltaic Solar Energy Conference*, Freiburg, Germany, 1989, pp. 1142–1145.
9. S. McCarthy, in *9th European Commission Photovoltaic Solar Energy Conference*, Freiburg, Germany, 1989, pp. 832–834.

10. I. B. Willer, in *9th European Commission Photovoltaic Solar Energy Conference*, Freiburg, Germany, 1989, pp. 795–798.
11. E. W. T. Horst, in *8th European Commission Photovoltaic Solar Energy Conference*, Florence, Italy, 1988, pp. 461–465.
12. D. J. Spiers and A. A. Rasinkoski, *Sol. Energ.*, 1996, **58** (4–6), 147–154.
13. H. Bode, *Lead-Acid Batteries*, Wiley Interscience, New York, 1977, p. 333.
14. M. A. Hamdy, *J. Power Sourc.*, 1993, **41**, 65–76.

CHAPTER 6

Case Studies of PV/T Systems

6.1 Introduction

In order to investigate the technical, operation and maintenance issues of PV/T systems, many studies have been carried out, aiming at:

1) demonstrating new energy technologies;
2) exploring local PV/T industries and markets;
3) investigating the environment protection actions of PV power supply in urban and rural areas of the country;
4) accumulating technical study and cost effective design experiences; and
5) providing a practical site for training local PV technicians and students.

This chapter presents some of the case studies on application of photovoltaic systems. They demonstrate the design and installation aspects, output power analysis, energy and emission savings and costs incurred. They also demonstrate the successful use of sustainable materials, conservation of resources and integration of renewable energy technologies. The examples are chosen from different climatic zones so as to present a wide variety of techniques.

6.2 Case Study I: Grid-connected Building Integrated Photovoltaic System (BIPV): Hong Kong

The rapid development in recent years of grid-connected building integrated photovoltaic (BIPV) systems is due to government-initiated renewable energy programs aiming at the development of renewable energy applications and reduction of greenhouse gas emissions. The first grid-connected BIPV system in Hong Kong was completed in 1999, funded by the Hong Kong SAR Government.[1] PV has been installed on the three walls and the roof of a plant room on a building, as shown in Figure 6.1. PV panels are integrated on the horizontal roof and the vertical east, west and south facades. An air gap was

RSC Energy Series No. 2
Fundamentals of Photovoltaic Modules and Their Applications
By G. N. Tiwari and Swapnil Dubey

Published by the Royal Society of Chemistry, www.rsc.org

Figure 6.1 The first grid-connected BIPV system in Hong Kong[1] (courtesy H. Yang, Hong Kong).

designed between the massive wall and the PV panels for the three vertical facades so that a natural ventilation effect can be obtained. The system consists of 100 PV panels (made by BP) each with 80 Wp and a TCG 4000/6 inverter, in which 20 panels face east, 22 south, 18 west and 40 on the top. The system was rated at 8 kW with output DC voltage of 75–105 V and output AC voltage of 220 V. The total PV power capacity of the project is 8 kWp and the integration area is 55 m^2; 11 m^2 is located on the vertical west facade, 11 m^2 on the vertical east facade and an additional 12 m^2 on the vertical south facade. The remaining 21 m^2 is located on the roof of the building. In order to increase the DC voltage, 7 PV modules are connected in series. The DC output is about 100 V. All the PV modules were involved in the remote system tests, but only 6.6 kWp was used for the actual grid-connected BIPV system test. Electricity generated from the BIPV system is used for daytime lighting of the building in an isolated lighting area for about 250 m^2 floor area in the building.

The overall energy efficiency of this system was found to be 9% while the energy efficiency of the inverter is 86–87%. Table 6.1 gives the results of monthly energy output from different PV facades for Hong Kong climatic conditions. The roof PV array has the maximum power output, since the annual average solar incident angle is the smallest compared with the solar incident angles of the other three facades. Depending on the local latitude, the roof is the best area for installing BIPV modules. However, the simulation indicated that the south facade generates nearly as much output as the west and east facades due to the lower local latitude. The total annual energy required for the lighting is 16,700 kWh. The total lighting energy supplied by the solar energy system is around 41%. The total harmonics from this BIPV system is less than 12% for most of the time, even when the incident solar irradiation is very weak.

Table 6.1 Monthly power output from PV arrays for different orientations for Hong Kong climatic conditions (kWh).[1]

Facade	*January*	*February*	*March*	*April*	*May*	*June*	*July*	*August*	*September*	*October*	*November*	*December*	*Total*
Roof	294	220	231	252	294	286	367	347	346	389	346	336	3708
South	156	90	72	48	36	31	36	54	90	150	163	180	1106
West	72	60	63	72	88	88	116	106	99	99	86	83	1032
East	72	60	63	72	88	88	116	106	99	99	86	83	1032
Total	594	430	429	444	506	493	635	613	634	737	681	682	6878

The power price of the BIPV system in Hong Kong is found to be HK\$1.5–2.0 kWh^{-1} (US\$ 0.19 0.25 kWh^{-1}) while the average price of electricity purchased from the two local power companies is about HK\$0.90 kWh^{-1} (US\$ 0.12 kWh^{-1}). The cost of BIPV systems with monocrystalline silicon PV modules is about HK\$40 Wp^{-1} (US\$ 5.16 Wp^{-1}) including installation and other component costs (inverter, safety control and cables). When the cost reduction of the building facade outer skins due to PV integration is considered, the pay back period for roofs is about 20–30 years, *i.e.* the lifetime period of PV modules. It is more advantageous to the BIPV if the environmental pollution costs are considered, *e.g.* greenhouse emission cost, business loss due to pollution in urban areas and medical cost increase caused by pollution from conventional power generation plants. Both the energy saving and environmental impact of BIPV application must be considered, which will make the use of BIPV technology applicable and economical. This project will play a very important role in education and deployment of renewable energy applications.

6.3 Case Study II: Simulation of an Existing BIPV System for Indian Climatic Conditions

The BIPV system shown in Figure 6.1 has been considered for Indian climatic conditions. Analysis of the system has been evaluated considering four weather conditions (a, b, c and d types) for five different cities (New Delhi, Bangalore, Mumbai, Srinagar and Jodhpur) in India.[2] The total area of building and integrated areas on the south, east, west and roof is considered to be the same as mentioned by Yang *et al.*[1] The 3D representation of a working model of a building integrated photovoltaic (BIPV) system installed in Hong Kong is shown in Figure 6.2. In total 35 PV modules are integrated over the roof and 14 different possible series and parallel combinations of duct have been considered for the calculation of thermal and electrical energy gain.

The hourly variation of beam and total radiation for a typical day in a summer month (May) for New Delhi conditions is shown in Figures 6.3 and 6.4. The beam radiation on the east and west facades (inclined at 90°) during the evening and morning hours has been found to be zero, as expected. Beam radiation on the south facade (inclined at 90°) at 8 a.m., 4 p.m. and 5 p.m. are also found to be zero due to the overhead motion of the Sun. The hourly variations of power output from the roof (inclined at 30°), south, east and west facades are shown in Figure 6.5. Table 6.2 gives the results of monthly energy output from different PV facades for New Delhi climatic conditions. Maximum power output has obtained for the inclined roof. The variation of annual thermal and electrical gain from different combinations of air duct considering the four types (a, b, c and d) of weather conditions of New Delhi (air velocity in duct is 2 $m\,s^{-1}$) is shown in Figure 6.6. Maximum thermal (13.42 MWh) and electrical (4.38 MWh) gain has been obtained for a sixth combination (11 ducts connected in parallel (1 set) and one duct having two PV modules; all are connected in parallel) because in a parallel connection outlet air temperature and losses were

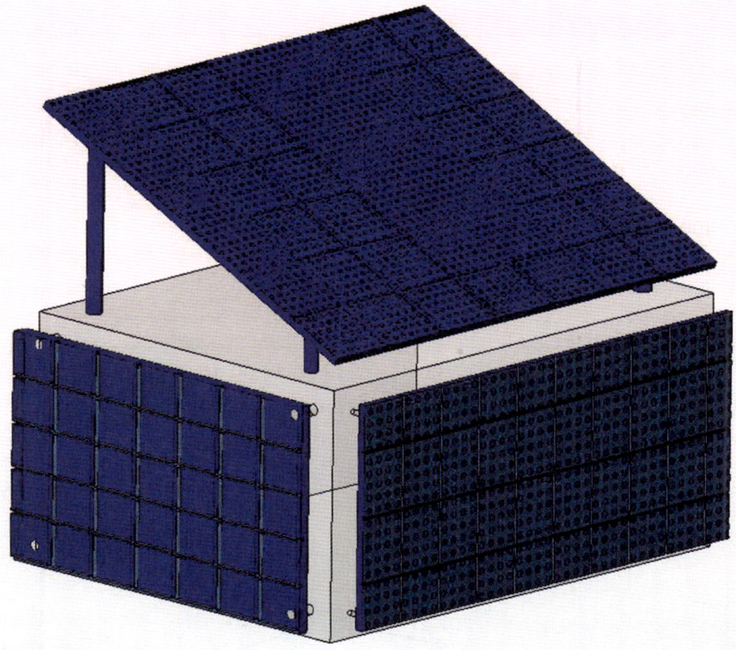

Figure 6.2 3D representation of the building integrated photovoltaic system (BIPV).[2]

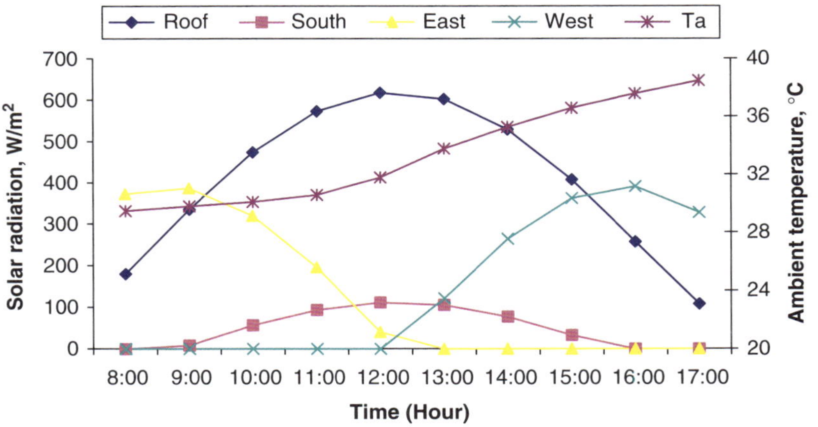

Figure 6.3 Hourly variation of beam radiation on roof (inclined), south, east and west facades for a typical day of summer (May) month (New Delhi conditions).[2]

less. The annual variation of thermal and electrical gain for five climatic conditions of India (New Delhi, Bangalore, Mumbai, Srinagar and Jodhpur) considering four weather conditions for the sixth combination is shown in Figures 6.7 and 6.8. The maximum thermal energy gain was obtained for Jodhpur city,

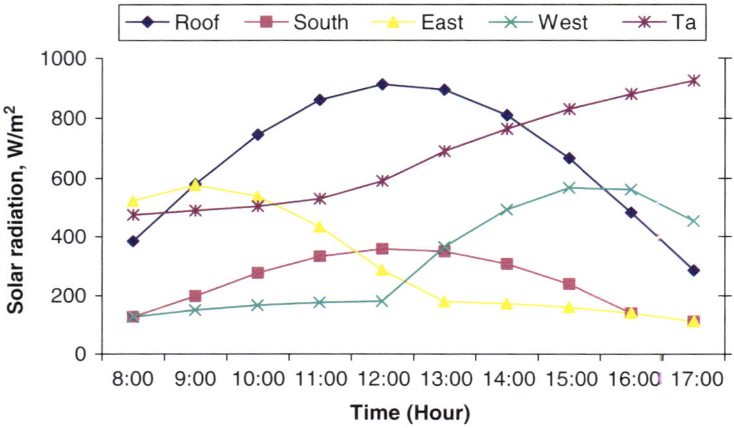

Figure 6.4 Hourly variation of total radiation on roof (inclined), south, east and west facades for a typical day of summer (May) month (New Delhi conditions).[2]

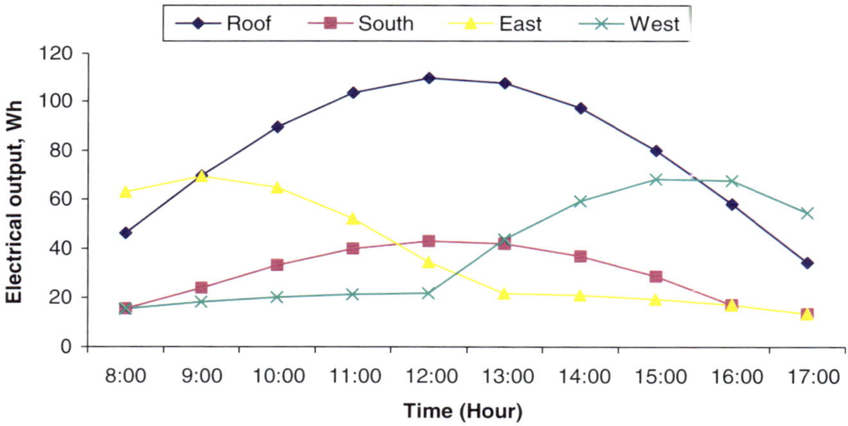

Figure 6.5 Hourly variation of power output from roof (inclined), south, east and west facades for a typical day of summer (May) month (New Delhi conditions).[2]

however maximum electrical energy gain was obtained for Bangalore city due to the lower annual ambient temperature. Minimum thermal and electrical energy gain was obtained for Srinagar city due to less availability of solar radiation. The percentage variation between Jodhpur and Srinagar city was 15.9% and 9.5% for thermal and electrical energy gain, respectively. The percentage variations between New Delhi, Mumbai and Bangalore with Srinagar were 8.1%, 13.3% and 14.1% on a thermal basis and 1.8%, 7.8% and 10.8% on an electrical basis, respectively. The average electrical efficiency of the system was found to be 9.64%.[2] The energy pay back time of the system considering overall energy and exergy gain is 1.4 and 7.6 years, respectively.

Table 6.2 Monthly power output from PV arrays for different orientations for New Delhi, India, climatic conditions (kWh).[2]

Facade	*January*	*February*	*March*	*April*	*May*	*June*	*July*	*August*	*September*	*October*	*November*	*December*	*Total*
Roof	265.7	237.5	332.6	357.5	377.9	357.2	359.9	344.5	357.3	315.9	247.4	298.3	3851.6
South	128.8	96.5	109.2	81.7	63.4	63.8	65.3	69.5	26.6	115.5	108.2	124.8	1053.1
West	56.8	53.2	75.5	81.1	83.0	82.3	77.5	71.0	24.4	67.1	55.3	57.7	784.7
East	56.6	53.5	75.1	78.2	79.4	80.2	76.7	70.2	24.4	62.9	51.6	54.8	763.6
Total	507.8	440.8	592.4	598.4	603.8	583.4	579.4	555.2	432.7	561.4	462.4	535.6	6452.9

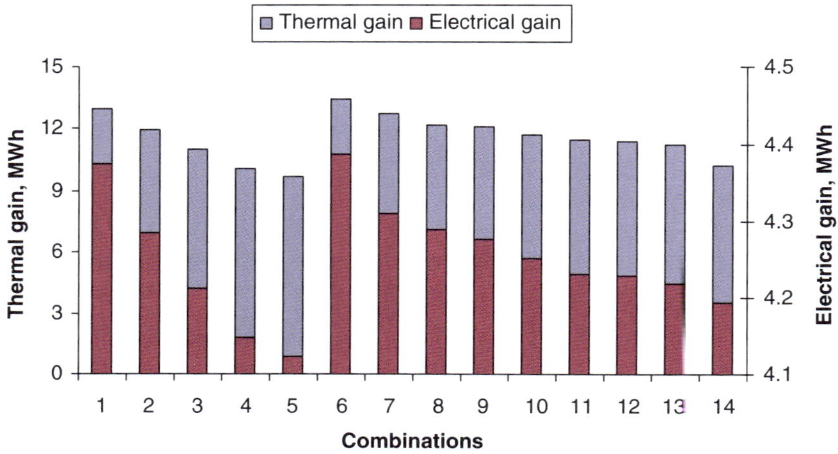

Figure 6.6 Variation of annual thermal and electrical gain from different combinations of air duct for New Delhi conditions, air velocity in duct is 2 m s^{-1}.[2]

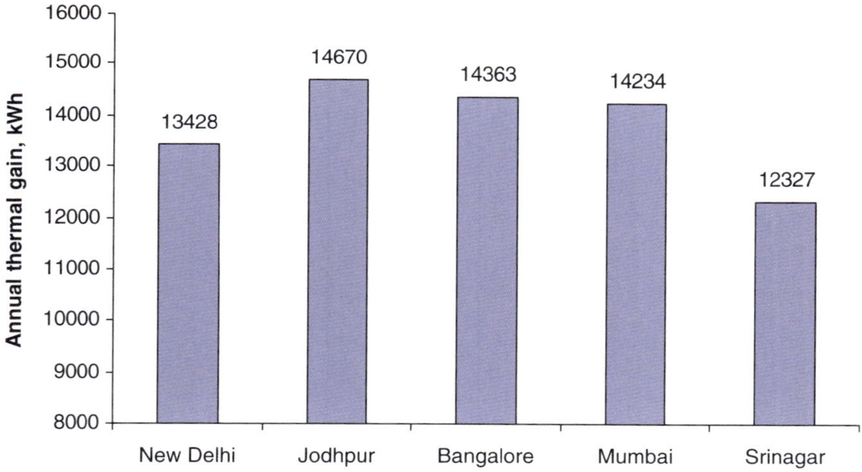

Figure 6.7 Annual variation of thermal gain for five climatic conditions of India by considering a sixth combination.[2]

6.4 Case Study III: PV-integrated Water-pumping Application in Nebraska

Water is an absolute necessity for human survival. Tapping groundwater with a dependable, economic and pollution-free energy source has become almost mandatory for rural development and agricultural self-reliance. Solar water pumping systems, in particular, are totally pollution-free and require very little

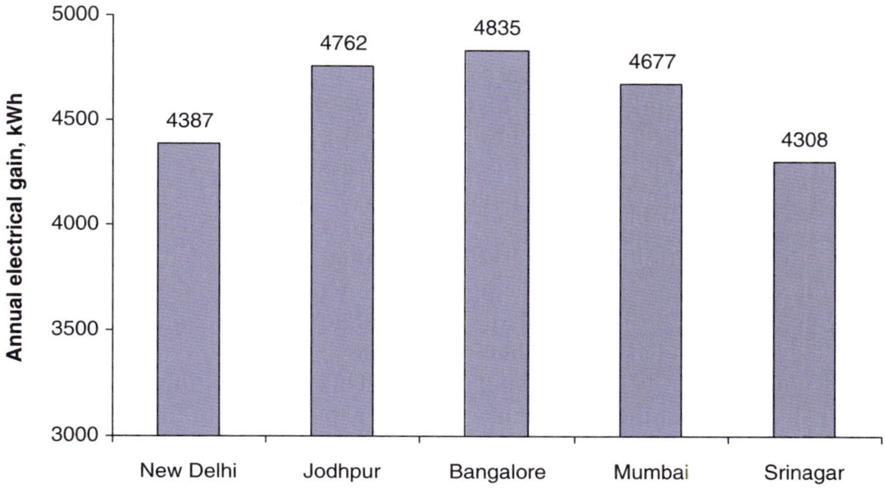

Figure 6.8 Annual variation of electrical gain for five climatic conditions of India by considering a sixth combination.[2]

maintenance as compared to the diesel-operated/AC-operated pump sets. The solar water pumping systems function only during the sunshine hours, thereby eliminating the use of costly battery banks. These pumping systems are ideal for small/marginal farmers to meet their irrigation requirements. Advantages of solar pump sets are as follows:

- No fuel cost – uses abundantly and freely available sunlight;
- Expensive transmission lines not required;
- Long operating life;
- Highly reliable and trouble-free performance;
- Easy to operate and maintain;
- Eco-friendly;
- Savings of conventional diesel fuel and electric energy.

In September 2001, a solar water-pumping system for all areas of the ranch in Bassett, Nebraska, was successfully developed and installed for uniform cattle grazing activity across the areas.[3] Because grid power was not available near the location, a PV system has been installed for pumping. The total design, procurement and installation period was about six months. The system pumps water at about 25 gallons min^{-1} ($1.57 \times 10^{-3} m^3 s^{-1}$). Including well drilling, the total installation cost was $5,510. The system installation has resulted in significant cost savings and emission reductions and has demonstrated the applicability and suitability of solar PV technology for remote locations. As the first step, Grundfos, a manufacturer of solar pumping systems, obtained the solar radiation information from the National Renewable Energy Laboratory. They then determined the amount of water to be pumped for grazing, which was about 5000 gallons day^{-1} (2.2×10^{-4} m^3 s^{-1}). They estimated the depth of the water level below ground and the height difference between ground and the water-storage tank/reservoir outlet. Based on the selected pipe height, diameter and variables, they calculated the total friction losses and total head required by the pump. Once the site parameters were defined, Grundfos selected their 25-SQF-3 pump and the GF-43 solar panel without any battery backup, as it was not required for this application. The GF-43 solar panel consists of eight modules, each capable of generating 43 watts (146.8 BTU hr^{-1}) for a total of 344 watts (1174 BTU hr^{-1}). The module is made of amorphous silicon thin-film type designed specifically for use with SQF systems and comes equipped with plugs and sockets enabling easy and simple installation. The installation required no major hardware components other than basic materials available from a hardware store.

The total project cost was about \$5510.[3] This cost is estimated to be significantly lower than the cost to run a traditional power line to the well. There is no maintenance required or suggested for the pump by the manufacturer. Suggested maintenance for the solar panels consists of cleaning the solar modules with clean water, cutting down plants that might shade the modules and tightening any loose bolts on the support structure, all as necessary depending on environmental conditions. The system originally included a flow meter to measure output from the pump.

6.4.1 Energy and Emission Savings

In the solar water pumping system all electrical production from the solar array is used by the pump. The panels produce power that is adequate for power usage and no 'excess' electricity is required. The pump control box consists of a simple on/off switch, and requires no power for the pump controller. The pump motor handles the DC-to-AC power conversion and pump condition monitoring. The PV pump system was sized to pump 5000 gallons day^{-1} of water at a head of 10 metres (33 feet). The installed GF-43 solar panels had a capacity of 344 watts. Solar panel ratings indicated that the annual energy production would be approximately 700 kWh.[3] Table 6.3 shows the estimated emission and energy savings from solar PV system installed at Nebraska.

6.4.2 Solar Water-pumping Systems in Punjab, India

Central Electronics Limited (CEL), Shahibabad, has developed two models for solar water pump sets suitable for shallow well applications *viz*. SW 900 and

Table 6.3 Energy and emission savings with a capacity of 344 watts at Nebraska.[3]

Annual generation (kWh)	*Energy savings (\$)*	*Sulfur dioxide offset (lbs)*	*Nitrogen oxides offset (lbs)*	*Carbon dioxide offset (lbs)*
700	42	3.1	2.41	1,572.1

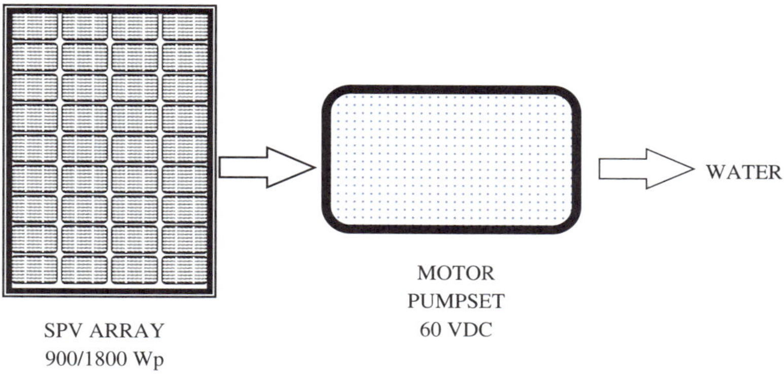

Figure 6.9 Block diagram of a typical SPV water-pumping system.

Figure 6.10 CEL's solar water-pumping systems at Block Adampur District Jalundhar, Punjab, India.

SW 1800 comprising 900 Wp and 1800 Wp SPV panels, respectively. A block diagram of the system is shown in Figure 6.9. The specifications of the SPV water pumping system (Figure 6.10) are as follows:

Model No.	SW 1800
Solar PV panel	1800 Wp
Motor pumpset type	Centrifugal DC monoblock
Motor capacity	2 HP
Operating voltage	60 V DC (nominal)

Max. suction head	7 metres
Max. total dynamic head	15 metres
Bore well size	100 mm dia. (Min.)
Required shadow free area	60 sq. metres
Module mounting structure	MS hot dipped galvanised
Facilities provided in the panel	Seasonal tilt angle adjustment Three times manual tracking facilities (east-south-west) in morning, noon and afternoon
Water output (at total head of 10 metres)	140,000 litres per day
Cost	Rs. 450,000/- (US$ 9000)

6.5 Case Study IV: Grid-interactive Photovoltaic Park on the Island of Crete

The favourable climate conditions of the island of Crete and the recent legislation for utilization of renewable energy sources provide a substantial incentive for installation of photovoltaic power plants. The pilot PV Park is located in Xirolimni, Sitia, Crete, and has been in operation since 2002. The PV Park is the largest operating PV Park in Greece with an installed capacity of 171.36 kWp, grid-connected with a 20 kV TEP transmission line, covering a total surface area of 3784 m^2 with an active area of 1142.4 m^2. The park is comprised of 1428 MSX 120 Solarex (now BP Solar) polycrystalline silicon PV modules. The PV modules are arranged in 120 parallel strings, with 12 modules in each, and connected to 60 Sunny Boy SB2500 inverters installed on the supporting structure, plus connection boxes, irradiance and temperature measurement instrumentation and a data logging system. The inverters are tied to the national grid *via* a 0.4/20 kV transformer and an electrical energy meter. The PV system was mounted on a stainless steel support structure facing south and tilted at 30°. Such a tilt angle was chosen to maximize yearly energy production, as shown in Figure 6.11.[4]

The PV park system was fully monitored to assess the performance of the system with the local power grid during 2007. To evaluate the PV park performance, the final yield (Y_F), reference yield (Y_R), performance ratio (P_R) and capacity factor (C_F) were calculated. The final yield is defined as the annual, monthly or daily net AC energy output of the system (kWh) (E_{AC}) divided by the peak power of the installed PV array (kW) (P_{DC}) at standard test conditions (STC) of 1000 W m^{-2} solar irradiance and 25 °C cell temperature,[4]

$$Y_F = \frac{E_{AC}}{P_{DC}} \tag{6.1}$$

Figure 6.11 View of the C. Rokas SA Photovoltaic Park. The PV modules are tilted at 30° and oriented south[4] (courtesy Emmanuel Kymakis, Greece).

The reference yield is the total in-plane solar insolation I_t (kWh m^{-2}) divided by the array reference irradiance (1 kW m^{-2}); therefore, the reference yield is the number of peak sun-hours,[4]

$$Y_R = \frac{I_t}{1\,k\text{Wh/m}^2} \tag{6.2}$$

The performance ratio is the final yield divided by the reference yield; it represents the total losses in the system when converting from name plate DC rating to AC output. The typical losses of a PV park include losses due to panel degradation (η_{deg}), temperature (η_{tem}), soiling (η_{soil}), internal network (η_{net}), inverter (η_{inv}), transformer (η_{tr}) and system availability and grid connection network (η_{ppc}). Therefore, P_R can be expressed as[4]

$$P_R = \frac{Y_F}{Y_R\eta_{deg} \cdot \eta_{tem} \cdot \eta_{soil} \cdot \eta_{net} \cdot \eta_{inv} \cdot \eta_{tr} \cdot \eta_{ppc}} \tag{6.3}$$

While the array yield (Y_A) is defined as the annual or daily energy output of the PV array divided by the peak power of the installed PV, the system losses (L_S) are gained from the inverter and transformer conversion losses, and the array capture losses (L_C) are due to the PV array losses,[4]

$$Y_A = \frac{E_A}{P_R} \tag{6.4a}$$

Table 6.4 Details of data obtained from field tests of different solar dryers at Nepal.[6]

Site of installation	*Type of dryer*	*Collector size, m^2*	*Drying Capacity (kg), Fresh*	*Airflow type*	*Date of experiment*	*Material*	*Available solar radiation, $W m^{-2}$*	*Time taken, hours*	*Thermal efficiency, %*
Bijulibazar, Kathmandu	STD	24	70	Forced (373 W)	Oct 22–23, 2002	Tomato	743	11	21.6
Bijulibazar, Kathmandu	STD	24	70	Forced (373 W)	Oct 12–13, 2002	Masyaura	602	10	19.7
IOE, Pulchowk, Lalitpur	SRD	1.65	4	Natural	Sept 02–03, 2002	Cauliflower	649	9	21.4
IOE, Pulchowk, Lalitpur	SRD	1.65	4	Natural	Aug 31 and Sept 01, 2002	Cauliflower	620	10	20.4
IOE, Pulchowk, Lalitpur	SRD	1.65	4	Natural	Aug 28–30, 2002	Cauliflower	564	11	20.5
CRT, Tripureshwor	SRDpg	1.6	4	Natural	Jan 31 and Feb 01, 2002	Radish	746	13	14.1
CRT, Tripureshwor	SCDgg	1	2	Natural	Jan 25–26, 2002	Radish	744	11	14.4
CRT, Tripureshwor	SCDpg	1	2	Natural	Jan 25–26, 2002	Radish	744	12	13.1
CRT, Tripureshwor	SCDgg	1	2	Natural	Jan 23–24, 2002	Radish	608	11	17.7
CRT, Tripureshwor	SCDpg	1	2	Natural	Jan 23–24, 2002	Radish	608	12	15.9
	SRD	12.5	40	Forced (15 W)	Apr 21–22, 2001	Cocoon	613	12	15.9

Kalyanpur VDC, Chitwan									
Malekhu, Dhading	HSBRD	4.2	10	Natural	March 8–9, 2001	Fish	265	11	9
RECAST, Kirtipur	ISRD	4.2	10	Natural	Jan 16–17, 2001	Apple	610	12	17.3
Kalyanpur VDC, Chitwan	SRD	12.5	40	Forced (15 W)	Dec 01–02, 2000	Cocoon	580	12	16.1
RECAST, Kirtipur	SRD	3	10	Forced (10 W)	Jan 07–08, 1999	Carrot	712	12	19.6
RECAST, Kirtipur	STD	24	70	Forced (75 W)	Aug 03–04, 1998	Onion	725	11	20.5
RECAST, Kirtipur	STD	24	70	Forced (75 W)	July 30–31, 1998	Radish	655	12	21.7
RECAST, Kirtipur	SRD	3	10	Natural	May 08–10, 1998	Banana	688	14	17.7
RECAST, Kirtipur	SCD	1.2	4	Natural	Apr 29–30, 1998	Cauliflower	688	12	21.1
RECAST, Kirtipur	SCD	1.2	4	Natural	Apr 21–22, 1998	Cauliflower	647	12	22.1

SRD = Solar radiant drying, SCD = Solar cabinet dryer, HSBRD = Hybrid solar/biomass rack dryer, STD = Solar tunnel dryer, SRD = Solar rack dryer, RECAST = Research Centre for Applied Science and Technology, CRT = Centre for Rural Technology, IOE = Institute of Engineers.

$$L_S = Y_A - Y_F \tag{6.4b}$$

$$L_C = Y_R - Y_A \tag{6.4c}$$

The capacity factor (C_F) is defined as the ratio of the actual annual energy output to the amount of energy the PV park would generate (E_G) if it operated at full rated power (P_r) for 24 h per day for a year,[4]

$$C_F = \frac{Y_F}{E_G} \tag{6.5}$$

The efficiency of a PV panel depends on the operation temperature and the power density of the solar radiation. As the temperature of the PV panels increases, the efficiency decreases linearly, since the peak power of the PV panels refers to STC conditions. At different temperatures, the output power of the PV panels depends on the difference of the panel temperature and the STC temperature and the power density of the incident solar radiation. The highest value of total in-plane insolation was in July with 224.66 kWh m^{-2} and the lowest, in December, was 92.35 kWh m^{-2}. The annual insolation was 1984.38 kWh m^{-2}, and the mean ambient temperature was 16.46 °C. The PV Park supplied 229 MWh to the grid during 2007, ranging from 335.48 to 869.68 kWh. The performance ratio was distributed within the range of 58–73%, and the annual mean value was 67.36%. The performance ratio was distributed within the range of 58–73%, and the annual mean value was 67.36%.[4]

6.6 Case Study V: Performance Study of Solar Drying Systems in Nepal

The developmental activities on solar dryers have been taking place in Nepal for over two decades. Their operations are based on direct drying (solar cabinet dryer), indirect drying (some versions of solar rack dryers) or mixed drying (solar tunnel dryers and some other versions of solar rack dryers). With the exception of solar tunnel dryers and large-size solar rack dryers, which are based on forced circulation of airflow, most of the dryers developed so far run on natural circulation of airflow.[5] Recently, a new concept of hybrid drying technology has also emerged. This case study describes the thermal efficiencies obtained for different types of solar dryers installed at different parts of the country under steady-state conditions. Data obtained from 20 laboratories as well as outdoor field tests have been used in the calculation of thermal efficiencies. Tests were carried out on 12 different solar dryers including 3 solar cabinet dryers, 6 solar rack dryers, 2 solar tunnel dryers and 1 hybrid solar/biomass rack dryer. The analytical part of this study provides an overview of how efficiently the food in a given mode of solar drying system uses the heat to warm up and evaporate the water. The efficiency of a solar drying system is affected by the properties of drying materials *e.g.* moisture content, size, shape and geometry as well as ambient conditions *e.g.* solar radiation and temperature, relative humidity, velocity and atmospheric pressure of ambient air. Due to the difference in the drying period for different drying materials, different values of efficiencies have been found for the same dryer. The details of data obtained from field tests of different solar dryers are shown in Table 6.4. The maximum value of thermal efficiency obtained from these tests is found as 22.1% for solar cabinet dryer, 21.4% for solar rack dryer and 21.7% for solar tunnel dryer.[6] For forced convection drying in the rural applications, PV/T solar dryers have been developed. In these, the electricity generated by PV is used for the circulation of air and removal of moisture in forced mode. Detailed analysis of PV/T dryers with results will be been discussed in Section 7.5.

References

1. H. Yang, G. Zheng, C. Lou, D. An and J. Burnett, *Sol. Energ.*, 2004, **76**, 55–59.
2. A. Ranjan, S. Dubey, B. Agarwal and G. N. Tiwari, *Open Renew. Energ. J.*, 2008, **1**, 1–9.
3. Nebraska Case Study, http://www.neo.ne.gov/publications/NebPVCaseStudy.pdf, accessed 10 November 2008.
4. E. Kymakis, S. Kalykakis and T. M. Papazoglou, *Energ. Convers. Manag.*, 2009, **50**, 433–438.
5. C. B. Joshi and M. B. Gewali, *Int. Energ. J.*, 2002, **3**(2), 53–74.
6. C. B. Joshi, M. B. Gewali and R. C. Bhandari, *IE(I) Journal-ID*, 2004, **85**, 53–57.

CHAPTER 7

Thermal Modelling of Hybrid Photovoltaic/Thermal (PV/T) Systems

7.1 Introduction

Solar Thermal Technology is employed for collecting the Sun's energy and converting it to heat energy for applications such as water and air heating, cooking and drying, steam generation, distillation, *etc*. Basically a solar thermal device consists of a solar energy collector – the 'absorber', a heating or heat transferring medium. Solar photovoltaic technology is employed for directly converting solar energy to electrical energy by the using 'solar silicon cell'.

Photovoltaic/thermal (PV/T) technology refers to the integration of a PV module and a conventional solar thermal system in a single piece of equipment. The rationale behind the hybrid concept is that a solar cell converts solar radiation to electrical energy with peak efficiency in the range of 9 to 12%, depending on specific solar-cell type and thermal energy through water heating. More than 80% of the solar radiation falling on photovoltaic (PV) cells is not converted to electricity, but is either reflected or converted to thermal energy. This leads to an increase in the PV cell's working temperature and, consequently, a drop of electricity conversion efficiency. In view of this, hybrid photovoltaic and thermal (PV/T) systems are introduced to generate electricity and thermal power simultaneously.

The collector is the heart of any solar energy collection system designed for operation in a low or medium temperature range. It is used to absorb solar energy, convert it into heat and transfer it into a stream of liquid or air. In a conventional solar thermal collector, electrical energy is required to circulate the working fluid through the collector and the required electrical energy is usually supplied by grid electricity or a DC battery as a power source. In the case of a hybrid photovoltaic/thermal (PV/T) system, the electrical power source is not required as the PV/T collector produces both

RSC Energy Series No. 2
Fundamentals of Photovoltaic Modules and Their Applications
By G. N. Tiwari and Swapnil Dubey

Published by the Royal Society of Chemistry, www.rsc.org

electrical and thermal energy. Kern and Russell,[1] give the main concepts of these systems with results, by the use of water or air as heat removal fluid. Hendrie[2] presents a theoretical model on PV/T systems using conventional thermal collector techniques. Florschuetz[3] suggests an extension of the Hottel–Whillier model for the analysis of PV/T systems and Raghuraman[4] presents numerical methods predicting the performances of liquid and air photovoltaic thermal flat-plate collectors. Lalovic[5] proposes a novel transparent type of a-Si cell as a low-cost improvement of hybrid systems and Loferski *et al.*[6] give results for a hybrid system with air circulation installed on a residential building, by using two separate one-dimensional analyses compared with test measurements. Bhargava *et al.*[7] and Prakash[8] present results regarding the effect of air mass flow rate, air channel depth, length and fraction of absorber plate area covered by solar cells (packing factor, PF) on a single pass.

Thermal energy has wider applications in our lives. It can be generally utilized in the form of either low grade (low temperature) or high grade (high temperature). Jones and Underwood[9] have studied the temperature profile of the photovoltaic (PV) module in a non-steady-state condition with respect to time. They conducted experiments for cloudy as well clear day conditions. They observed that the PV module temperature varies in the range of 300–325 K (27–52 °C) for an ambient air temperature of 297.5 K ($\sim$ 24.5 °C). The main reasons for reduction of the electrical efficiency of the PV module are the packing factor (PF) of the PV module, ohmic losses between two consecutive solar cells and the temperature of the module. The overall electrical efficiency of the PV module can be increased by increasing the packing factor (PF) and reducing the temperature of the PV module by withdrawing the thermal energy associated with the PV module.[10,11] The packing factor is the ratio of the total area of solar cells to the area of the PV module. The carrier of thermal energy associated with the PV module may be either air or water. Once thermal energy withdrawal is integrated with the photovoltaic (PV) module, it is referred to as a hybrid PV/T system.

The hybrid photovoltaic/thermal (PV/T) system has the following applications:

1. Air heating system;[7–8,12–20] and
2. water heating system.[11,14,20–27]

Solar thermal technology is now a mature technology. Widespread utilization of solar thermal technology can reduce a significant portion of the conventional energy. Internationally the market for solar technology has expanded significantly during the last decade. Though the initial investment for these technologies is high compared to available conventional alternatives, the return on investment has become increasingly attractive with the increase in prices of conventional energy. The pay back period depends on the site of installation, utilization pattern and fuel replaced.

7.2 PV/T Air Collectors

PV/T air collectors are used for heating the air and electricity generation simultaneously. The hot air is used for space heating and drying purposes. The various designs of air duct have been studied by the earlier researchers; some of the designs and results are discussed in this section. Bhargava *et al.*[7] have studied the solar air heater combined with solar cells. Hagazy[12] and Sopian *et al.*[28] investigated glazed photovoltaic/thermal air (PV/T air) system for a single and a double pass air heater for space heating and drying purposes. Hegazy[7] has studied four configurations of a photovoltaic/thermal solar air collector and observed that the configuration with air flow between the top glass cover and a solar cell gives an overall (electrical and thermal) efficiency of about 55% at $0.04\,\text{kg}\,\text{m}^{-2}\,\text{s}$ mass flow rate of air. Radziemska[29] reviewed the thermal performance of Si- and GaAs-based solar cells and modules including work on air- and water-cooled hybrid PV/thermal solar air collectors. Sandnes and Rekstad[26] have studied the behaviour of a combined PV/T collector which was constructed by pasting single-crystal silicon cells onto a black plastic solar heat absorber (unglazed PV/T system). They recommended that the combined PV/T concept must be used for low-temperature thermal application for increasing the electrical efficiency of PV system *e.g.* space heating of a building. Zakharchenko *et al.*[25] have also studied unglazed hybrid PV-thermal systems with a suitable thermal contact between the panel and the collector. They have proved that the areas of the PV panel and a collector in the PV/T system need not be equal for higher overall efficiency. Tripanagnostopoulos *et al.*[14] suggested that a PV/T system with reflector gives higher electrical and thermal output. Coventry[30] studied the performance of a concentrating PV/T solar collector and reported that the overall thermal and electrical efficiency of a PV/T concentrating system were 58% and 11%, respectively which gives a total efficiency of 69%.

The electrical efficiency (η_{el}), as a function of temperature is given by (Radziemska,[29])

$$\eta_{el} = \eta_0\left[1 - \beta_o\left(\overline{T_c} - \overline{T_a}\right)\right] \tag{7.1}$$

where $\eta_{el} = \eta_{ec}$; η_0 is the standard efficiency of a PV module at a temperature of 298 K and solar intensity of $1000\,\text{W}\,\text{m}^{-2}$; β_0 is the silicon efficiency temperature coefficient ($0.0045\,\text{K}^{-1}$ or $0.0064\,\text{K}^{-1}$) and T_c is the cell temperature (K).

Thus, the electrical efficiency (η_{el}) reduces with an increase in PV temperature as shown in Figure 7.1. It is clear from Figure 7.1 that the cell efficiency decreases with an increase of temperature as expressed by eqn (7.1). It is further to be noted from Figure 7.1(a) that an unglazed PV module gives better electrical efficiency than a glazed PV module due to the low operating temperature of the solar cell.[31,32]

It is clear from eqn (7.1) that the decrease in PV module temperature will enhance the electrical efficiency of the PV module. This can be achieved by removing the thermal energy associated with the PV module. This is done by flowing fluid (air/water) below the PV module as mentioned above.

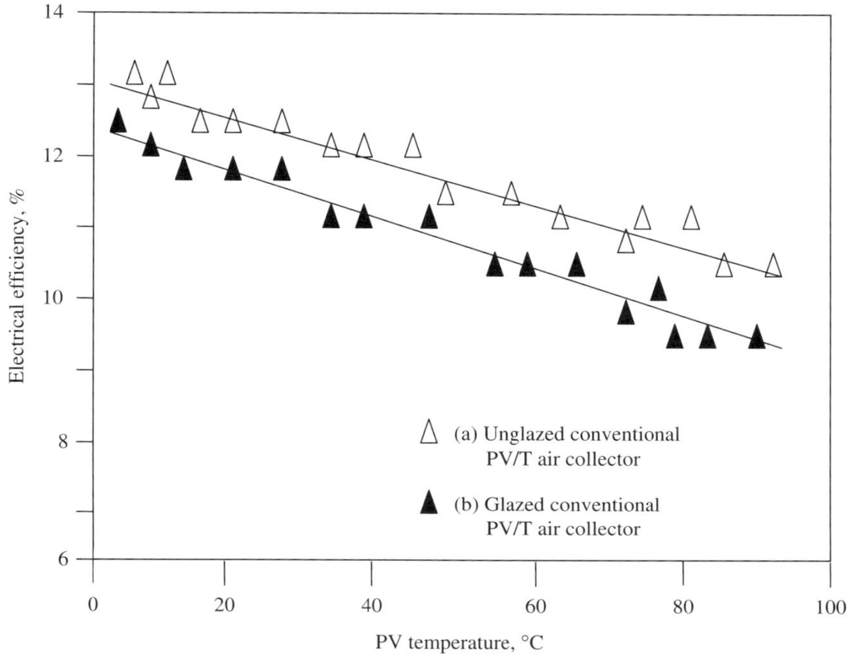

Figure 7.1 Variation of electrical efficiency of (a) an unglazed and (b) a glazed collector with PV module temperature.

There may be several combinations of PV/T solar collector for its performance improvement and some of them are given below.

7.2.1 Hybrid Air Collector

A conventional unglazed PV/T air collector is one in which a PV module is used as the absorber plate. The heat is transferred from the back surface of the PV module to the flowing air. A conventional single pass unglazed PV/T air collector is shown in Figure 7.2(a–c).

Figure 7.3(a) represents the unglazed PV/T air collector with thin metallic sheet (TMS). A suspended thin metallic sheet has been used at the middle of the air duct, which doubles the heat extraction surface and the air flow is as shown in Figure 7.3(b). The glazed PV/T air collector with thin metallic sheet (TMS) is shown in Figure 7.3(c).

The PV/T air collector is glazed to reduce the top heat loss from the PV module to the ambient. Figure 7.4(a and b) shows the typical glazed PV/T air collector. The various heat transfer modes for a single-pass glazed PV/T air collector are shown in Figure 7.4(c).

Figure 7.5(a and b) represents the PV/T air collector with fins, unglazed and glazed, respectively. The fins (height and spacing distance each 4 cm) with

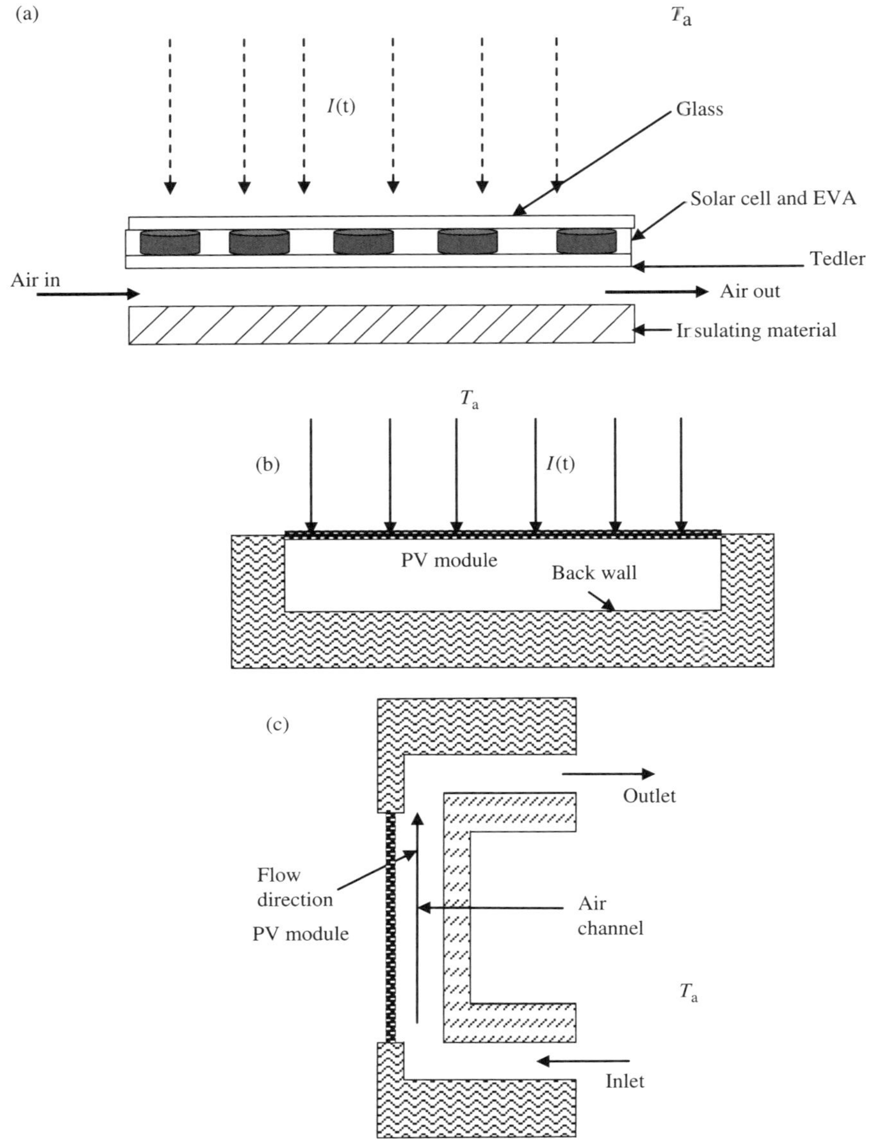

Figure 7.2 Unglazed PV/T air collector (a) with tedlar, (b) unglazed and air flow direction perpendicular to the page and (c) another view of unglazed.

rectangular profiles are attached to the back wall of the air duct and parallel to the air-flow direction.

The effect of fins on the heat transfer coefficient from the absorber to the flowing air for different channel depth has been shown in Figure 7 6. The results of the heat transfer coefficient with fin, TMS and without fin indicate that the

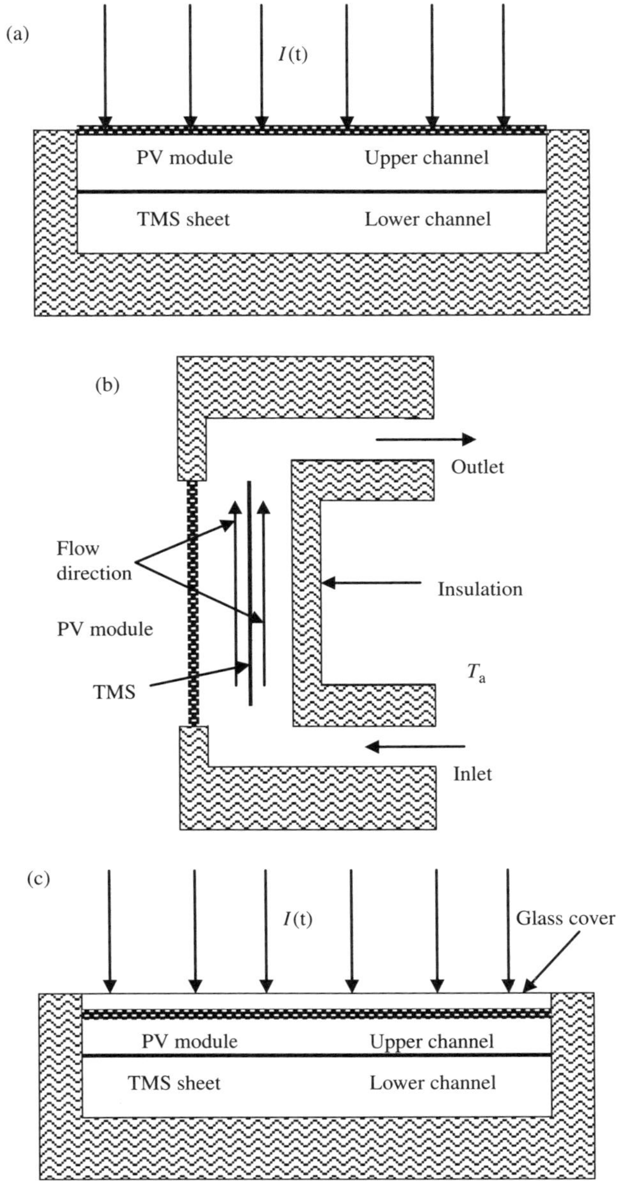

Figure 7.3 Conventional PV/T air collector with thin metallic sheet (TMS) (a) unglazed, (b) showing air flow direction and (c) glazed.

heat transfer coefficient with fin and TMS are the same for all the channel depths under study. However, the values of heat transfer coefficients are higher than without fin (conventional) as shown in Figure 7.6. One can also observe that TMS and fin has a significant effect for channel depth less than 10 cm.

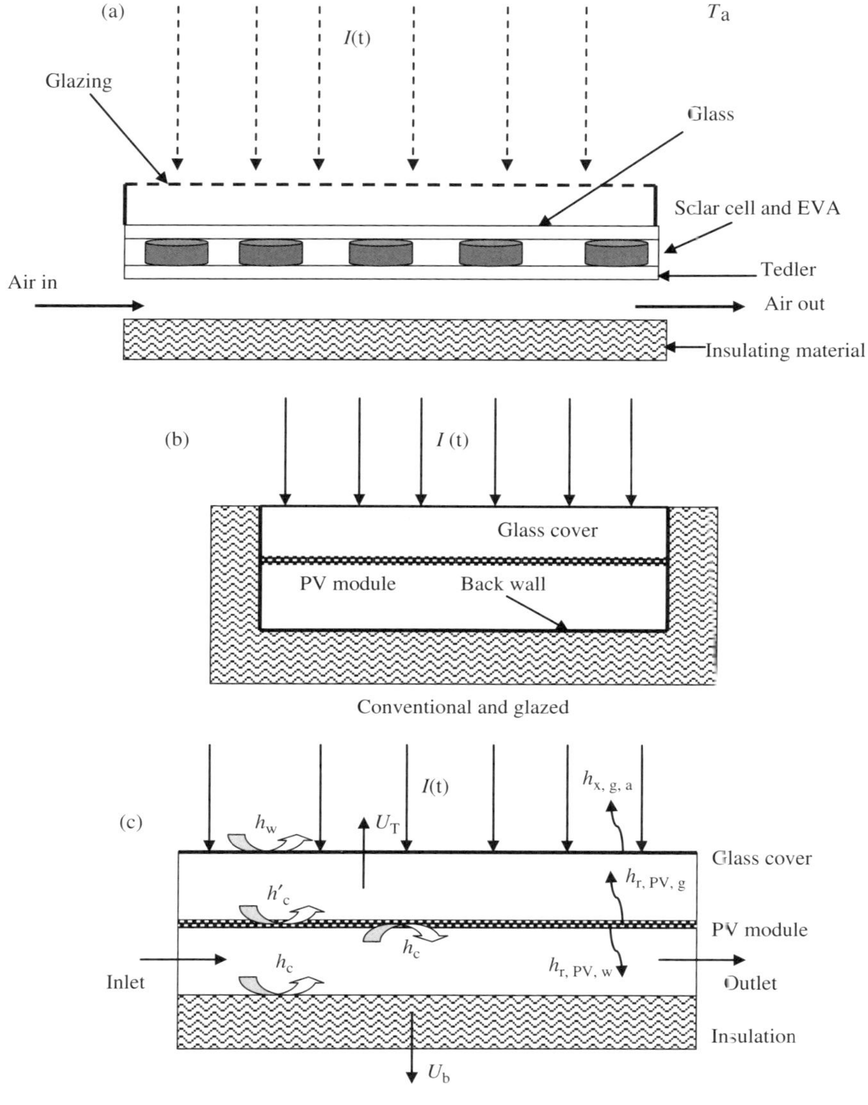

Figure 7.4 Conventional PV/T air collector (a) glazed with tedlar, (b) glazed and (c) various heat transfer coefficients.

Due to the introduction of the fin, the flow of air under natural mode becomes difficult, hence the air pump/blower is required for a smooth flow of air across the fin. The capacity of the blower depends on the length of the collector as shown in Figure 7.7. The capacity of the blower in the cases of fin, TMS and conventional is more or less the same for collector length less than 5 m as indicated in Figure 7.7. For higher collector length, the capacity of the blower required is maximum for TMS.

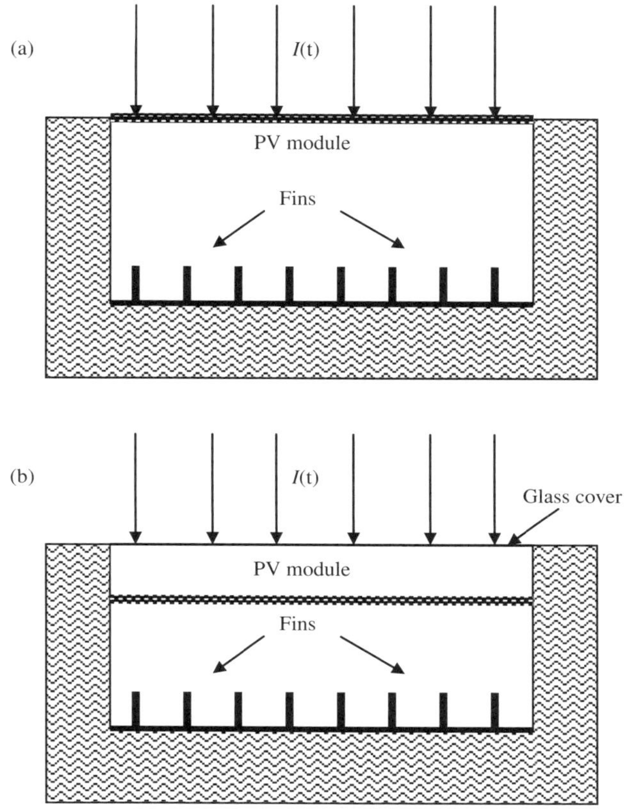

Figure 7.5 Conventional PV/T air collector with fins (a) unglazed and (b) glazed.

Figure 7.8(a) represents the unglazed single-pass PV/T air collector without tedlar. In this case, the air flows below the PV module and receives the solar radiations transmitted through its non-packing area. Figure 7.8(b) represents the glazed single-pass PV/T air collector without tedlar.

7.2.2 Double-pass PV/T Solar Air Collector

Figure 7.9(a) shows the double-pass PV/T solar air collector, in which air flows through the upper channel, *i.e.* between the glass cover and the PV panel, and then through the lower channel, *i.e.* between the absorber plate and back plate.

Othman *et al.*[33] studied the performance of a double-pass PV/T solar collector with compound parabolic concentrator (CPC) and fins (Figure 7.9(b)). The absorber of the collector consists of an array of solar cells to generate electricity, CPC and fins attached to the back side of the absorber plate. Air enters through the upper channel (between the glass cover and the PV panel) and is heated directly by the Sun and then it enters through the lower channel

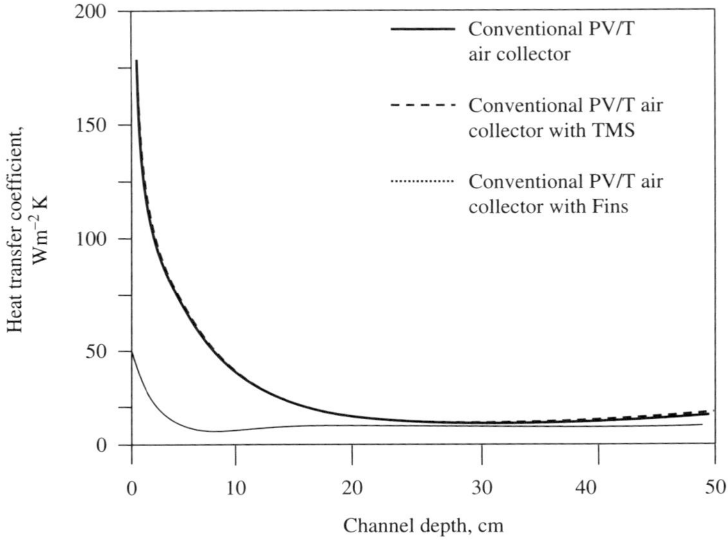

Figure 7.6 Variation of convective heat transfer coefficient with channel depth for different PV/T air collectors.

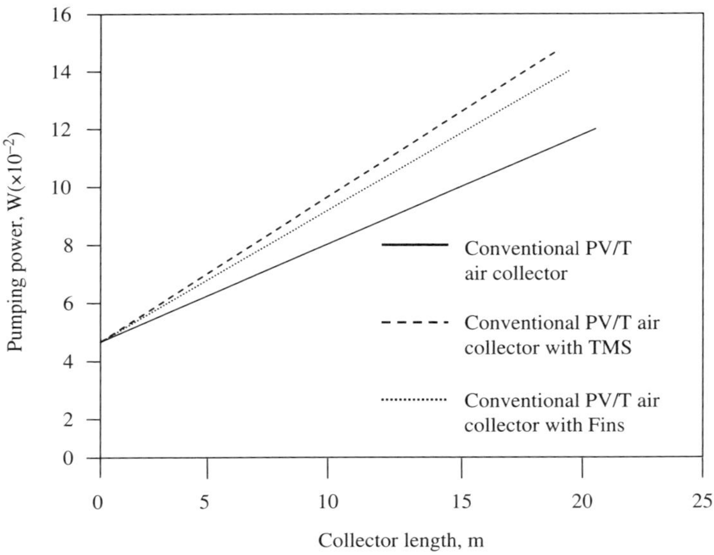

Figure 7.7 Effect of channel depth on pumping power.

(between the back plate and the photovoltaic panel). The compound parabolic concentrators concentrate solar radiation onto the PV cells. The fins on the back of the photovoltaic panel increase the heat transfer to the air and enhance the efficiency of the system.

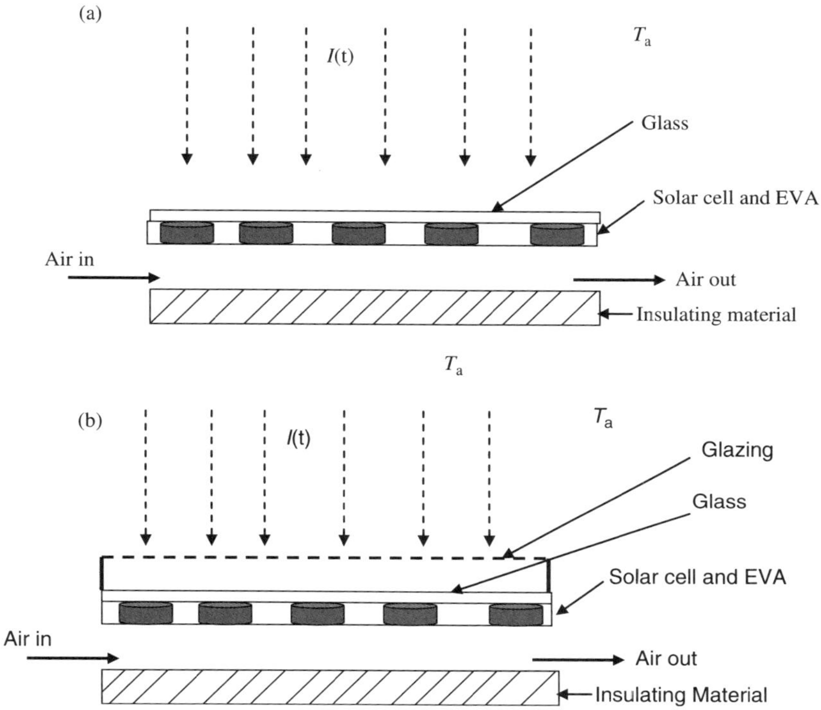

Figure 7.8 PV/T air collector without tedlar, (a) unglazed and (b) glazed.

Othman *et al.*[34] studied the double-pass finned PV/T solar air heater in which fins are attached parallel to the length of the collector at the back side of the absorber surface *i.e.* in the lower channel (Figure 7.9(c)). Air is made to flow through the upper channel and is heated directly by the Sun. The air then enters through the lower channel of the collector. The fins, provided on the back of the photovoltaic panel, increase the heat transfer to the air and thus enhance the efficiency of the system. Turbulence is introduced as the airflow is interrupted and the heat transfer area is increased due to the fins provided. The heat transfer from the absorber plate to the flowing air is increased due to the combined effect of these two phenomena. The air extracts heat from the PV cells and hence the electrical efficiency of a PV cell is improved by the reduction of its operating temperature. Figure 7.10 and Figure 7.11 show the cross section of the PV module with air duct of hybrid air collector with tedlar and glass at the back of the PV module.

7.2.3 Thermal Modelling of PV/T Air Collector Covered by Glass-to-Tedlar Type PV Module

Figures 7.10 and Figure 7.12 show the cross-sectional view and elemental length 'dx', respectively, of a PV/T air collector (with tedlar). The working fluid *i.e.* air is used to flow below the tedlar. A thermal resistance circuit diagram and a photograph of a PV/T air heater are shown in Figures 7.13 and 7.14.

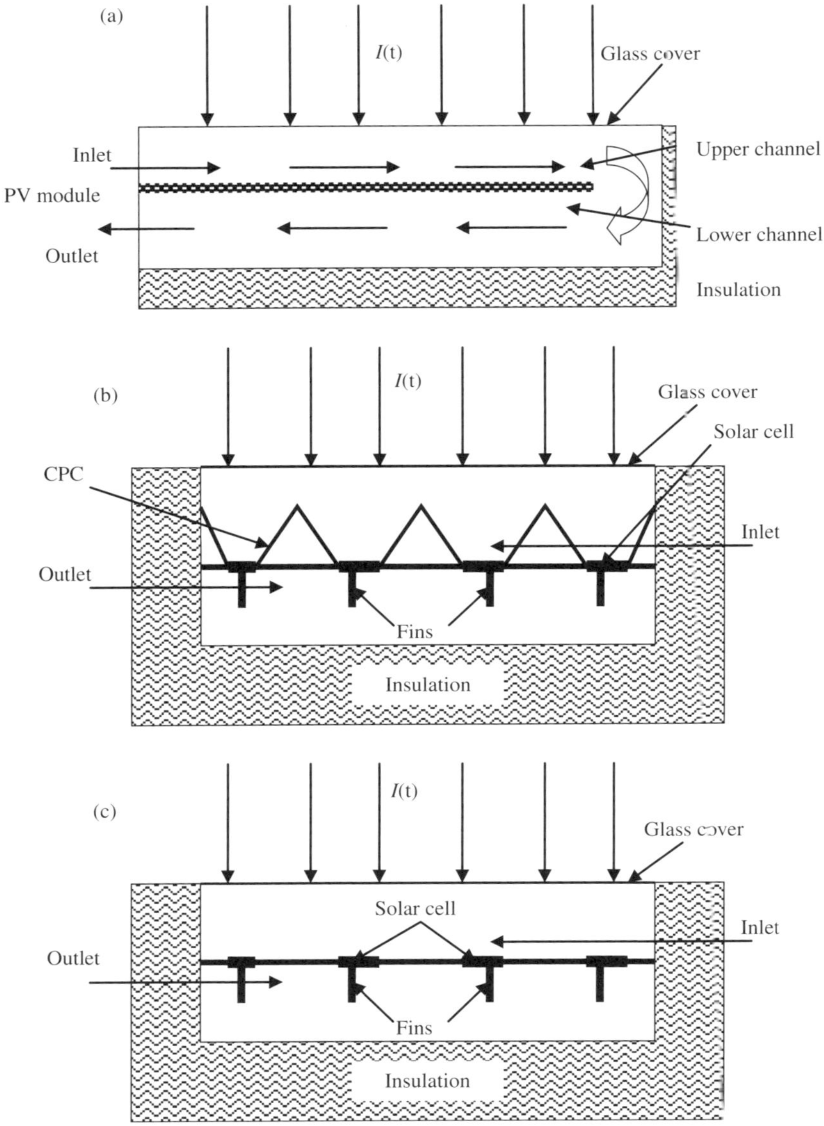

Figure 7.9 Schematic double-pass PV/T solar air collector (a) with air cooling, (b) with CPC and fins and (c) with fins.

In order to write the energy balance equation for each component of the PV/T air collector, the following assumptions are made:

- The system is in a quasi-steady state;
- The transmissivity of EVA is approximately 100%;

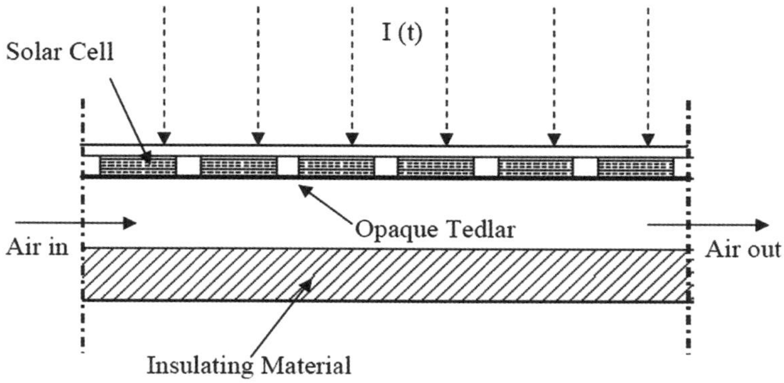

Figure 7.10 Cross sectional view of glass-to-tedlar PV module with duct.

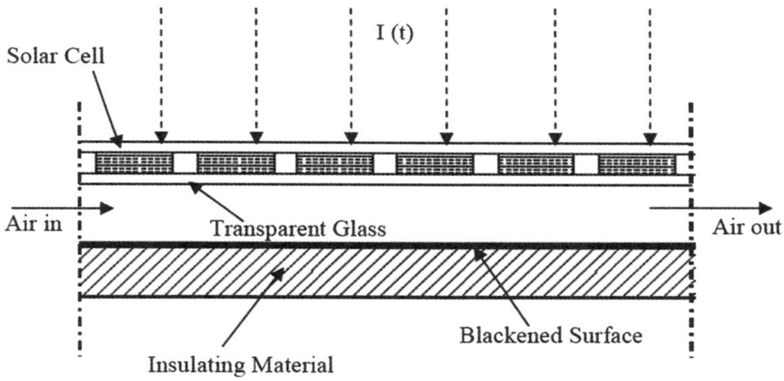

Figure 7.11 Cross sectional view of glass-to-glass PV module with duct.

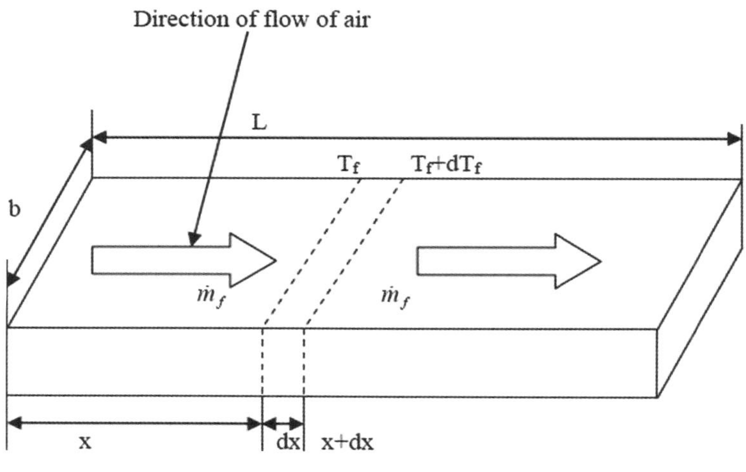

Figure 7.12 Elemental length 'dx' shows flow pattern of air below tedlar.[17]

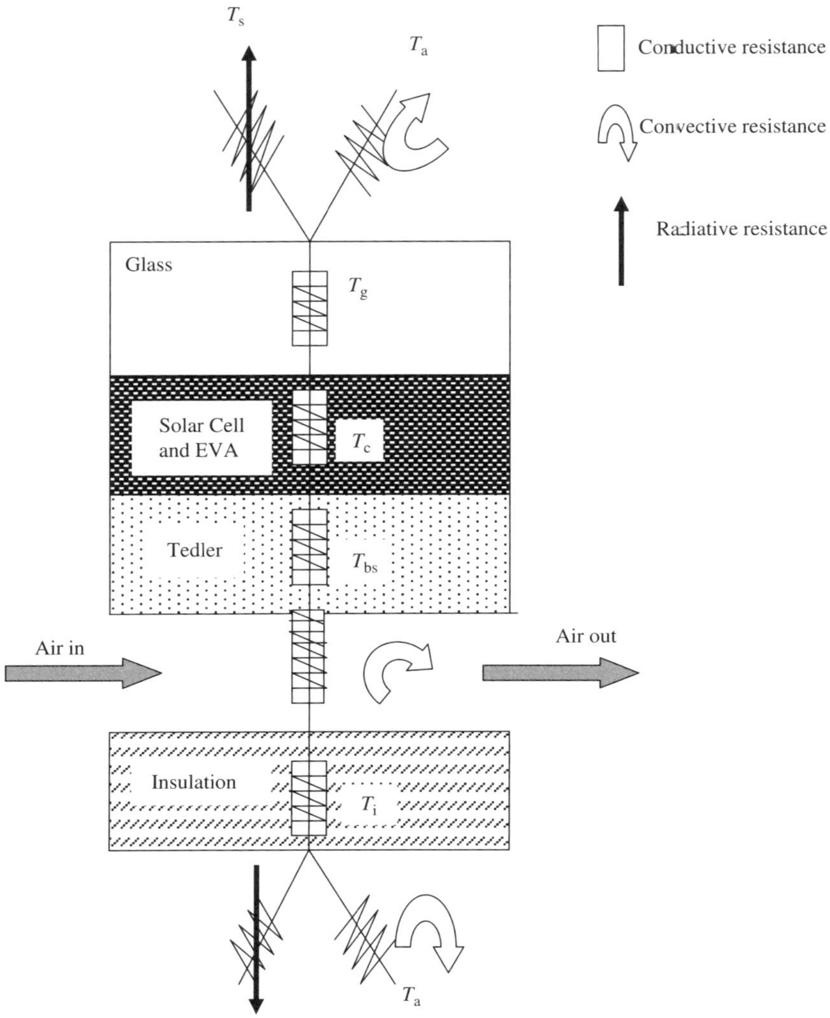

Figure 7.13 Thermal resistance circuit diagram for unglazed PV/thermal air with tedlar.

- The temperature variation along the thickness as well as along the width is negligible;
- The air flow between the tedlar and the wood structure is uniform for the forced mode of operation; and
- The ohmic losses in the solar cell are negligible.

7.2.3.1 Energy Balance

Following Figures 7.10, 7.12 and 7.13, the energy balance equations for each component in watts (W) are written as follows:

Figure 7.14 Photograph of a PV/T hybrid air-heating system.

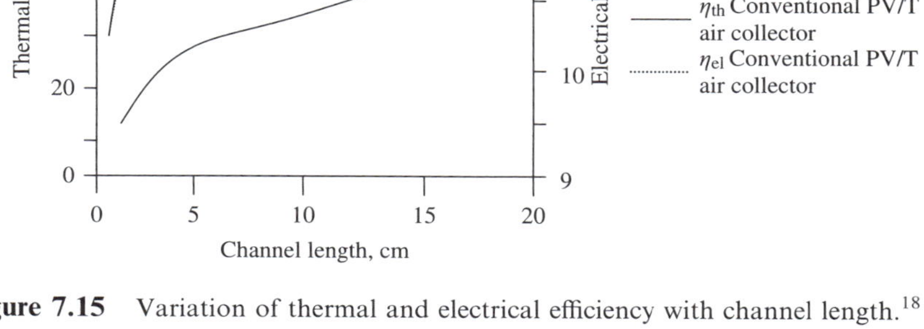

Figure 7.15 Variation of thermal and electrical efficiency with channel length.[18]

(i) *PV module:*

$$\tau_G[\alpha_c I(t)\beta_c + (1-\beta_c)\alpha_T I(t)]bdx = bdx[U_T(T_c - T_a) + h_T(T_c - T_{bs})] + bdx \times \eta_c \tau_G I(t)\beta_c \tag{7.2}$$

$$\begin{bmatrix}\text{The rate of solar}\\ \text{energy available}\\ \text{on PV module}\end{bmatrix} = \begin{bmatrix}\text{An overall heat}\\ \text{loss from top}\\ \text{surface of cell}\\ \text{to ambient}\end{bmatrix} + \begin{bmatrix}\text{An overall heat}\\ \text{transfer from cell to}\\ \text{back surface of tedlar}\end{bmatrix} + \begin{bmatrix}\text{The rate of}\\ \text{electrical energy}\\ \text{produced}\end{bmatrix}$$

where $U_T = \left[\frac{L_G}{k_G} + \frac{1}{h_0}\right]^{-1}$ = overall heat transfer coefficient between solar cell to ambient through glass cover; $h_T = \left[\frac{L_T}{k_T}\right]^{-1}$ = conductive heat transfer coefficient through the tedlar, T_c = temperature of solar cell, T_a = ambient temperature, T_{bs} = temperature of back surface of tedlar, τ_G is the transmissivity of glass of the PV module, α_c and α_T are the absorptivities of the solar cell and tedlar respectively, β_c is the packing factor of the solar cell, b is the width of the PV module and η_c is the solar cell efficiency.

(ii) *Back surface of the tedlar*:

$$bdxh_T(T_c - T_{bs}) = bdxh_t(T_{bs} - T_{air}) \tag{7.3}$$

$$\begin{bmatrix}\text{An overall heat}\\ \text{transfer from cell to}\\ \text{back surface of tedlar}\end{bmatrix} = \begin{bmatrix}\text{The rate of heat transfer}\\ \text{from back surface of the}\\ \text{tedlar to flowing fluid}\end{bmatrix}$$

where h_t = convective heat transfer coefficient from the tedlar back surface to the working fluid (air) and T_{air} = temperature of air flowing below the tedlar (air duct).

(iii) *Air flowing below the tedlar (air duct)*:

$$bdxh_t(T_{bs} - T_{air}) = \dot{m}_a c_a \frac{dT_{air}}{dx} dx + bdxU_b(T_{air} - T_a) \tag{7.4}$$

$$\begin{bmatrix}\text{The rate of heat transfer}\\ \text{from back surface of the}\\ \text{tedlar to flowing fluid}\end{bmatrix} = \begin{bmatrix}\text{The mass flow}\\ \text{rate of flowing}\\ \text{fluid}\end{bmatrix} + \begin{bmatrix}\text{An overall heat transfer}\\ \text{from flowing fluid to}\\ \text{ambient}\end{bmatrix}$$

where $U_b = \left[\frac{L_i}{k_i} + \frac{1}{h_i}\right]^{-1}$.

From eqn (7.2), an expression for a solar-cell temperature in terms of a back surface temperature of the PV module and climatic parameters can be written as

$$T_c = \frac{(\alpha\tau)_{eff} I(t) + U_T T_a + h_T T_{bs}}{U_T + h_T} \tag{7.5}$$

where $(\alpha\tau)_{eff} = \tau_G\{\alpha_c\beta_c + \alpha_T(1-\beta_c) - \eta_c\beta_c\}$.

Now, by substituting an expression for T_c from eqn (7.5) in eqn (7.3), an expression for the back surface temperature of a PV module is given by

$$T_{bs} = \frac{h_{p1}(\alpha\tau)_{eff} I(t) + U_{tT} T_a + h_t T_{air}}{U_{tT} + h_i} \tag{7.6}$$

where $h_{p1} = \frac{h_T}{U_T + h_T}$, the penalty factor due to the glass of a PV module; $U_{tT} = \frac{U_T h_T}{U_T + h_T}$.

With the help of eqns (7.5) and (7.6), eqn (7.4) can be rewritten as

$$b\left[h_{p1} h_{p2} (\alpha\tau)_{eff} I(t)\right] = \dot{m}_a c_a \frac{dT_{air}}{dx} + bU_L (T_{air} - T_a) \tag{7.7}$$

where $h_{p2} = \frac{h_i}{U_{tT} + h_i}$, the penalty factor due to the tedlar of a PV module; $U_L = U_{tair} + U_b$; $U_{tair} = \left[\frac{1}{U_{tT}} + \frac{1}{h_t}\right]^{-1}$.

By integrating eqn (7.7) with an initial condition at $x = 0$, $T_{air} = T_{airin}$, we get an expression for the temperature of the flowing air below the tedlar which is given by

$$T_{air} = \left[\frac{h_{p1} h_{p2} (\alpha\tau)_{eff} I(t)}{U_L} + T_a\right]\left(1 - e^{\frac{-bU_L}{\dot{m}_a c_a} x}\right) + T_{airin} e^{\frac{-bU_L}{\dot{m}_a c_a} x} \tag{7.8}$$

Now, the outlet air temperature (T_{airout}) of the flowing air below the tedlar can be obtained from the above equation as

$$T_{airout} = T_{air}\Big|_{x=L} = \left[\frac{h_{p1} h_{p2} (\alpha\tau)_{eff} I(t)}{U_L} + T_a\right]\left(1 - e^{\frac{-bU_L}{\dot{m}_a c_a} L}\right) + T_{airin} e^{\frac{-bU_L}{\dot{m}_a c_a} L} \tag{7.9}$$

The average air temperature of the flowing air below the tedlar over the length of air duct below the PV module is obtained as:

$$\bar{T}_{air} = \frac{1}{L}\int_0^L T_{air}\, dx = \left[\frac{h_{p1} h_{p2} (\alpha\tau)_{eff} I(t)}{U_L} + T_a\right]\left(1 - \frac{1 - e^{\frac{-bU_L}{\dot{m}_a c_a} L}}{\frac{bU_L}{\dot{m}_a c_a}}\right) + T_{airin} \frac{1 - e^{\frac{-bU_L}{\dot{m}_a c_a} L}}{\frac{bU_L}{\dot{m}_a c_a}} \tag{7.10}$$

From knowing an average air temperature of the flowing air below the tedlar from the above equation, the back surface temperature of a PV module can be obtained from eqn (7.6). Once the back surface temperature of a PV module is known, the solar-cell temperature can be evaluated from eqn (7.5) for given climatic parameters of a solar intensity and an ambient air temperature.

The rate of useful thermal energy obtained from the PV/T air collector is obtained as

$$\begin{aligned}\dot{q}_{\mathrm{u}} &= \dot{m}_{\mathrm{a}} c_{\mathrm{a}} (T_{\mathrm{airout}} - T_{\mathrm{airin}}) \\ &= \frac{\dot{m}_{\mathrm{a}} c_{\mathrm{a}}}{U_{\mathrm{L}}} \left\{ h_{\mathrm{p1}} h_{\mathrm{p2}} (\alpha\tau)_{\mathrm{eff}} I(t) - U_{\mathrm{L}} (T_{\mathrm{airin}} - T_{\mathrm{a}}) \right\} \left(1 - e^{\frac{-bU_{\mathrm{L}}}{\dot{m}_{\mathrm{a}} c_{\mathrm{a}}} L} \right)\end{aligned} \tag{7.11}$$

The overall efficiency of the PV/T collector is

$$\eta_o = \frac{\sum_{i=1}^{T} \left[\eta_{\mathrm{c}} I(t) bL \, \tau_g \, \beta_c + \dot{q}_{\mathrm{u}} \right]}{\sum_{i=1}^{T} I(t) bL} \tag{7.12a}$$

or

$$\eta_o = \eta_{\mathrm{E}} + \frac{\sum_{i=1}^{T} \dot{q}_{\mathrm{u}}}{\sum_{i=1}^{T} I(t) bL} = \eta_{\mathrm{E}} + \eta_{\mathrm{TH}} \tag{7.12b}$$

If the conversion factor of the thermal power plant (0.38) is taken into account, then an overall thermal efficiency of the PV/T air collector becomes

$$\eta_0 = (\eta_{\mathrm{E}}/0.38) + \eta_{\mathrm{TH}} \tag{7.13}$$

which is same as eqn (4.5)

7.2.3.2 Analytical Results

Case (i): For $\dot{m}_a = 0$, eqn (7.11) gives $\dot{q}_u = 0$, which indicates no withdrawal of thermal energy from the PV module, and eqn (7.12b) reduces to $\eta_{\mathrm{o}} = \eta_E$ as expected.

Case (ii): For a very large value of $\dot{m}_a$ (a large value of either air-duct velocity or air-duct depth), eqn (7.11) reduces to

$$\begin{aligned}\dot{q}_{\mathrm{u}} &= \dot{m}_{\mathrm{a}} c_{\mathrm{a}} (T_{\mathrm{airout}} - T_{\mathrm{airin}}) \\ &= \left\{ h_{\mathrm{p1}} h_{\mathrm{p2}} (\alpha\tau)_{\mathrm{eff}} I(t) - U_{\mathrm{L}} (T_{\mathrm{airin}} - T_{\mathrm{a}}) \right\} b\mathrm{L}\end{aligned} \tag{7.14}$$

and eqn (7.12b) becomes

$$\eta_{\mathrm{o}} = \eta_{\mathrm{E}} + \frac{\sum_{i=1}^{T} \dot{q}_{\mathrm{u}}}{\sum_{i=1}^{T} I(t) bL} = \eta_{\mathrm{E}} + h_{\mathrm{p1}} h_{\mathrm{p2}} (\alpha\tau)_{\mathrm{eff}} - U_{\mathrm{L}} \frac{(T_{\mathrm{airin}} - T_{\mathrm{a}})}{\bar{I}} \tag{7.15}$$

with $\bar{I} = \frac{\sum_{i=1}^{T} I(t)}{T}$.

It can be seen from eqn (7.15) that an overall efficiency of a PV/T collector is significantly increased due to withdrawal of the thermal energy from the back of the PV module at very large flow rate of the hot air withdrawal. However, the overall efficiency of the PV/T air collector is further increased by using eqn (7.13). In this case an electrical efficiency is also increased due to lowering of the temperature of the PV module.

Case (iii): For a very large value of L, eqn (7.11) reduces to

$$\begin{aligned} \dot{q}_u &= \dot{m}_a c_a (T_{airout} - T_{airin}) \frac{\dot{m}_a c_a}{U_L} \\ &= \left\{ h_{p1} h_{p2} (\alpha\tau)_{eff} I(t) - U_L (T_{airin} - T_a) \right\} \end{aligned} \tag{7.16a}$$

and eqn (7.12b) becomes

$$\begin{aligned} \eta_o &= \eta_E + \frac{\sum_{i=1}^{T} \dot{q}_u}{\sum_{i=1}^{T} I(t) bL} \\ &= \eta_E + \frac{\dot{m}_a c_a}{U_L bL} \left[h_{p1} h_{p2} (\alpha\tau)_{eff} - U_L \frac{(T_{airin} - T_a)}{\bar{I}} \right] \end{aligned} \tag{7.16b}$$

Case (iv): For $L = 0$, eqn (7.11) gives $\dot{q}_u = 0$, similar to case (i).

7.2.3.3 Temperature-dependent Electrical Efficiency[35]

If $T_{fi} = T_a$ and $T_f = \bar{T}_f$, then from eqns (7.1), (7.5), (7.6) and (7.10) the expression for temperature-dependent electrical efficiency can be obtained as

$$\eta = \frac{\eta_0 \left[1 - \frac{\beta_o \tau_g [\alpha_c \beta_c + \alpha_T (1 - \beta_c)] I(t)}{U_T + h_T} \left\{ 1 + \frac{h_T h_{p1}}{h_t + U_{tT}} + \frac{h_T h_t h_{p1} h_{p2}}{(h_T + U_{tT}) U_L} \left(1 - \frac{1 - \exp(-X_o)}{X_o} \right) \right\} \right]}{1 - \frac{\beta_o \eta_0 \tau_g \alpha_c \beta_c I(t)}{U_T + h_T} \left[1 + \frac{h_T h_{p1}}{h_t + U_{tT}} + \frac{h_T h_t h_{p1} h_{p2}}{(h_T + U_{tT}) U_L} \left(1 - \frac{1 - \exp(-X_o)}{X_o} \right) \right]} \tag{7.17}$$

where $X_o = \frac{b U_L L}{\dot{m}_a C_a}$.

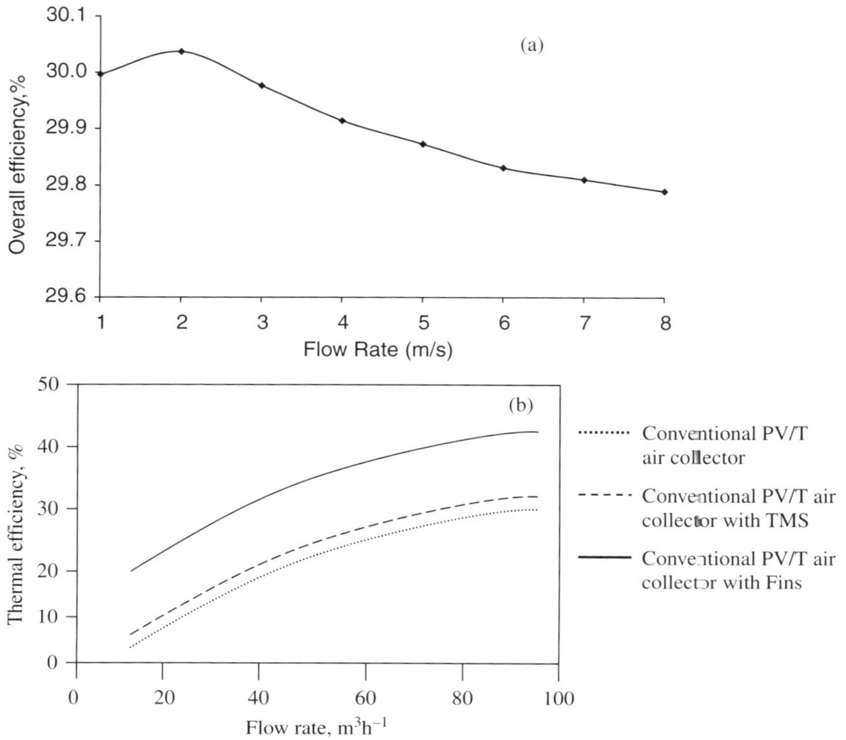

Figure 7.16 Variation of (a) overall efficiency with flow velocity and (b) thermal efficiency with flow rate.[18]

7.2.3.4 Discussion

From Figure 7.15, it is clear that with an increase in collector or channel length the thermal efficiency increases and becomes almost constant with further increase in length while the electrical efficiency decreases as the PV temperature increases with increase in collector length (eqn (7.1)).

Figure 7.16(a) shows the variation of overall efficiency with flow velocity (m s^{-1}) and it is clear that the optimum value of flow velocity is about 2 m s^{-1}. Figure 7.16(b) shows the variation of thermal efficiency with flow rate (m^3 h^{-1}) and it is clear that the value for the PV/T air collector with fins is higher than for other PV/T air collectors.

Example 7.1

Calculate the outlet air temperature for an air duct having cross sectional area 1 m × 0.45 m × 0.04 m. Air is flowing at the rate of 2 m s^{-1}, the penalty factor is 0.7, gain and loss are 0.8 and 6.2 W m^{-2} K^{-1}, respectively, $T_a = 25\,°C$ and $T_{airin} = T_a + 1\,°C$, $I(t) = 750$ W m^{-2}.

Solution

Using eqn (7.9), we have

$$\begin{aligned}\dot{m} &= \rho A V \\ &= 1 \times 0.45 \times 0.04 \times 2 = 0.036\,\mathrm{kg\,s^{-1}} \\ T_{\mathrm{out}} &= \left[\frac{0.7 \times 0.8 \times 750}{6.2} + 25\right]\left[1 - \exp\left(\frac{-0.45 \times 6.2 \times 1}{0.036 \times 1005}\right)\right] \\ &\quad + 26 \times \exp\left(\frac{-0.45 \times 6.2 \times 1}{0.036 \times 1005}\right) \\ &= 30.95\,^{\circ}\mathrm{C}\end{aligned}$$

Example 7.2

Calculate the useful heat gain for Example 7.1.

Solution

Using eqn (7.11), we get

$$\begin{aligned}\dot{Q}_{\mathrm{u}} &= 0.036 \times 1005\{0.7 \times 0.8 \times 750 - 6.2 \times 1\} \\ &\quad \times \left[1 - \exp\left(\frac{-0.45 \times 6.2 \times 1}{0.036 \times 1005}\right)\right] \\ &= 1.1\,\mathrm{kW}\end{aligned}$$

7.2.4 Thermal Modelling of PV/T Air Collector Covered by Glass-to-Glass Type PV Module

7.2.4.1 Energy Balance[35]

(i) *For solar cells of PV module:*

The energy balance equation for a solar cell of a PV module can be written as

$$\alpha_{\mathrm{c}}\tau_{\mathrm{g}}\beta_{\mathrm{c}}I(t)b\,dx = [U_{\mathrm{tc,a}}(T_{\mathrm{c}} - T_{\mathrm{a}}) + U_{\mathrm{Tc,f}}(T_{\mathrm{c}} - T_{\mathrm{f}})]b\,dx + \tau_{\mathrm{g}}\eta\alpha_{\mathrm{c}}\beta_{\mathrm{c}}I(t)b\,dx \tag{7.18a}$$

$$\begin{bmatrix}\text{The rate of solar}\\ \text{energy available}\\ \text{on solar cell}\end{bmatrix} = \begin{bmatrix}\text{An overall heat}\\ \text{loss from top}\\ \text{surface of cell}\\ \text{to ambient}\end{bmatrix} + \begin{bmatrix}\text{The rate of heat}\\ \text{transfer from cell}\\ \text{to flowing fluid}\end{bmatrix} + \begin{bmatrix}\text{The rate of electrical}\\ \text{energy produced}\end{bmatrix}$$

From eqn (7.18a), the expression for cell temperature is

$$T_c = \frac{\tau_g \alpha_c \beta_c (1-\eta) I(t) + U_{tc,a} T_a + U_{tc,f} T_f}{U_{tc,a} + U_{bc,f}} \tag{7.18b}$$

(ii) *For blackened absorber plate:*

$$\left[\alpha_p (1-\beta_c)\tau_g^2 I(t)\right] bdx = \left[h_{p,f}(T_p - T_f) + U_{bp,a}(T_p - T_a)\right] bdx \tag{7.19a}$$

$$\begin{bmatrix}\text{The rate of solar energy}\\ \text{available on blackened}\\ \text{surface from non packing}\\ \text{area of PV module}\end{bmatrix} = \begin{bmatrix}\text{The rate of heat}\\ \text{transfer from}\\ \text{blackened plate}\\ \text{to flowing fluid}\end{bmatrix} + \begin{bmatrix}\text{An overall heat}\\ \text{loss from plate}\\ \text{to ambient}\end{bmatrix}$$

From eqn (7.19a), the expression for plate temperature is

$$T_p = \frac{\alpha_p (1-\beta_c)\tau_g^2 I(t) + h_{p,f} T_f + U_{bp,a} T_a}{U_{bp,a} + h_{p,f}} \tag{7.19b}$$

(iii) *For air flowing through the duct:*

The energy balance of flowing water through an absorber pipe is given by

$$\dot{m}_a c_a \frac{dT_f}{dx} dx = \left[h_{p,f}(T_p - T_f) + U_{Tc,f}(T_c - T_f)\right] bdx \tag{7.20}$$

$$\begin{bmatrix}\text{The mass flow}\\ \text{rate of flowing}\\ \text{fluid}\end{bmatrix} = \begin{bmatrix}\text{The rate of heat}\\ \text{transfer from}\\ \text{blackened plate to}\\ \text{flowing fluid}\end{bmatrix} + \begin{bmatrix}\text{An overall heat}\\ \text{transfer from cell}\\ \text{to flowing fluid}\end{bmatrix}$$

The solution of eqn (7.20) with the help of eqns (7.18b) and (7.19b) and initial conditions, namely at $T_f\mid_{x=0}$, $T_f = T_{fi1}$ and at $T_f|_{x=L}$, $T_f = T_{fo1}$, we get

$$T_{fo} = \left[\frac{(\alpha\tau)_{eff} I(t)}{U_L} + T_a\right]\left[1 - \exp\left(-\frac{bU_L L}{\dot{m}_a C_a}\right)\right] + T_{fi}\exp\left(-\frac{bU_L L}{\dot{m}_a C_a}\right) \tag{7.21a}$$

The average air temperature over the length of air duct below the PV module is obtained as

$$\overline{T_f} = \frac{1}{L}\int_0^L T_{fo} dx = \left[\frac{(\alpha\tau)_{eff} I(t)}{U_L} + T_a\right]\left[1 - \frac{1-\exp\left(-\frac{bU_L L}{\dot{m}_a C_a}\right)}{\frac{bU_L L}{\dot{m}_a C_a}}\right] + T_{fi}\frac{1-\exp\left(-\frac{bU_L L}{\dot{m}_a C_a}\right)}{\frac{bU_L L}{\dot{m}_a C_a}} \tag{7.21b}$$

For a number of collectors connected in series, the outlet temperature of the first collector will be the inlet of the second collector, the outlet temperature of the second will be the inlet of the third and so on. Hence, for a system of N collectors connected in series, the outlet fluid temperature from the Nth collector can be expressed in terms of the inlet temperature of the first collector.

The outlet fluid temperature at the Nth collector fully covered with a PV module is derived as

$$T_{\mathrm{fo}\,N} = \left[\frac{(\alpha\tau)_{\mathrm{eff}} I(t)}{U_{\mathrm{L}}} + T_{\mathrm{a}}\right]\left[1 - \exp\left(-\frac{N\,bU_{\mathrm{L}}L}{\dot{m}_{\mathrm{a}}C_{\mathrm{a}}}\right)\right] + T_{\mathrm{fi}} \exp\left(-\frac{N\,bU_{\mathrm{L}}L}{\dot{m}_{\mathrm{a}}C_{\mathrm{a}}}\right) \tag{7.22a}$$

The useful heat output of the Nth collector is derived as

$$\dot{Q}_{\mathrm{u,N}} = \left[F_{\mathrm{R}}(\alpha\tau)_{\mathrm{eff}}\left\{\frac{1-(1-K_{\mathrm{K}})^{N}}{N\,K_{\mathrm{K}}}\right\}\right] I(t) - \left[F_{\mathrm{R}}U_{\mathrm{L}}\left\{\frac{1-(1-K_{\mathrm{K}})^{N}}{N\,K_{\mathrm{K}}}\right\}\right](T_{\mathrm{fi}} - T_{\mathrm{a}}) \tag{7.22b}$$

where

$$K_{\mathrm{K}} = \left[\frac{b.\,L.\,F_{\mathrm{R}}U_{\mathrm{L}}}{\dot{m}_{\mathrm{a}}C_{\mathrm{a}}}\right] \quad \text{and} \quad F_{\mathrm{R}} = \frac{\dot{m}_{\mathrm{a}}C_{\mathrm{a}}}{A_{\mathrm{c}}U_{\mathrm{L}}}\left[1 - \exp\left(-\frac{A_{\mathrm{c}}U_{\mathrm{L}}F'}{\dot{m}_{\mathrm{a}}C_{\mathrm{a}}}\right)\right]$$

7.2.4.2 Temperature-dependent Electrical Efficiency[35]

If $T_{\mathrm{fi}} = T_{\mathrm{a}}$ and $T_{\mathrm{f}} = \bar{T}_f$ then, from eqns (7.1), (7.18b) and (7.21b), the expression for temperature-dependent electrical efficiency can be obtained as

$$\eta = \frac{\eta_0\left[1 - \dfrac{\tau_{\mathrm{g}}\beta_{\mathrm{o}}}{U_t\mathrm{c,a} + U_{T,\mathrm{c,f}}}\left\{\alpha_{\mathrm{c}}\beta_{\mathrm{c}} + \dfrac{U_{\mathrm{Tc,f}}}{U_{\mathrm{L}}}\left(h_{\mathrm{p1}}\alpha_{\mathrm{c}}\beta_{\mathrm{c}} + h_{\mathrm{p2}}\alpha_{\mathrm{p}}(1-\beta_{\mathrm{c}})\tau_{\mathrm{g}}\right)\left(1 - \dfrac{1-\exp(-X_{\mathrm{o}})}{X_{\mathrm{o}}}\right)\right\} I(t)\right]}{1 - \dfrac{\eta_0\beta_{\mathrm{o}}\tau_{\mathrm{g}}\beta_{\mathrm{c}}\alpha_{\mathrm{c}}\,I(t)}{U_{\mathrm{tc,a}} + U_{T\mathrm{c,f}}}\left(1 + \dfrac{U_{\mathrm{Tc,f}}h_{\mathrm{p1}}}{U_{\mathrm{L}}}\left(1 - \dfrac{1-\exp(-X_{\mathrm{o}})}{X_{\mathrm{o}}}\right)\right)} \tag{7.23}$$

where

$$X_{\mathrm{o}} = \frac{bU_{\mathrm{L}}L}{\dot{m}_{\mathrm{a}}C_{\mathrm{a}}}$$

$$(\alpha\tau)_{\text{eff}} = h_{\text{p1}}(\alpha\tau)_{1,\text{eff}} + h_{\text{p2}}(\alpha\tau)_{2,\text{eff}}$$

$$(\alpha\tau)_{1,\text{eff}} = \tau_g \alpha_c \beta_c (1-\eta) \text{ and } (\alpha\tau)_{2,\text{eff}} = \alpha_p (1-\beta_c)\tau_g^2$$

h_{p1} and h_{p2} are the penalty factors due to the glass cover of the PV module, which are defined as

$$h_{\text{p1}} = \frac{U_{\text{Tc,f}}}{U_{\text{tc,a}} + U_{T\text{c,f}}} \text{ and } h_{\text{p2}} = \frac{h_{\text{p,f}}}{U_{\text{p,a}} + h_{\text{p,f}}}$$

$$U_{\text{tc,a}} = \left[\frac{L_g}{K_g} + \frac{1}{h_o}\right]^{-1} \qquad U_{tT} = \frac{U_{\text{Tc,f}} \cdot U_{\text{tc,a}}}{U_{\text{Tc,f}} + U_{\text{tc,a}}}$$

$$h_o = 5.7 + 3.8\,V, \quad V = 0.5\,\text{m/s} \qquad U_T = \frac{U_{\text{bp,a}} \cdot h_{\text{p f}}}{U_{\text{bp,a}} + h_{\text{p,f}}}$$

$$U_{\text{Tc,f}} = \left[\frac{L_g}{K_g} + \frac{1}{h_i}\right]^{-1} \qquad U_L = U_{tT} + U_T$$

$$h_{\text{p,f}} = h_i = 2.8 + 3v, \; v = 2\,\text{m/s}$$

7.2.4.3 Discussion

The hourly variations of outlet air temperature over the length of duct and electrical efficiency of PV modules are shown in Figures 7.17 and 7.18, respectively. The figures show that higher temperature and efficiency are obtained by using glass-to-glass type PV modules due to the solar radiation falling on the non-packing area of the glass-to-glass module being transmitted through the glass and absorbed by the blackened plate, so that the heat is convected to the flowing air in two ways: from the back surface of the PV

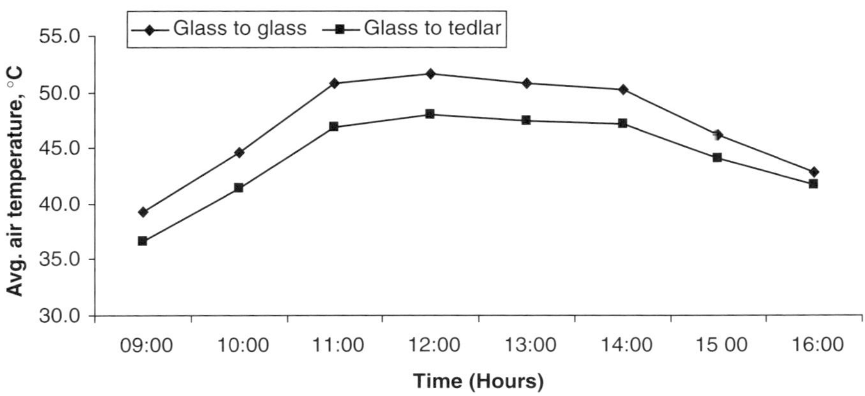

Figure 7.17 Hourly variation of average air temperature over the length of the duct below glass-to-glass and glass-to-tedlar type PV modules.[35]

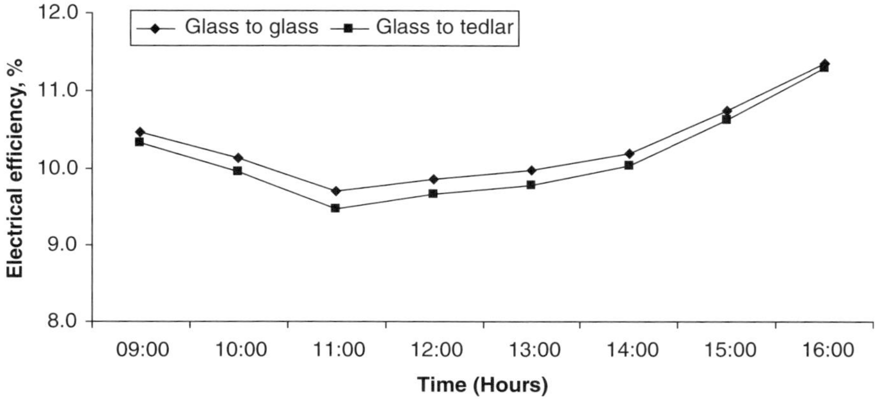

Figure 7.18 Hourly variation of electrical efficiency of glass-to-glass and glass-to-tedlar type PV modules.[35]

module as well as from the top surface of the blackened plate. However, in the case of glass to tedlar all the radiation is absorbed by the tedlar and then carried away by the conduction. This increases the temperature of the solar cell and its efficiency decreases.

7.2.5 Testing of the Solar Air Collector

A schematic view of a solar air collector test facility is shown in Figure 7.19. This test facility has been developed at the University of Waterloo. The air-flow circuit operates as an open loop. The air at ambient temperature passes through a temperature-sensing thermocouple array. The volume flow rate of air is kept at 7.5 and 10 litres m^{-2} s respectively. To keep the inlet air temperature constant up to 80 °C, the inlet air is further allowed to pass through a thermostatically controlled electric heater (2 kW). A pyranometer is fixed at the top of an inclined solar air collector to measure solar intensity (I(t)) during the testing period. The inlet air at constant temperature is then passed through air collector between 11 a.m. and 1 p.m. for air heating. There is less variation in the value of solar intensity between 11 a.m. and 1 p.m.. Further, the heated outlet air is passed to the temperature-sensing array and a calibrated flow measuring orifice through a mixer duct.

The instantaneous efficiency (η_i) can be calculated as

$$\eta_i = \frac{\dot{m}C_{air}(T_{fo} - T_{fi})}{A_p I(t)} \tag{7.24}$$

The plot of η_i with $\left(\frac{T_{fi}-T_a}{I(t)}\right)$ for 7.5 and 10 litres m^{-2} s for a conventional air collector (characteristic curve) is shown in Figure 7.20.[37] It is clear from the figure that the air collector is more efficient at high flow rate.

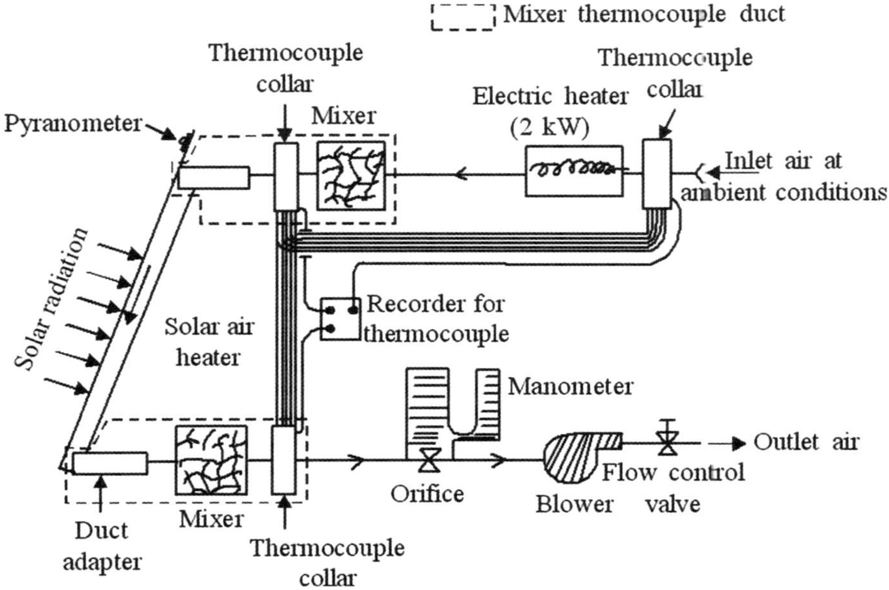

Figure 7.19 Schematic view of air heating collector testing apparatus.

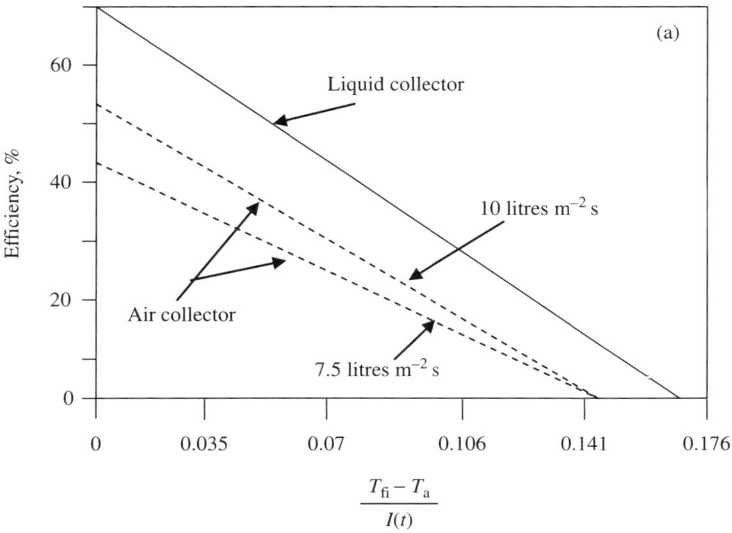

Figure 7.20 Characteristic curve of liquid and air flat-plate collectors.

Figure 7.21a shows the characteristic curve for a hybrid PV/T air collector with TMS and fins.[32] The results for a conventional PV/T air collector have also been shown in the same figure. It is seen that the characteristic curve for a hybrid PV/T air collector with fins gives the best results in comparison to the other two cases.

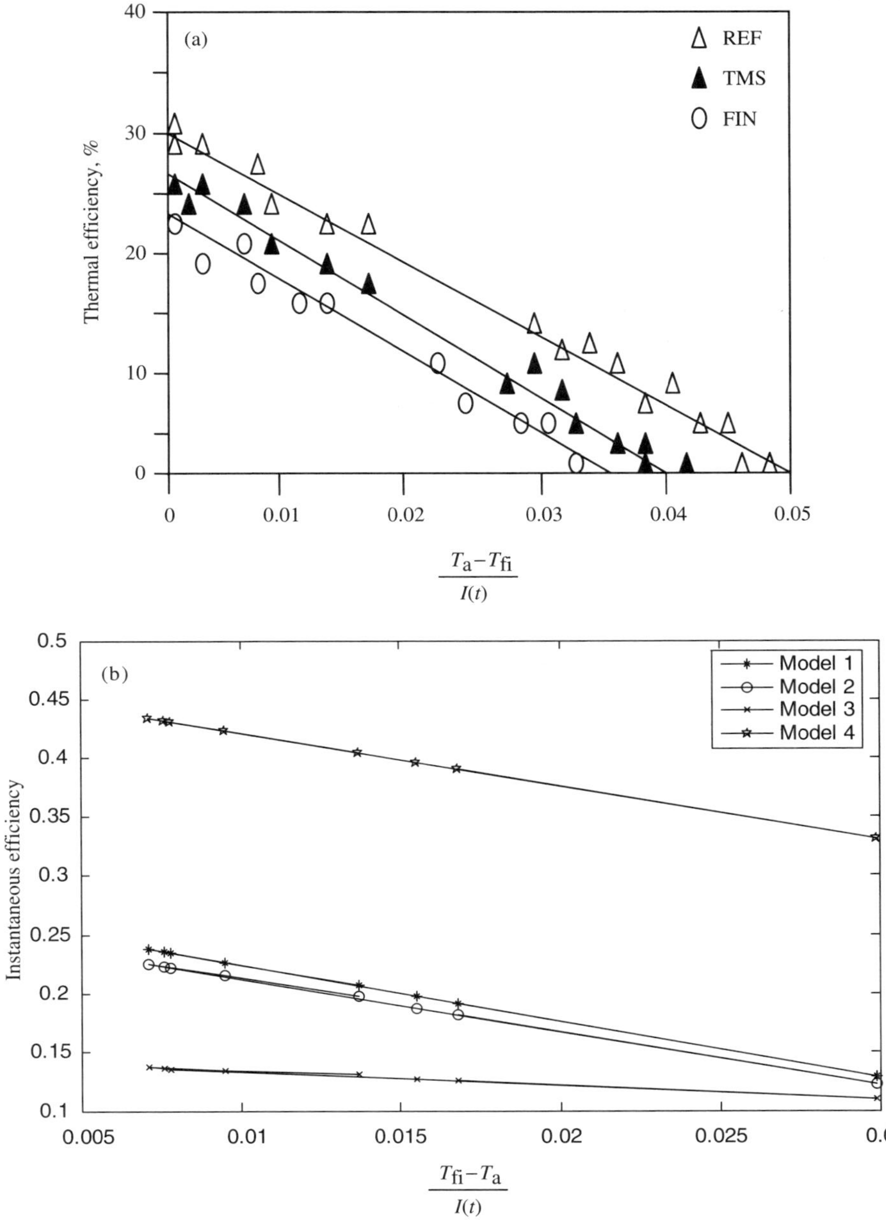

Figure 7.21 Comparison of various hybrid PV/T air collectors (a) REF, TMS and FIN (Figures 7.2(b), 7.3(a), 7.5(a)) and[32] (b) Model 1, 2, 3, and 4 (Figures 7.2(a), 7.4(a), 7.8(a) and (b).[27]

Figure 7.21(b) gives the results of the characteristic curve for four configurations as shown in Figures 7.2(a), 7.4(a), 7.8(a) and (b). This figure shows that a glazed PV/T air collector without tedlar (Figure 7.8b) gives the best results due to maximum thermal energy utilization.

It is further to be noted that the characteristic curves of an unglazed PV/T air collector shown in Figure 7.21(a and b) have exactly the same nature. This validates the results obtained by Tonui and Tripanagnostopoulos[32] and Tiwari and Sodha.[38]

Further, the characteristic equation for a hybrid PV/T air collector for different conditions of Figure 7.21(a) has been developed and the results are reported in Table 7.1.[39] From this table, the PV/T air collector with fins gives maximum efficiency (30%) and minimum U value 6.14 W m^{-2} K^{-1}.

7.3 PV/T Solar Water Heater

A solar water heater (SWH) is a device that uses solar energy to heat water. A solar water-heating system consists of a collector, an insulated storage tank and connecting pipelines. The solar panel of the solar water heater collects the Sun's energy with a black absorber, facing the Sun to catch as much solar radiation as possible. The heat collected by the absorber is transferred to the water flowing through the absorber and is stored in the storage tank. The storage tank is insulated so the water stays hot and can be used later in the day or even the following day. There are two modes by which the heated water is circulated between the collector and storage tank:

- the thermosyphon mode, in which the circulation of heated water is accomplished by natural convection; and
- the forced circulation mode, where a small pump is required for the flow of water.

In the case of forced circulation, a water pump at the inlet of the collector is used to transfer the hot water available at the upper header of the collector to the insulated storage tank. The collector can also be connected in series for higher operating temperatures. The stratification problem can be avoided in the case of forced circulation, unlike in natural circulation. A DC pump can be used for forced circulation of water and the pump is run by a PV module. The PV module is integrated on the collector. The integration area of the PV module depends upon the requirement of hot water or electricity generation. This type of collector is called a PV/T water collector. In the case of partial integration of a PV module, an equal area can be integrated on each collector.

For the PV/T water-heating system, two types of combi-panel (hybrid PV/T) have been considered, namely:

a) *The parallel plate configuration*;[8,17,18,40] and
b) *The tube-in-plate configuration*.[11,21,22,24,27,40]

Table 7.1 Efficiency equation for different hybrid PV/T air collectors (Figure 7.21(a)).

PV/T system type	*Equation*
Conventional PV/T air collector	$\eta_{th} = 0.25 - 7.31\left(\frac{T_{fi}-T}{I(t)}\right)$
Conventional PV/T air collector with TMS	$\eta_{th} = 0.28 - 7.14\left(\frac{T_{fi}-T}{I(t)}\right)$
Conventional PV/T air collector with fins	$\eta_{th} = 0.30 - 6.14\left(\frac{T_{fi}-T}{I(t)}\right)$

Zondag *et al.*[21] developed a model of a hybrid PV/T water collector and performed experimental studies of such systems for varying sizes. Chow *et al.*[24] have concluded that the tube-in-plate absorber collector with single glazing is regarded as one of the most promising designs and that the flat-plate collector partially covered by a PV module gives better thermal and electrical output. They have concluded their findings on the basis of indoor simulation. Robles-Ocampo *et al.*[41] have designed and made an original water-heating planar collector and a set of reflecting planes and concluded that the estimated overall solar energy utilization efficiency for the system related to the direct radiation flux is of the order of 60%, with an electric efficiency of 16.4%. Recently, Zondag[20] carried out a rigorous review of research work on a PV-thermal collector and system, carried out by various scientists up to 2006. His review includes the history and importance of photovoltaic hybrid systems and their application in various sectors. It also includes characteristic equations, a study of design parameters and marketing, *etc.*

7.3.1 Integration of a PV Module on a Collector

A PV module can be integrated on the lower or upper portion of the collector. The analytical expression has been developed for calculating the overall thermal and electrical efficiency of a PV/T collector by varying the position of the PV module on the collector, which is derived by using basic energy balance equations. The following assumptions have been made:

- One-dimension heat conduction is a good approximation for the present study;
- The system is in quasi-steady state;
- The ohmic losses in the solar cell are negligible.

An instantaneous thermal efficiency of a flat-plate collector can be obtained as[42,43]

$$\eta_i = \frac{\dot{Q}_u}{N_c \times A_c \times I(t)} \tag{7.25a}$$

or

$$\eta_i = F_R\left[(\alpha\tau) - U_L\frac{T_{fi} - T_a}{I(t)}\right] \tag{7.25b}$$

Equation (7.25b) is known as the Hottel–Whiller–Bliss equation of a flat-plate collector. This is also known as the characteristic equation of a flat-plate collector.

7.3.1.1 The Lower Portion of the Absorber is Partially Covered by the PV Module

In this case the lower portion of the absorber is covered by the PV module and the upper portion is covered by a glass cover. The outlet of water at the end of the PV module-absorber combination becomes the inlet to the glass-absorber combination. Following Dubey and Tiwari,[44] an expression for the rate of thermal energy available from the PV-integrated flat-plate collector can be given as

$$\dot{Q}_{u,(m+c)} = \dot{m}_f C_f(T_{fo2} - T_{fi})$$

then the total thermal energy available from the PV-integrated (bottom side) flat-plate collector can be derived as

$$\begin{aligned}\dot{Q}_{u,(m+c)} = {} & A_m F_{Rm}\left[h_{p2}(\alpha\tau)_{m,eff} I(t) - U_{L,m}(T_{fi} - T_a)\right]\\ & + A_c F_{Rc}\left[(\alpha\tau)_{c,eff} I(t) - U_{L,c}(T_{fo1} - T_a)\right]\end{aligned}$$

Here

$$T_{fo1} = T_{fi} + \frac{\dot{Q}_{u,m}}{\dot{m}_f C_f}$$

On simplifying the above equation we get

$$\begin{aligned}\dot{Q}_{u,(m+c)} = {} & \left[A_m F_{Rm} h_{p2}(\alpha\tau)_{m,eff}\left(1 + \frac{A_c F_{Rc} U_{L,c}}{\dot{m}_f C_f}\right) A_c F_{Rc}(\alpha\tau)_{c,eff}\right] I(t)\\ & - \left[A_m F_{Rm} U_{Lm}\left(1 + \frac{A_c F_{Rc} U_{L,c}}{\dot{m}_f C_f}\right) + A_c F_{Rc} U_{Lc}\right](T_{fi} - T_a)\end{aligned} \tag{7.26a}$$

In this case an instantaneous efficiency can be obtained by using eqns (7.25a) and (7.26) as

$$\eta_i = 0.56 - 0.42\frac{T_{fi} - T_a}{I(t)} \tag{7.26b}$$

The expression for $(\alpha\tau)_{m,eff}$ and $(\alpha\tau)_{c,eff}$ have been given after eqn (7.32a).

7.3.1.2 The Upper Portion of the Absorber is Partially Covered by the PV Module

In this case the upper portion of the absorber is covered by the PV module and the lower portion is covered by a glass cover. The outlet of water at the end of the glass-absorber combination becomes the inlet to the PV module-absorber combination.

An expression for the rate of thermal energy available from the flat-plate collector can be evaluated as

$$\dot{Q}_{u,(c+m)} = \dot{m}_f C_f (T_{fo2} - T_{fi})$$

An expression for the total thermal energy available from the PV integrated (upper side) flat-plate collector can be evaluated as

$$\begin{aligned}\dot{Q}_{u,(c+m)} = {} & A_c F_{Rc}\left[(\alpha\tau)_{c,eff} I(t) - U_{L,c}(T_{fi} - T_a)\right] \\ & + A_m F_{Rm}\left[h_{p2}(\alpha\tau)_{m,eff} I(t) - U_{L,m}(T_{fo1} - T_a)\right]\end{aligned}$$

Here

$$T_{fo1} = T_{fi} + \frac{\dot{Q}_{u,c}}{\dot{m}_f C_f}$$

On simplifying the above equation we get

$$\begin{aligned}\dot{Q}_{u,(c+m)} = {} & \left[A_c F_{Rc}(\alpha\tau)_{c,eff}\left(1 - \frac{A_m F_{Rm} U_{L,m}}{\dot{m}_f C_f}\right) + A_m F_{Rm} h_{p2}(\alpha\tau)_{m,eff}\right] I(t) \\ & - \left[A_c F_{Rc} U_{Lc}\left(1 - \frac{A_m F_{Rm} U_{L,m}}{\dot{m}_f C_f}\right) + A_m F_{Rm} U_{Lm}\right](T_{fi} - T_a)\end{aligned} \qquad (7.27a)$$

In this case an instantaneous efficiency can be obtained by using eqns (7.25a) and (7.27) as

$$\eta_i = 0.55 - 3.63\frac{T_{fi} - T_a}{I(t)} \qquad (7.27b)$$

7.3.2 Overall Thermal and Electrical Efficiency

- Thermal efficiency: An expression for the overall thermal efficiency can be obtained as[27]

$$\eta_{overall,\,thermal} = \eta_{thermal} + \frac{\eta_{electrical}}{0.38} \qquad (7.28a)$$

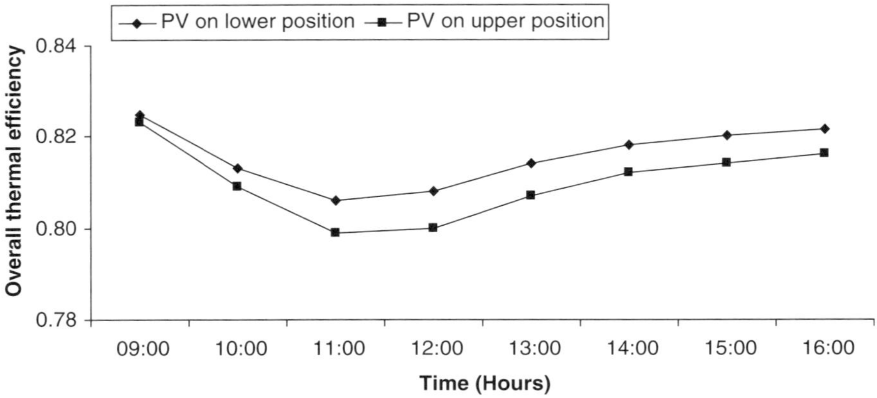

Figure 7.22 Hourly variation of overall thermal efficiency when the upper and lower portions of the absorber are partially covered by a PV module.

- Electrical efficiency: An expression for the overall electrical efficiency can be obtained as[27]

$$\eta_{\text{overall, electrical}} = \eta_{\text{electrical}} + \eta_{\text{thermal}}\left(1 - \frac{T_a + 273}{T_f + 273}\right) \tag{7.28b}$$

The overall thermal and electrical efficiency have been calculated using eqns (7.28a) and (7.28b). Higher overall thermal and electrical efficiency are obtained in the case of PV on the lower position than PV on the upper position because, in the case of the upper position, the water gets pre-heated from the glass-absorber combination of the collector area and then heated water goes into the PV-absorber combination of the collector area, which decreases the heat transfer from the PV module and increases the cell temperature. A higher cell temperature decreases the cell efficiency and consequently the overall efficiency of the collector system. The variations in overall thermal and electrical efficiency are shown in Figures 7.22 and 7.23), respectively. The figures show that the overall thermal efficiency varies from 80.6% to 82.5% and 79.9% to 82.1% and electrical efficiency varies from 10.4% to 11.7% and 10.1% to 11.2% when the PV module is in the lower and upper position, respectively.

7.3.3 Hybrid PV/T Water-heating System

In this section, thermal modelling and performance of a PV-integrated solar water-heating system, which is installed at Solar Energy Park, IIT Delhi, have been discussed. A hybrid PV/T solar water-heating system consists of two flat-plate collectors connected in series, each having an effective area of 2 m^2. The embedded design of an absorber is shown in Figure 7.24. Specifications of a flat-plate collector are given in Table 7.2. The whole absorber and glass cover is encased in an aluminium metallic box with 0.1 m glass wool insulation below the absorber to reduce bottom losses.

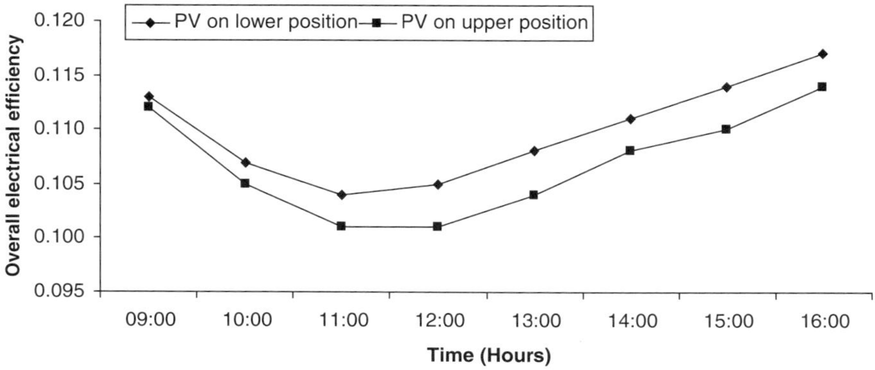

Figure 7.23 Hourly variation of overall electrical efficiency when the upper and lower portions of the absorber are partially covered by a PV module.

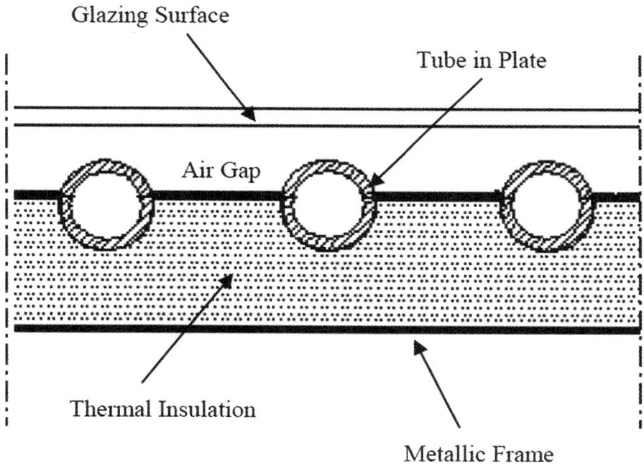

Figure 7.24 Cross sectional front view of an embedded design of a flat-plate collector.

A glass-to-glass photovoltaic (PV) module with an effective area of 0.605 m^2 is integrated at the bottom of one of the collectors. The flow pattern of water in such a configuration has also been depicted in Figure 7.25. In this case; solar radiation is transmitted through the non-packing area of a PV module and finally absorbed by the blackened absorber. Further, the thermal energy associated with the PV module is transferred to the absorber by convection for further heating of the absorber. Water below the absorber gets heated and moves in the upward direction. The outlet of water at the end of the absorber, which is covered with the PV module (T_{fo1}), becomes the inlet to the glass-absorber combination. Such a collector is referred to as a photovoltaic/thermal (PV/T) water collector.

The outlet of the photovoltaic/thermal (PV/T) water collector (T_{fo2}) is further connected to the inlet of a conventional flat-plate collector for higher

Table 7.2 Dimensions of photovoltaic/thermal (PV/T) solar water-heating system.

Sr. No.	*Components*	*Specifications*
1.	Capacity of storage tank	200 litres
2.	Collectors	Flat-plate, tube-in-plate type
3.	Area of collector	$2\,m^2$
4.	Tube diameter	0.0125 m
5.	Tube material	Copper tubes
6.	Plate thickness	0.002 m
7.	Air gap	0.01 m
8.	Thickness of insulation	0.1 m
9.	Thickness of glass	0.004 m
10.	Angle of collector	30°
11.	PV module	Glass-to-glass type
12.	Area of module	$0.605\,m^2$
13.	Area of solar cell	$0.015\,m^2$
14.	Total area of solar cell	$0.54\,m^2$
15.	Non-packing area	$0.065\,m^2$
16.	No. of solar cells	36
17.	PV module	75 W
18.	DC motor	18 V, 60 W, 2800 rpm

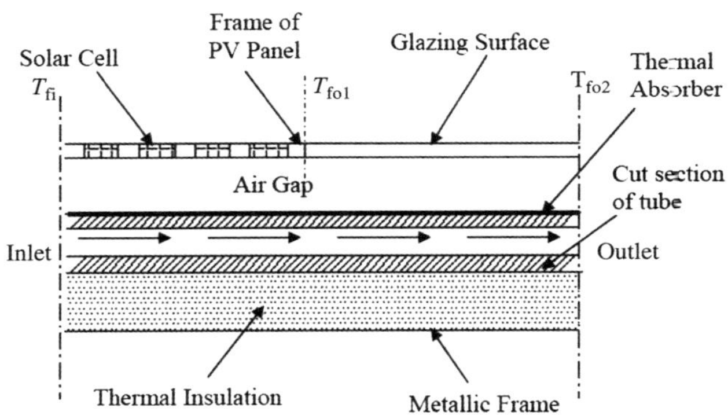

Figure 7.25 Cross sectional side view of a PV-integrated flat-plate collector.

operation temperature. Both collectors are connected to an insulated storage tank of 200 litres capacity. There is a provision of a DC water pump (18 V, 60 W, 2800 rpm) connected to the PV module to circulate the water between collectors and storage tank in a forced mode. A photograph of the complete experimental set-up is shown in Figure 7.26.[44]

7.3.3.1 Energy Balance Equations[44]

In order to write the energy balance equation for each component of a combined system of photovoltaic/thermal (PV/T) solar water heater, the following assumptions have been made:

Figure 7.26 Photograph of a combined photovoltaic/thermal (PV/T) glass-to-glass solar water-heating system.[44]

- The heat capacity of the photovoltaic/thermal (PV/T) collector has been neglected in comparison with the heat capacity of water in the storage tank;
- There is no temperature stratification in the water of the storage tank due to forced mode of operation;
- One-dimension heat conduction is a good approximation for the present study;
- The system is in a quasi-steady state;
- The ohmic losses in the solar cell are negligible.

The energy balance equations for each component of a PV/T solar water heating system are as follows:

(i) *For solar cells of PV module (glass-glass):*

$$\alpha_c \tau_c \beta_c I(t) W dx = \left[U_{tc,a}(T_c - T_a) + h_{c,p}(T_c - T_p)\right] W dx + \tau_g \eta_c \beta_c I(t) \cdot W dx \tag{7.29a}$$

$$\begin{bmatrix}\text{The rate of solar}\\\text{energy available}\\\text{on solar cell}\end{bmatrix} = \begin{bmatrix}\text{An overall heat}\\\text{loss from top}\\\text{surface of cell}\\\text{to ambient}\end{bmatrix} + \begin{bmatrix}\text{The rate of heat}\\\text{transfer from cell}\\\text{to flowing fluid}\end{bmatrix} + \begin{bmatrix}\text{The rate of}\\\text{electrical energy}\\\text{produced}\end{bmatrix}$$

From eqn (7.29a), the expression for the cell temperature is

$$T_c = \frac{(\alpha\tau)_{1,\text{eff}} I(t) + U_{tc,a} T_a + h_{c,p} T_p}{U_{tc,a} + h_{c,p}} \tag{7.29b}$$

(ii) *For blackened absorber plate temperature below the PV module (glass-glass):*

$$\alpha_p (1 - \beta_c) \tau_g^2 I(t) + h_{c,p}(T_c - T_p) = h_{p,f}(T_p - T_f) \tag{7.30a}$$

$$\begin{bmatrix}\text{The rate of solar energy}\\ \text{available on blackened}\\ \text{surface from non packing}\\ \text{area of PV module}\end{bmatrix} + \begin{bmatrix}\text{The rate of heat}\\ \text{transfer from cell}\\ \text{to absorber}\end{bmatrix} = \begin{bmatrix}\text{The rate of heat transfer}\\ \text{from blackened plate to}\\ \text{flowing fluid}\end{bmatrix}$$

From eqn (7.30a), the expression for plate temperature is

$$T_p = \frac{(\alpha\tau)_{2,\text{eff}} I(t) + h_{p1}(\alpha\tau)_{1,\text{eff}} I(t) + U_{L1} T_a + h_{p,f} T_f}{U_{L1} + h_{p,f}} \tag{7.30b}$$

(iii) *For water flowing through an absorber pipe below the PV module (glass-glass):*

The energy balance of flowing water through the absorber pipe is given by

$$\dot{m}_f C_f \frac{dT_f}{dx} dx = F' h_{p,f}(T_p - T_f) W dx \tag{7.31}$$

$$\begin{bmatrix}\text{The rate of heat}\\ \text{withdrawal}\end{bmatrix} = \begin{bmatrix}\text{The rate of heat}\\ \text{transfer from}\\ \text{blackened plate to}\\ \text{flowing fluid}\end{bmatrix} + \begin{bmatrix}\text{The rate of heat}\\ \text{transfer from cell}\\ \text{to flowing fluid}\end{bmatrix}$$

The solution of eqn (7.31) with the help of eqns (7.29b) and (7.30b) and initial conditions, namely at $T_f|_{x=0}$, $T_f = T_{fi1}$ and at $T_f|x_{=L}$, $T_f = T_{fo1}$, we get

$$T_{fo1} = \left[\frac{h_{p2}(\alpha\tau)_{m,\text{eff}} I(t)}{U_{L,m}} + T_a\right]\left[1 - \exp\left(-\frac{F' A_m U_{L,m}}{\dot{m}_f C_f}\right)\right] + T_{fi1} \exp\left(-\frac{F' A_m U_{L,m}}{\dot{m}_f C_f}\right) \tag{7.32a}$$

Here, T_{fo1} is the outlet temperature of the water from the absorber PV module and T_{fo1} becomes the inlet temperature for the remaining part of the collector,

where

$$(\alpha\tau)_{m,eff} = h_{p1}(\alpha\tau)_{1,eff} + (\alpha\tau)_{2,eff}$$

and

$$(\alpha\tau)_{1,eff} = (\alpha_c - \eta_c)\beta_c\tau_c \; and \; (\alpha\tau)_{2,eff} = \alpha_p(1 - \beta_c)\tau_g^2$$

The penalty factors due to the glass cover of the PV module (h_{p1}) and absorber below the PV module (h_{p2}) are defined as

$$h_{p1} = \frac{h_{c,p}}{U_{tc,a} + h_{c,p}} \text{ and } h_{p2} = \frac{h_{p,f}}{U_{L1} + h_{p,f}}$$

$$U_{tc,a} = 5.7 + 3.8\,V, \; U_{L1} = \frac{U_{tc,a} \cdot h_{c,p}}{U_{tc,a} + h_{c,p}},$$

$$U_{Lm} = \frac{U_{L1} \cdot h_{p,f}}{U_{L1} + h_{p,f}}$$

$$h_{c,p} = 5.7 + 3.8\,V, \; V = 0\,m/s$$

The rate of thermal energy available at the end of the absorber PV module (glass-glass) is evaluated as

$$\dot{Q}_{u,m} = \dot{m}_f C_f (T_{fo1} - T_{fi})$$

After substituting the expression for T_{fo1} from eqn (7.32a), we get

$$\dot{Q}_{u,m} = A_m F_{Rm}\left(h_{p2}(\alpha\tau)_{m,eff} I(t) - U_{L,m}(T_{fi} - T_a)\right) \tag{7.32b}$$

Following Duffie and Beckman[42] and Tiwari,[43] the flat-plate collector efficiency is given by

$$F' = \frac{1}{\dfrac{W \times U_L}{\pi D h} + \dfrac{W}{D + (W - D)F}} \tag{7.32c}$$

where

$$F = \frac{\tanh[m(W - D)]/2}{[m(W - D)]/2} \text{ and } m = \sqrt{\frac{U_L}{K\delta}}$$

Now, the flow rate factor (F_R) is given by

$$F_R = \frac{\dot{m}C_f}{A_c U_L}\left[1 - \exp\left(-\frac{A_c U_L F'}{\dot{m}C_f}\right)\right] \tag{7.32d}$$

(iv) *The outlet water temperature at the end of first collector (Figure 7.25):*

Following Duffie and Beckman[42] and Tiwari,[43] an expression for the outlet water temperature at the end of the first collector will be

$$T_{fo2} = \left[\frac{(\alpha\tau)_{c1,eff} I(t)}{U_{L,c1}} + T_a\right]\left[1 - \exp\left(-\frac{F' A_{c1} U_{L,c1}}{\dot{m}_f C_f}\right)\right] + T_{fi2} \exp\left(-\frac{F' A_{c1} U_{L,c1}}{\dot{m}_f C_f}\right) \tag{7.33a}$$

Here, $T_{fi2} = T_{fo1}$ can be evaluated from eqn (7.32a).

The rate of thermal energy available from the first flat-plate collector can be evaluated as

$$\dot{Q}_{u,c1} = \dot{m}_f C_f (T_{fo2} - T_{fo1})$$

After substituting the expression for T_{fo2} from eqn (7.33a), we get

$$\dot{Q}_{u,c1} = A_{c1} F_{Rc1}\left((\alpha\tau)_{c1,eff} I(t) - U_{L,c1}(T_{fo1} - T_a)\right) \tag{7.33b}$$

Here,

$$T_{fo1} = T_{fi} + \frac{\dot{Q}_{u,m}}{\dot{m}_f C_f}$$

(v) *The outlet temperature from the second collector:*

Similarly, an expression of outlet water temperature at the end of the second flat-plate collector can be written as a function of the outlet water temperature ($T_{fi3} = T_{fo2}$), which is inlet to the second collector as

$$T_{fo3} = \left[\frac{(\alpha\tau)_{c2,eff} I(t)}{U_{L,c2}} + T_a\right]\left[1 - \exp\left(-\frac{F' A_{c2} U_{L,c2}}{\dot{m}_f C_f}\right)\right] + T_{fi3} \exp\left(-\frac{F' A_{c2} U_{L,c2}}{\dot{m}_f C_f}\right) \tag{7.34a}$$

Here, $T_{fi3} = T_{fo2}$ can be evaluated from eqn (7.33a).

The above equations can be rearranged to get the final outlet water temperature at the end of collectors connected in series (T_{fo3}),

$$\begin{aligned} T_{fo3} = {} & \left(\frac{(\alpha\tau)_{c2,eff} I(t)}{U_{L,c2}} + T_a\right)\left(1 - \exp\left(-\frac{F' A_{c2} U_{L,c2}}{\dot{m}_f C_f}\right)\right) + \\ & \left[\left(\frac{(\alpha\tau)_{c1,eff} I(t)}{U_{L,c1}} + T_a\right)\left(1 - \exp\left(-\frac{F' A_{c1} U_{L,c1}}{\dot{m}_f C_f}\right)\right) + \right. \\ & \left. \left\{\begin{array}{l} \left(\frac{h_{p2}(\alpha\tau)_{m,eff} I(t)}{U_{L,m}} + T_a\right)\left(1 - \exp\left(-\frac{F' A_m U_{L,m}}{\dot{m}_f C_f}\right)\right) + \\ T_{fi1} \exp\left(-\frac{F' A_m U_{L,m}}{\dot{m}_f C_f}\right) \end{array}\right\} \exp\left(-\frac{F' A_{c1} U_{L,c1}}{\dot{m}_f C_f}\right)\right] \\ & \exp\left(-\frac{F' A_{c2} U_{L,c2}}{\dot{m}_f C_f}\right) \end{aligned} \tag{7.34b}$$

(vi) *The rate of thermal energy available at the end of the second collector:*
An expression for the rate of thermal energy available at the end of the second collector will be as follows:

$$\dot{Q}_{\mathrm{u(m+c1+c2)}} = \dot{m}_{\mathrm{f}} C_{\mathrm{f}} (T_{\mathrm{fo3}} - T_{\mathrm{fi}})$$

The rate of thermal energy available at the end of the second flat-plate collector in terms of outlet temperature from the first collector (T_{fo2}) can be evaluated as

$$\dot{Q}_{\mathrm{u,c2}} = \dot{m}_{\mathrm{f}} C_{\mathrm{f}} (T_{\mathrm{fo3}} - T_{\mathrm{fo2}})$$

After substituting the expression for T_{fo3} from eqn (7.34a), we get

$$\dot{Q}_{\mathrm{u,c2}} = A_{\mathrm{c2}} F_{\mathrm{Rc2}} \Big((\alpha\tau)_{\mathrm{c2,eff}} I(t) - U_{\mathrm{L,c2}} (T_{\mathrm{fo2}} - T_{\mathrm{a}}) \Big) \tag{7.35}$$

Here,

$$T_{\mathrm{fo2}} = T_{\mathrm{fi}} + \frac{\dot{Q}_{\mathrm{u,m}}}{\dot{m}_{\mathrm{f}} C_{\mathrm{f}}} + \frac{\dot{Q}_{\mathrm{u,c1}}}{\dot{m}_{\mathrm{f}} C_{\mathrm{f}}}$$

On solving eqns (7.32b), (7.33b) and (7.35) we get

$$\begin{aligned} Q_{\mathrm{u(m+c1+c2)}} = & \left[\begin{array}{l} A_{\mathrm{m}} F_{\mathrm{Rm}} h_{\mathrm{p2}} (\alpha\tau)_{\mathrm{m,eff}} (1 - K_1) \\ + A_{\mathrm{c1}} F_{Rc1} (\alpha\tau)_{\mathrm{c1,eff}} (1 - K_2) + A_{\mathrm{c2}} F_{\mathrm{Rc2}} (\alpha\tau)_{\mathrm{c2,eff}} \end{array} \right] I(t) \\ & - \left[\begin{array}{l} A_{\mathrm{m}} F_{\mathrm{Rm}} U_{\mathrm{L,m}} (1 - K_1) \\ + A_{\mathrm{c1}} F_{\mathrm{Rc1}} U_{\mathrm{L,c1}} (1 - K_2) + A_{\mathrm{c2}} F_{\mathrm{Rc2}} U_{\mathrm{L,c2}} \end{array} \right] (T_{\mathrm{fi1}} - T_{\mathrm{a}}) \end{aligned} \tag{7.36}$$

where

$$K_1 = \left[\frac{A_{\mathrm{c1}} F_{\mathrm{Rc1}} U_{\mathrm{L,c1}}}{\dot{m}_{\mathrm{f}} C_{\mathrm{f}}} + \frac{A_{\mathrm{c2}} F_{\mathrm{Rc2}} U_{\mathrm{L,c2}}}{\dot{m}_{\mathrm{f}} C_{\mathrm{f}}} - \frac{A_{\mathrm{c1}} F_{\mathrm{Rc1}} U_{\mathrm{L,c1}} A_{\mathrm{c2}} F_{\mathrm{Rc2}} U_{\mathrm{L,c2}}}{(\dot{m}_{\mathrm{f}} C_{\mathrm{f}})^2} \right]$$

and

$$K_2 = \left[\frac{A_{\mathrm{c2}} F_{\mathrm{Rc2}} U_{\mathrm{L,c2}}}{\dot{m}_{\mathrm{f}} C_{\mathrm{f}}} \right]$$

7.3.3.2 Instantaneous Thermal Efficiency

An instantaneous thermal efficiency of a flat-plate collector can be obtained as

$$\eta_{\mathrm{i}} = (\alpha\tau)_{\mathrm{eff}} - U_{\mathrm{L}} \frac{T_{\mathrm{fi}} - T_{\mathrm{a}}}{I(t)} \tag{7.37}$$

where

$$(\alpha\tau)_{\text{eff}} = \left[\frac{A_{\text{m}}F_{\text{Rm}}h_{\text{p2}}(\alpha\tau)_{\text{m,eff}}(1-K_1) + A_{\text{c1}}F_{\text{Rc1}}(\alpha\tau)_{\text{c1,eff}}(1-K_2) - A_{\text{c2}}F_{\text{Rc2}}(\alpha\tau)_{\text{c2,eff}}}{(A_{\text{m}} + A_{\text{c1}} + A_{\text{c2}})}\right]$$

and

$$U_{\text{L}} = \left[\frac{A_{\text{m}}F_{\text{Rm}}U_{\text{L,m}}(1-K_1) + A_{\text{c1}}F_{\text{Rc1}}U_{\text{L,c1}}(1-K_2) + A_{\text{c2}}F_{\text{Rc2}}U_{\text{L,c2}}}{(A_{\text{m}} + A_{\text{c1}} + A_{\text{c2}})}\right]$$

Example 7.3

A flat-plate collector system has an aluminium absorber plate ($K_{\text{p}} = 211$ W m^{-1} °C^{-1}) of thickness 0.35 mm and area 1.5 m^2 and it has two riser tubes of diameter 0.025 m each. The length of the tubes being 1 m, find out the collector efficiency factor F' for this collector, if the convective heat transfer coefficient from the inner tube surface to water is 50,100 and 500 W m^{-2} °C. The overall loss coefficient is 7.2 W m^{-2} °C^{-1}.

Solution

The width of the spacing between the two riser tubes is

$$W = (1.5 - 0.025 \times 10)/10 = 0.125 \text{ m}.$$

The value of m and the fin efficiency factor (F) can be obtained as

$$m = \left(\frac{7.2}{211 \times 0.35 \times 10^{-3}}\right)^{1/2} = 9.87$$

$$F = \frac{\tanh[9.87(0.125 - 0.025)/2]}{9.87 \times (0.125 - 0.025)/2}$$

$$= \frac{\tanh 0.4935}{0.4935} = 0.926$$

The collector efficiency factor (F') (eqn (7.32c)), for $h = 50$ W m^{-2} °C^{-1} and $b = D = 0.025$ m is:

$$F' = \frac{1/7.2}{0.125\left[\frac{1}{7.2(0.125-0.025)\times 0.926+0.025} + \frac{1}{3.14\times 0.025\times 50}\right]}$$

$$= \frac{1}{\left[\frac{0.125}{(0.125-0.025)\times 0.926+0.025} + \frac{0.125\times 7.2}{3.14\times 0.025\times 50}\right]}$$

$$= \frac{1}{1.0629 + 0.2293} = 0.774$$

Similarly, for $h = 100\,\mathrm{W\,m^{-2}\,^{\circ}C^{-1}}$, $F' = 0.849$ and for $h = 500\,\mathrm{W\,m^{-2}\,^{\circ}C^{-1}}$, $F' = 0.921$ and for $h = 1000\,\mathrm{W\,m^{-2}\,^{\circ}C^{-1}}$, $F' = 0.931$.

It can be seen from the above calculation that there is no significant variation in the value of F' for $h \geq 500\,\mathrm{W\,m^{-2}\,^{\circ}C^{-1}}$.

Example 7.4

Calculate the fin efficiency factor and the collector efficiency factor for the following data:

Overall loss coefficient $= 6\,\mathrm{W\,m^{-2}\,^{\circ}C^{-1}}$, tube spacing $= 100$ mm, tube diameter $= 8$ mm, plate thickness $= 0.45$ mm, thermal conductivity $= 385\,\mathrm{W\,m\,^{\circ}C^{-1}}$, heat transfer coefficient inside tubes $= 100\,\mathrm{W\,m^{-2}\,^{\circ}C^{-1}}$, bond resistance $= 0$.

Also, calculate the collector efficiency factor for the value of the heat transfer coefficient inside the tubes as $300\,\mathrm{W\,m^{-2}\,^{\circ}C^{-1}}$ and $1000\,\mathrm{W\,m^{-2}\,^{\circ}C^{-1}}$.

Solution

The value of m and the fin efficiency factor (F) can be obtained as

$$m = \left(\frac{6}{385 \times 4.5 \times 10^{-4}}\right)^{1/2} = 5.88$$

$$\text{and, } F = \frac{\tanh\,[5.88(0.10 - 0.008)/2]}{5.88 \times (0.10 - 0.008)/2} = 0.976$$

The collector efficiency factor F' (from eqn (7.32c)) is

$$F' = \frac{1/6}{0.10\left[\frac{1}{6[(0.10-0.008)0.976+0.008]} + \frac{1}{\pi(0.008)\times 100}\right]} = 0.800$$

The collector efficiency for $h_{fi} = 300\,\mathrm{W\,m^{-2}\,^{\circ}C}$ is given as

$$F' = \frac{1/6}{0.10\left[\frac{1}{6[(0.10-0.008)0.976+0.008]} + \frac{1}{\pi(0.008)\times 300}\right]} = 0.91$$

Similarly for $h_{fi} = 1000\,\mathrm{W\,m^{-2}\,^{\circ}C^{-1}}$, we have $F' = 0.96$ and for $h_{fi} = 2000\,\mathrm{W\,m^{-2}\,^{\circ}C^{-1}}$, $F' = 0.97$.

We see that as the heat transfer coefficient inside the tube is increased, the collector efficiency factor increases. However, not much increase in efficiency is observed when the value of h_{fi} is increased beyond $1000\,\mathrm{W\,m^{-2}\,^{\circ}C^{-1}}$.

Example 7.5

Calculate the fin efficiency factor and the collector efficiency factor for the data given below:

Tube spacing = 100 mm, tube diameter (inside) = 8 mm
Plate thickness = 0.45 mm, plate thermal conductivity = 385 W m °C^{-1}
Heat transfer coefficient inside tubes = 300 W m^{-2} °C^{-1}
U = 2, 4 and 8 W m^{-2} °C^{-1}.

Solution

For $U = 2$ W m^{-2} °C, we have

$$m = \left(\frac{2}{385 \times 4.5 \times 10^{-4}}\right)^{1/2} = 3.40$$

$$F = \frac{\tanh\,[3.40(0.10 - 0.008)/2]}{3.40 \times (0.10 - 0.008)/2} = 0.99$$

Further, F' is given by eqn (7.32c) and its value will be

$$F' = \frac{1/2}{0.10\left[\frac{1}{2[(0.10-0.008)0.99+0.008]} + \frac{1}{\pi(0.008)\times 300}\right]} = 0.96$$

For $U = 4$, m, F and F' are given by

$$m = \left(\frac{4}{385 \times 4.5 \times 10^{-4}}\right)^{1/2} = 4.81$$

$$F = \frac{\tanh\,[4.81(0.10 - 0.008)/2]}{4.81 \times (0.10 - 0.008)/2} = 0.98$$

$$\text{and, } F' = \frac{1/4}{0.10\left[\frac{1}{4[0.099]} + \frac{1}{\pi(0.008)\times 300}\right]} = 0.94$$

For $U = 8$, m, F and F' are given by

$$m = \left(\frac{8}{385 \times 4.5 \times 10^{-4}}\right)^{1/2} = 6.795$$

$$F = \frac{\tanh\,[6.795(0.10 - 0.008)/2]}{4.81 \times (0.10 - 0.008)/2} = 0.968$$

$$\text{and, } F' = \frac{1/8}{0.10\left[\frac{1}{8[0.097]} + \frac{1}{\pi(0.008)\times 300}\right]} = 0.879$$

Hence, we see that with an increase in the overall loss coefficient the collector efficiency factor F' decreases.

7.3.3.3 Energy Balance for Complete Water-heating System without Withdrawal[44]

The rate of thermal energy available at the outlet of the second collector is fed into an insulated storage tank, and then the energy balance of whole system will be

$$\dot{Q}_{u,(m+c1+c2)} = M_w C_w \frac{dT_w}{dt} + (UA)_{tk}(T_w - T_a) \tag{7.38a}$$

The above equation can be solved by assuming $T_{fi} = T_w$ due to perfectly insulating connecting pipes. Here it is assumed that there is no withdrawal of hot water from the storage tank. Using eqn (7.36) the tank water temperature can be obtained as

$$(\alpha\tau)_{eff} I(t) - (UA)_{eff}(T_w - T_a) = M_w C_w \frac{dT_w}{dt} + (UA)_{tk}(T_w - T_a) \tag{7.38b}$$

or

$$\frac{dT_w}{dt} + aT_w = f(t)$$

In order to obtain an approximate solution of the above equation, the following assumptions have been made:

a) The time interval Δt $(0 < t < \Delta t)$ is small.
b) The function $f(t)$ is constant, *i.e.* $f(t) = \overline{f(t)}$ for the time interval Δt.
c) a is constant during the time interval Δt

where $a = \frac{\left[(UA)_{eff} + (UA)_{tk}\right]}{M_w C_w}$ and $f(t) = \frac{(\alpha\tau)_{eff} I(t) + \left[(UA)_{eff} + (UA)_{tk}\right] T_a}{M_w C_w}$

On solving the above differential equation the expression for the tank water temperature can be obtained as

$$T_w = \frac{\overline{f(t)}}{a}\left(1 - e^{-at}\right) + T_{w0} e^{-at} \tag{7.39}$$

where T_{w0} is the temperature of the storage tank water at $t=0$ and $\overline{f(t)}$ is the average value of $f(t)$ for the time interval between 0 and t.

The thermal energy output from the tank is given as

$$\dot{Q}_{u,thermal} = M_w C_w (T_w - T_a) \tag{7.40}$$

7.3.3.4 Energy Balance for Complete Water Heating System with Withdrawal

The energy balance of a PV/T water heating system, considering withdrawal from the tank is given as

$$\dot{Q}_{u,(m+c1+c2)} = M_w C_w \frac{dT_w}{dt} + (UA)_{tk}(T_w - T_a) + \dot{m}_w C_w (T_w - T_a) \tag{7.41a}$$

or

$$(\alpha\tau)_{eff} I(t) - (UA)_{eff}(T_w - T_a) = M_w C_w \frac{dT_w}{dt} + (UA)_{tk}(T_w - T_a) + \dot{m}_w C_w (T_w - T_a) \tag{7.41b}$$

or

$$\frac{dT_w}{dt} + aT_w = f(t)$$

where $a = \frac{[(UA)_{eff} + (UA)_{tk} + \dot{m}_w C_w]}{M_w C_w}$ and $f(t) = \frac{(\alpha\tau)_{eff} I(t) + [(UA)_{eff} + (UA)_{tk} + \dot{m}_w C_w] T_a}{M_w C_w}$.

On solving the above differential equation the expression for the tank water temperature can be obtained as

$$T_w = \frac{\overline{f(t)}}{a}(1 - e^{-at}) + T_{w0} e^{-at} \tag{7.42}$$

where T_{w0} is the temperature of the storage tank water at $t=0$ and $\overline{f(t)}$ is the average value of $f(t)$ for the time interval between 0 and t.

The thermal energy output from the tank is given as

$$\dot{Q}_{u,thermal} = \dot{m}_w C_w (T_w - T_a) \tag{7.43}$$

To compare the results of the calculations with the experimental results, the correlation coefficient (r) and root mean square percent deviation (e) have been evaluated by using the following expressions:

$$r = \frac{N \sum X_i Y_i - (\sum X_i)(\sum Y_i)}{\sqrt{N \sum X_i^2 - (\sum X_i)^2} \sqrt{N \sum Y_i^2 - (\sum Y_i)^2}} \tag{7.44a}$$

and

$$e = \sqrt{\frac{\sum (e_i)^2}{N}} \tag{7.44b}$$

where

$$e_i = \left[\frac{X_i - Y_i}{X_i}\right] \times 100$$

Example 7.6(a)

Calculate the net rate of useful energy per m^2 for the following parameters:

(i) The overall heat loss coefficient (U_L) = 6.0 W m^{-2} °C^{-1} and F' = 0.8 (Example 7.5);
(ii) $\dot{m}$ = 0.35 kg s^{-1} and C_f = 4190 J^{-1} kg °C^{-1};
(iii) $I(t)$ = 500 W m^{-2} and $(\alpha\tau)$ = 0.8;
(iv) T_{fi} = 60 °C and T_a = 40 °C.

Solution

The flow rate factor is given by

$$\begin{aligned} F_R &= [\dot{m}\, C_f/(A_c\, U_L)][1 - \exp(-A_c\, U_L\, F'/(\dot{m}C_f))] \\ &= [0.35 \times 4190/(1 \times 6)][1 - \exp(-6 \times 1 \times 0.8/(0.35 \times 4190))] = 0.7986. \end{aligned}$$

The net rate of useful energy per m^2 can be calculated as

$$\begin{aligned} \dot{q}_u &= F_R[\alpha_0\tau_0\, I(t) - U_L(T_{fi} - T_a)] \\ &= 0.7986[0.8 \times 500 - 6(60 - 40)] = 223.6 \text{ W/m}^2 \end{aligned}$$

Example 7.6(b)

Determine the rate of useful energy per m^2 for Example 7.6(a) with the mass flow rate of 0.035 kg s^{-1}.

Solution

The flow rate factor can be evaluated as

$$F_R = \frac{\dot{m}\, C_f}{A_c\, U_L}\left[1 - \exp\left(-\frac{A_c\, U_L\, F'}{\dot{m} C_f}\right)\right]$$

$$F_R = \frac{(0.035 \times 4190)}{(1 \times 6)}\left[1 - \exp\left(-\frac{1 \times 6 \times 0.8}{0.035 \times 4190}\right)\right] = 0.787$$

The net rate of useful energy per m^2 will be

$$\dot{q}_u = 0787[0.8 \times 500 - 6(60 - 40)] = 220.36 \text{ W m}^{-2}$$

It is clear that the change in flow rate has no effect on $\dot{q}_u$ for a given design and climatic parameters of a collector.

Example 7.7

Find out the threshold radiation flux for $(\alpha\tau) = 0.80$, 0.60, 0.40 and 0.20, given $T_p = 100\,°C$, $T_a = 16\,°C$ and $U_L = 6\text{ W m}^{-2}\,°C^{-1}$.

Solution

The threshold radiation flux levels are

$$I_{th} = \frac{6(100 - 16)}{0.8} = 630\text{ W m}^{-2} \text{ for } (\alpha\tau) = 0.8$$
$$= 840\text{ W m}^{-2} \text{ for } (\alpha\tau) = 0.6$$
$$= 1260\text{ W m}^{-2} \text{ for } (\alpha\tau) = 0.4$$
$$= 2520\text{ W m}^{-2} \text{ for } (\alpha\tau) = 0.2.$$

This indicates that solar radiation can not be used for thermal heating for $(\acute{a}\tau) = 0.2$ and 0.4 due to the higher value of I_{th}.

7.3.3.5 Overall Thermal Energy Gain

The energy analysis is based on the first law of thermodynamics, and the expression for total thermal gain can be defined as

$$\sum \dot{Q}_{u,total} = \sum \dot{Q}_{u,thermal} + \frac{\sum \dot{Q}_{u,electrical}}{0.38} \tag{7.45}$$

Overall thermal output from a PV/T system = thermal energy collected by the PV/T system + (Electrical output/e_{power}), where e_{power} is the electric power generation efficiency of a conventional power plant for India.

This is so because electrical energy is a high-grade form of energy which is required for the operation of a DC motor. This electrical energy has been converted to equivalent thermal energy by using an electric power generation efficiency of 0.38 for a conventional power plant.[40]

7.3.3.6 Discussion

Equation (7.34b) has been computed using MATLAB software for evaluating the outlet water temperature for typical days during the month of February, 2007, for a given design and climatic parameters. The hourly variations of theoretical and experimental results are shown in Figure 7.27. Similarly, eqn (7.37) has been computed for evaluating the instantaneous efficiency during the month of February, 2007. Theoretical and experimental variations of instantaneous efficiency *vs.* $\frac{T_{fi}-T_a}{I(t)}$ are shown in Figure 7.28. Equation (7.39) was used for evaluating the storage tank water temperature for a given design and climatic parameters and the results are shown in Figure 7.29. The correlation coefficient (r) and root mean square percent deviation (e) evaluated using eqns (7.44a) and (7.44b), respectively, are shown in the same figures. It is observed that there is a good agreement between theoretical values and experimental values of experimental set-up. Using eqn (7.45), the monthly variation of thermal energy gain for New Delhi weather conditions in the case of without withdrawal from the tank is evaluated and shown in Figure 7.30. The annual thermal gain obtained is 2877.9 kWh.

The combined system of photovoltaic/thermal (PV/T) solar water heater presented in this section is a self-sustainable system. This system can be installed at remote areas for fulfilment of hot-water requirements and the electrical energy saved by this system can be utilized for other purposes.

7.3.4 Collectors Connected in Series

Collectors are connected in series for obtaining a higher outlet water temperature. The expression for outlet water temperature and useful heat gain for five different combinations of collectors connected in series is derived in the following sections.

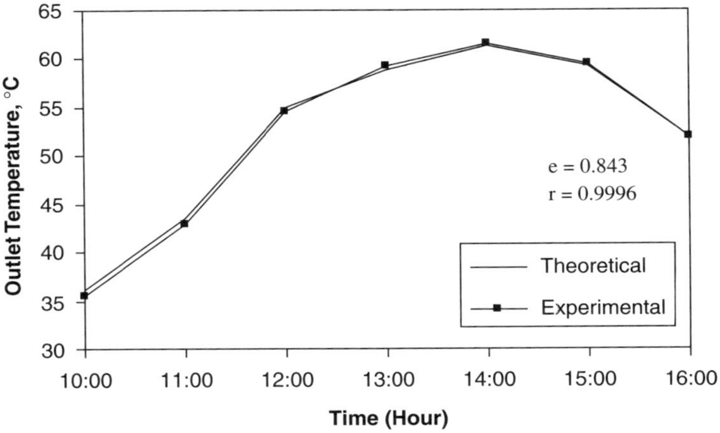

Figure 7.27 Hourly variation of outlet temperature in the month of February, 2007.[44]

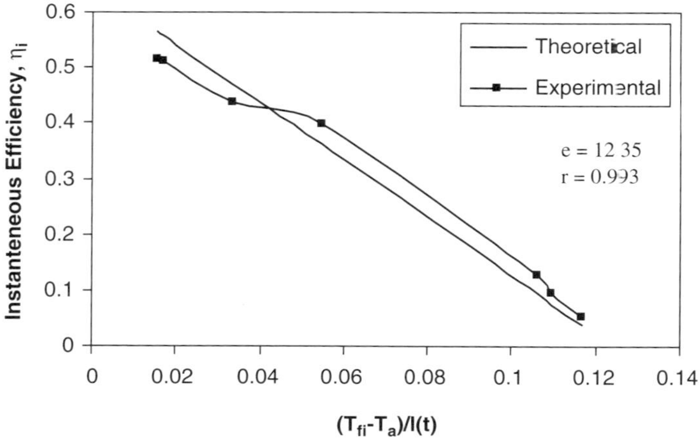

Figure 7.28 Hourly variation of instantaneous efficiency *vs.* $(T_{fi} - T_a)/I(t)$ in the month of February, 2007.[44]

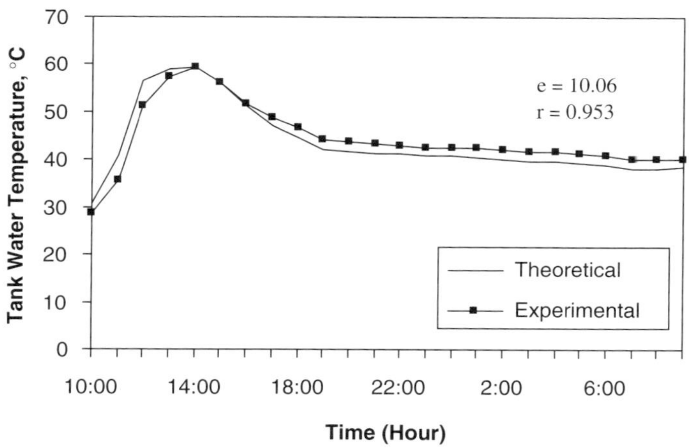

Figure 7.29 Hourly variation of tank water temperature in the month of February, 2007.[44]

7.3.4.1 Fully Covered by Transparent Glass

Following Duffie and Beckman[42] and Tiwari,[43] the energy balance on the flowing fluid along the x-direction through a single tube of length Δx can be written as

$$\dot{m}_f C_f \frac{dT_f}{dx} - n_0 W F' \left[(\alpha\tau)_{c,eff} - U_{L,c}(T_f - T_a) \right] = 0 \qquad (7.46a)$$

$$\begin{bmatrix} \text{Rate of heat} \\ \text{withdrawal} \end{bmatrix} - \begin{bmatrix} \text{Rate of} \\ \text{heat gain} \end{bmatrix} - \begin{bmatrix} \text{Heat loss} \\ \text{to ambient} \end{bmatrix} = 0$$

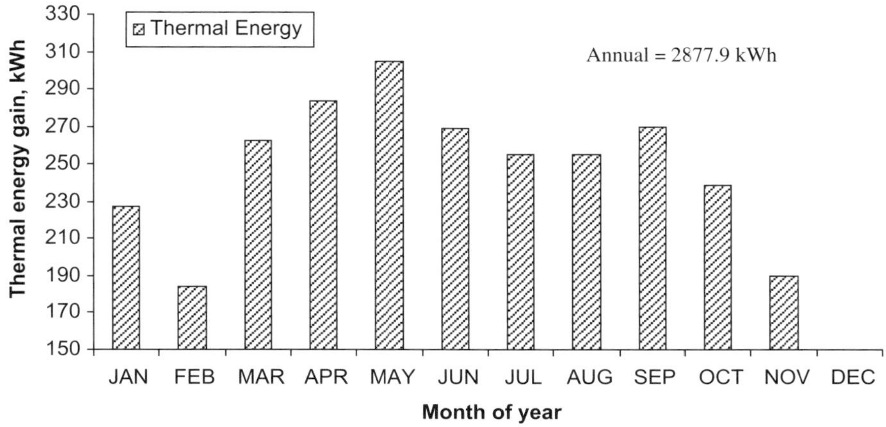

Figure 7.30 Monthly variation of overall thermal energy gain for New Delhi weather conditions in the case of without withdrawal from the tank.[44]

The outlet fluid temperature (T_{fo}) at $x = L$, using boundary conditions $T_f = T_{fi}$ at $x = 0$, can be obtained as

$$T_{f0} = \left[\frac{(\alpha\tau)_{c,eff}}{U_{L,c}} + T_a\right] + \left[T_{fi} - T_a - \frac{(\alpha\tau)_{c,eff}}{U_{L,c}}\right] \exp\left[-\frac{A_c U_{L,c} F'}{\dot{m}_f c_f}\right] \quad (7.46b)$$

Similarly, the outlet fluid temperature (T_{foN}) for the Nth collector, if all the collectors are identical and connected in series, can be given as

$$\begin{aligned} T_{fo\,N} = &\left[\frac{(\alpha\tau)_{c,eff} I(t)}{U_{L,c}} + T_a\right]\left[1 - \exp\left(-\frac{NF'AU_{L,c}}{\dot{m}_f C_f}\right)\right] \\ &+ T_{fi} \exp\left(-\frac{NF'AU_{L,c}}{\dot{m}_f C_f}\right) \end{aligned} \quad (7.46c)$$

and the useful heat output for N identical collectors is defined as

$$\begin{aligned} \dot{Q}_{u,N} = &\left[NAF_R(\alpha\tau)_{c,eff}\left\{\frac{1-(1-K_K)^N}{NK_K}\right\}\right] I(t) \\ &- \left[NAF_R U_{L,c}\left\{\frac{1-(1-K_K)^N}{NK_K}\right\}\right](T_{fi} - T_a) \end{aligned} \quad (7.47a)$$

where

$$K_K = \left[\frac{AF_R U_{L,c}}{\dot{m}_f C_f}\right].$$

The gain factor and loss factor can be defined as

$$(\alpha\tau)_{\mathrm{eff}} = F_{\mathrm{R}}(\alpha\tau)_{\mathrm{c,eff}}\left[\frac{1-(1-K_{\mathrm{K}})^{N}}{NK_{\mathrm{K}}}\right]$$
$$U_{\mathrm{L}} = F_{\mathrm{R}}U_{\mathrm{L,c}}\left[\frac{1-(1-K_{\mathrm{K}})^{N}}{NK_{\mathrm{K}}}\right] \quad (7.47\mathrm{b})$$

7.3.4.2 *Fully Covered by PV Module (Glass-to-Glass)*

In eqn (7.32a), T_{fo1} is the outlet temperature of the water from the first collector covered by the PV module and becomes the inlet temperature for the second collector. The outlet fluid temperature of the second collector can be given as

$$T_{\mathrm{fo2}} = \left[\frac{h_{\mathrm{p2}}(\alpha\tau)_{\mathrm{m,eff}}I(t)}{U_{\mathrm{L,m2}}} + T_{\mathrm{a}}\right]\left[1-\exp\left(-\frac{F_2' A_2 U_{\mathrm{L,m2}}}{\dot{m}_{\mathrm{f}}C_{\mathrm{f}}}\right)\right] + T_{\mathrm{fi2}}\exp\left(-\frac{F' A_2 U_{\mathrm{L,m2}}}{\dot{m}_{\mathrm{f}}C_{\mathrm{f}}}\right) \quad (7.48\mathrm{a})$$

as $T_{\mathrm{fi2}} = T_{\mathrm{fo1}}$.

For a number of collectors connected in series, the outlet temperature of the first collector will be the inlet of the second collector, the outlet temperature of the second will be the inlet of the third and so on. Hence, for a system of N collectors connected in series, the outlet fluid temperature ($T_{\mathrm{fo}N}$) from the Nth collector can be expressed in terms of the inlet temperature of the first.

If all the collectors are identical, *i.e.*

$$U_{L,\mathrm{m1}} = U_{\mathrm{L,m2}} = \ldots\ldots\ldots = U_{\mathrm{L,mN}} = U_{\mathrm{L,m}}$$
$$A_1 = A_2 = \ldots\ldots\ldots = A_N = A$$
$$F_1' = F_2' = \ldots\ldots\ldots = F_N' = F'$$

The outlet fluid temperature ($T_{\mathrm{fo}N}$) for N collectors fully covered by PV is derived as[45]

$$T_{\mathrm{fo}N} = \left[\frac{h_{\mathrm{p2}}(\alpha\tau)_{\mathrm{m,eff}}I(t)}{U_{\mathrm{L,m}}} + T_{\mathrm{a}}\right]\left[1-\exp\left(-\frac{NF' A U_{\mathrm{L,m}}}{\dot{m}_{\mathrm{f}}C_{\mathrm{f}}}\right)\right] + T_{\mathrm{fi}}\exp\left(-\frac{NF' A U_{\mathrm{L,m}}}{\dot{m}_{\mathrm{f}}C_{\mathrm{f}}}\right) \quad (7.48\mathrm{b})$$

The useful heat output of the combination is

$$\dot{Q}_{\mathrm{u,1}} + \dot{Q}_{\mathrm{u,2}} = A_1 F_{\mathrm{R1}}\left[h_{\mathrm{p2}}(\alpha\tau)_{\mathrm{m,eff}}I(t) - U_{L,m1}(T_{\mathrm{fi}} - T_{\mathrm{a}})\right] + A_2 F_{\mathrm{R2}}\left[h_{\mathrm{p2}}(\alpha\tau)_{\mathrm{m,eff}}I(t) - U_{\mathrm{L,m2}}(T_{\mathrm{fo1}} - T_{\mathrm{a}})\right]$$

Here

$$T_{fo1} = T_{fi} + \frac{\dot{Q}_{u,1}}{\dot{m}_f C_f}$$

$$\dot{Q}_{u,1+2} = \Big[A_1 F_{R1} h_{p2}(\alpha\tau)_{m,eff}(1 - K_K) + A_2 F_{R2} h_{p2}(\alpha\tau)_{m,eff}(1 - K_K)\Big] I(t) - [A_1 F_{R1} U_{L,m1}(1 - K_K) + A_2 F_{R2} U_{L,m2}(1 - K_K)](T_{fi} - T_a) \quad (7.49a)$$

where

$$K_K = \left[\frac{A_2 F_{R2} U_{L,m2}}{\dot{m}_f C_f}\right]$$

If the two sets of collectors are identical, the gain factor and loss factor can be defined as

$$(\alpha\tau)_{eff} = F_{R1} h_{p2}(\alpha\tau)_{m,eff}\left[1 - \frac{K_K}{2}\right]$$

$$U_L = F_{R1} U_{L,m1}\left[1 - \frac{K_K}{2}\right]$$

For N identical sets of collectors in series

$$(\alpha\tau)_{eff} = F_{R1} h_{p2}(\alpha\tau)_{m,eff}\left[\frac{1 - (1 - K_K)^N}{N K_K}\right]$$

$$U_L = F_{R1} U_{L,m1}\left[\frac{1 - (1 - K_K)^N}{N K_K}\right] \quad (7.49b)$$

Example 7.8

Calculate the outlet fluid temperature at the outlet of two and four collectors connected in series for the same configuration as in Example 7.6 with the following climatic and design parameters:

$$I(t) = 500\ \mathrm{W\,m^{-2}},\ T_a = 40°\mathrm{C}\ \text{and}\ (\alpha\tau) = 0.8.$$

Solution

(a) For two identical collectors connected in series
From eqn (7.46c)

$$T_{f02} = \left(\frac{0.8 \times 500}{6.0} + 40\right)\left\{1 - \exp\left(-\frac{2 \times 1 \times 6 \times 0.8}{0.35 \times 4190}\right)\right\} + 60\exp\left\{-\frac{2 \times 1 \times 6 \times 0.8}{0.35 \times 4190}\right\} = 60.30°\mathrm{C}$$

(b) For four identical collectors connected in series

$$T_{f04} = 60.6°C.$$

This indicates that the outlet temperature at the end of the fourth collector is higher than two collectors connected in series. However, the rise in temperature is insignificant due to the large value of $\dot{m}$. In this case, the water does not get sufficient time for thermal heating.

7.3.4.3 Partially Covered by PV Module (Glass-to-Glass) (Figure 7.31)

From eqn (7.26), PV on the lower portion, the useful heat output from the N collectors connected in series can be derived as

$$\dot{Q}_{u,N} = N.A_c\left[(\alpha\tau)_{eff,N} I(t) - U_{L,N}(T_{fi} - T_a)\right] \tag{7.50a}$$

where

$$(\alpha\tau)_{eff,N} = (F_R(\alpha\tau))_1\left[\frac{1-(1-K_K)^N}{NK_K}\right]$$
$$\text{and } U_{L,N} = (F_R U_L)_1\left[\frac{1-(1-K_K)^N}{NK_K}\right] \tag{7.50b}$$

where

$$K_K = \left[\frac{(AF_R U_L)_1}{\dot{m}_f C_f}\right]$$

and

$$(AF_R(\alpha\tau))_1 = \left[A_m F_{Rm} h_{p2}(\alpha\tau)_{m,eff}\left(1 - \frac{A_c F_{Rc} U_{L,c}}{\dot{m}_f C_f}\right) + A_c F_{Rc}(\alpha\tau)_{c,eff}\right]$$
$$(AF_R U_L)_1 = \left[A_m F_{Rm} U_{L,m}\left(1 - \frac{A_c F_{Rc} U_{L,c}}{\dot{m}_f C_f}\right) + A_c F_{Rc} U_{L,c}\right]$$

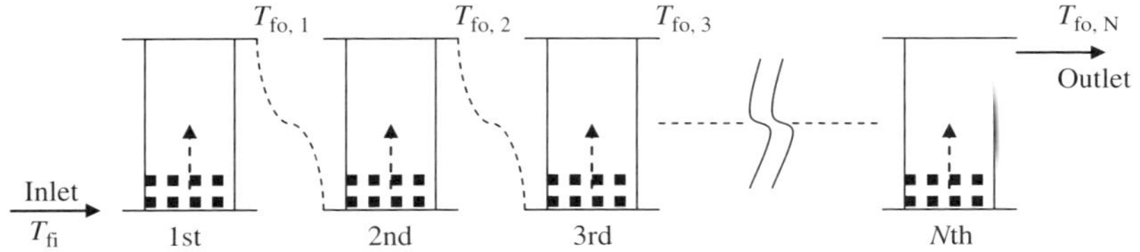

Figure 7.31 Collectors partially covered by PV connected in series.

The rate of thermal energy available at the end of first collector is given as

$$\dot{Q}_{u,1(m+c)} = \dot{m}_f C_f (T_{fo1} - T_{fi}) \tag{7.51}$$

or

$$\dot{Q}_{u,1(m+c)} = (AF_R(\alpha\tau))_1 I(t) - (AF_R U_L)_1 (T_{fi} - T_a) \tag{7.52}$$

From eqns (7.51) and (7.52), the outlet fluid temperature at the end of the first collector can be evaluated as

$$T_{fo1} = \frac{(AF_R(\alpha\tau))_1}{\dot{m}_f C_f} I(t) + \frac{(AF_R U_L)_1}{\dot{m}_f C_f} T_a + T_{fi}\left(1 - \frac{(AF_R U_L)_1}{\dot{m}_f C_f}\right)$$

Similarly, the outlet fluid temperature at the end of the second collector can be evaluated as

$$T_{fo2} = \frac{(AF_R(\alpha\tau))_2}{\dot{m}_f C_f} I(t) + \frac{(AF_R U_L)_2}{\dot{m}_f C_f} T_a + T_{fi2}\left(1 - \frac{(AF_R U_L)_2}{\dot{m}_f C_f}\right)$$

as $T_{fi2} = T_{fo1}$.

For a number of collectors connected in series, the outlet fluid temperature (T_{foN}) from the Nth collector can be expressed in terms of the inlet temperature of the first collector.

For N identical sets of collectors connected in series, the outlet fluid temperature at the end of the Nth collector can be defined as[45]

$$T_{foN} = \frac{(AF_R(\alpha\tau))_1}{\dot{m}_f C_f}\left(\frac{1 - K_K^N}{1 - K_K}\right) I(t) + \frac{(AF_R U_L)_1}{\dot{m}_f C_f}\left(\frac{1 - K_K^N}{1 - K_K}\right) T_a + T_{fi} K_K^N \tag{7.53}$$

where

$$K_K = \left[1 - \frac{(AF_R U_L)_1}{\dot{m}_f C_f}\right]$$

7.3.4.4 Fully Covered by PV Module and Fully Covered by Glass Cover (Figure 7.32)

An identical set of collectors fully covered by a PV module and fully covered by a glass cover are connected in series (PV-glass combination; Figure 7.32). The

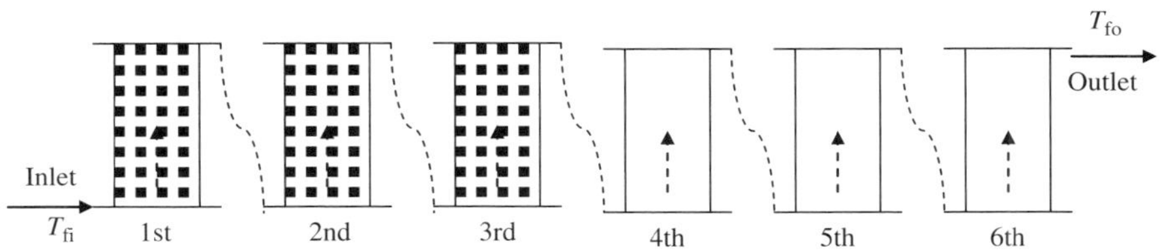

Figure 7.32 Collectors fully covered by PV module and fully covered by glass cover are connected in series (PV-glass combination).

expression for the outlet fluid temperature from a mixed combination is derived as

$$T_{\mathrm{foN}} = \left[\frac{(\alpha\tau)_{\mathrm{c,eff}}I(t)}{U_{\mathrm{L,c}}} + T_{\mathrm{a}}\right]\left[1 - \exp\left(-\frac{F'A_{\mathrm{c}}U_{\mathrm{L,c}}}{\dot{m}_{\mathrm{f}}C_{\mathrm{f}}}\right)\right]\left[\frac{1 - \left\{\exp\left(-\frac{F'A_{\mathrm{c}}U_{\mathrm{L,c}}}{\dot{m}_{\mathrm{f}}C_{\mathrm{f}}}\right)\right\}^{N_{\mathrm{c}}}}{1 - \exp\left(-\frac{F'A_{\mathrm{c}}U_{\mathrm{L,c}}}{\dot{m}_{\mathrm{f}}C_{\mathrm{f}}}\right)}\right]$$
$$+ \left[\left(\frac{h_{p2}(\alpha\tau)_{\mathrm{m,eff}}I(t)}{U_{\mathrm{L,m}}} + T_{\mathrm{a}}\right)\left\{1 - \exp\left(-\frac{N_{\mathrm{m}}F'A_{\mathrm{m}}U_{\mathrm{L,m}}}{\dot{m}_{\mathrm{f}}C_{\mathrm{f}}}\right)\right\}\right.$$
$$\left. + T_{\mathrm{fi}}\exp\left(-\frac{N_{\mathrm{m}}F'A_{\mathrm{m}}U_{\mathrm{L,m}}}{\dot{m}_{\mathrm{f}}C_{\mathrm{f}}}\right)\right]\left[\exp\left(-\frac{F'A_{\mathrm{c}}U_{\mathrm{L,c}}}{\dot{m}_{\mathrm{f}}C_{\mathrm{f}}}\right)\right]^{N_{\mathrm{c}}} \tag{7.54a}$$

where N_{c} is the number of collectors covered by the glass cover and N_{m} is the number of collectors covered by the PV module.

The expression for the useful heat gain from a mixed combination is derived as

$$\dot{Q}_{\mathrm{u,N}} = \dot{m}_{\mathrm{f}}C_{\mathrm{f}}\left[\begin{array}{l}\left[\frac{(\alpha\tau)_{\mathrm{c,eff}}I(t)}{U_{\mathrm{L,c}}} + T_{\mathrm{a}}\right]\left[1 - \exp\left(-\frac{F'A_{\mathrm{c}}U_{\mathrm{L,c}}}{\dot{m}_{\mathrm{f}}C_{\mathrm{f}}}\right)\right] \\ \left[\frac{1 - \left\{\exp\left(-\frac{F'A_{\mathrm{c}}U_{\mathrm{L,c}}}{\dot{m}_{\mathrm{f}}C_{\mathrm{f}}}\right)\right\}^{N_{\mathrm{c}}}}{1 - \exp\left(-\frac{F'A_{\mathrm{c}}U_{\mathrm{L,c}}}{\dot{m}_{\mathrm{f}}C_{\mathrm{f}}}\right)}\right] \\ + \left[\frac{h_{p2}(\alpha\tau)_{\mathrm{m,eff}}I(t)}{U_{\mathrm{L,m}}} + T_{\mathrm{a}} + T_{\mathrm{fi}}\right]\left[1 - \exp\left(-\frac{N_{\mathrm{m}}F'A_{\mathrm{m}}U_{\mathrm{L,m}}}{\dot{m}_{\mathrm{f}}C_{\mathrm{f}}}\right)\right] \\ \left[\exp\left(-\frac{F'A_{\mathrm{c}}U_{\mathrm{L,c}}}{\dot{m}_{\mathrm{f}}C_{\mathrm{f}}}\right)\right]^{N_{\mathrm{c}}}\end{array}\right] \tag{7.54b}$$

7.3.4.5 *Series and Parallel Combination of Collectors Fully Covered by PV (Figure 7.33)*

The expression for the outlet fluid temperature from a mixed combination is derived as

$$T_{\mathrm{fo}N_{\mathrm{S}}} = \left(\frac{h_{p2}(\alpha\tau)_{\mathrm{m,eff}}I(t)}{U_{\mathrm{L,m}}} + T_{\mathrm{a}}\right)\left\{1 - \exp\left(-\frac{N_{\mathrm{s}}F'A_{\mathrm{m}}U_{\mathrm{L,m}}}{N_{\mathrm{S}}\dot{m}_{\mathrm{f}}C_{\mathrm{f}}}\right)\right\} + T_{\mathrm{fi}}\left\{\exp\left(-\frac{N_{\mathrm{s}}F'A_{m}U_{\mathrm{L,m}}}{N_{\mathrm{S}}\dot{m}_{\mathrm{f}}C_{\mathrm{f}}}\right)\right\} \tag{7.55a}$$

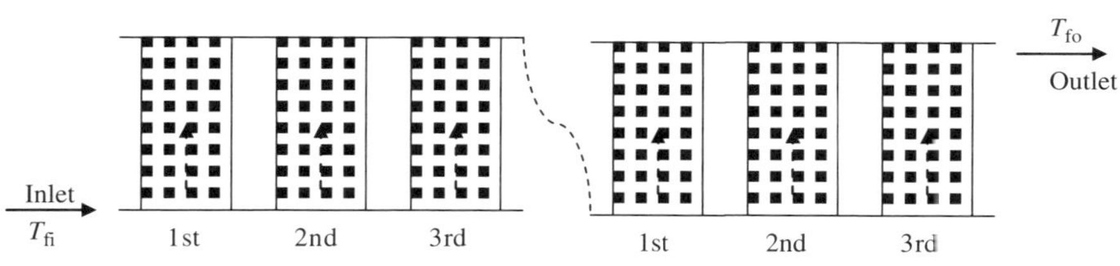

Figure 7.33 Series and parallel combination of collectors (two panels) fully covered by PV (mixed combination).

Here, N_m is the number of collectors covered by PV modules (connected in parallel), N_S is the number of identical set of panels (connected in series), and N_c is the number of collectors.

The useful heat output from N identical sets of panels is derived as

$$\dot{Q}_{u,N_S} = \frac{N_c A_m F_{Rm}}{N_S} \times \left[h_{p2}(\alpha\tau)_{m,eff} \left[\frac{1-(1-K_K)^{N_s}}{N_s K_K} \right] I(t) - U_{L,m} \left[\frac{1-(1-K_K)^{N_s}}{N_s K_K} \right] (T_{fi} - T_a) \right] \tag{7.55b}$$

where

$$K_K = \left[\frac{A_m F_{Rm} U_{L,m}}{\dot{m}_f C_f} \right]$$

For two sets of panels each having three collectors, $K_K = 0.0886$ and $F_{Rm} = 0.8404$.

Using eqn (7.45), the monthly variation of thermal energy gain for New Delhi weather conditions when the collector is fully and partially covered by a PV module is evaluated for six collectors connected in series and at constant mass flow rate of $0.04\,\mathrm{kg\,s^{-1}}$. The variation is shown in Figures 7.34 and 7.35. The annual thermal gain obtained is 21172.1 kWh and 1996.4 kWh for fully and partially covered collectors, respectively. Higher thermal gain is obtained in the case of fully covered collectors for two reasons: one is the lower outlet temperature, hence fewer losses and thermal energy is higher, and the second is higher electrical energy gain.

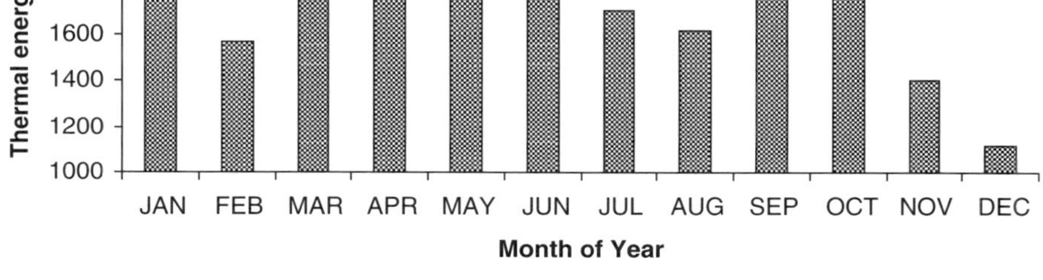

Figure 7.34 Monthly variation of overall thermal energy gain for New Delhi weather conditions, when the collector is fully covered by a PV module.

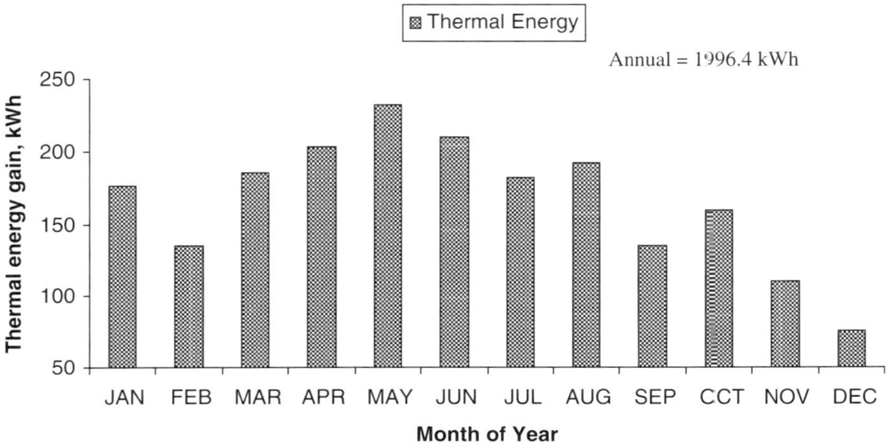

Figure 7.35 Monthly variation of overall thermal energy gain for New Delhi weather conditions, when the collector is partially covered by a PV module.

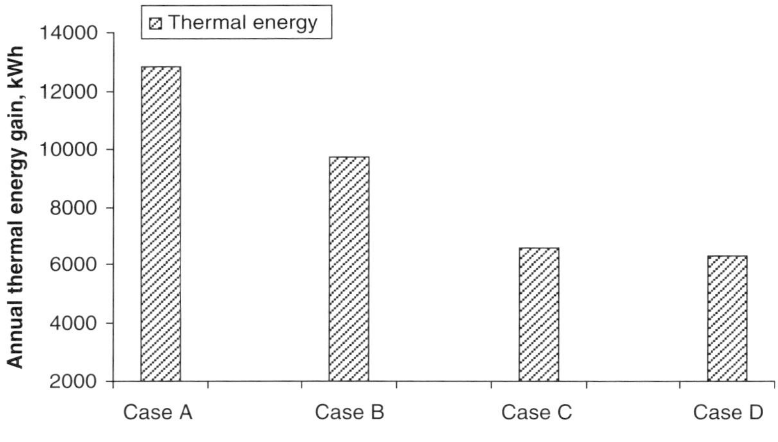

Figure 7.36 Variation of annual thermal energy gain for A, B, C and D cases considering six collectors and $\dot{m} = 0.04\,\text{kg}\,\text{s}^{-1}$ for New Delhi conditions.

Variation of annual thermal and electrical energy gain for A, B, C and D cases (Case A: fully covered by PV module, Case B: partially covered by PV module, Case C: PV-glass combination, Case D: mixed combination) considering six collectors and $\dot{m} = 0.04\,\text{kg}\,\text{s}^{-1}$ for New Delhi conditions is shown in Figures 7.36 and 7.37. Results shows that Case A is better from a thermal point of view and Case D is better from an electrical point of view. Depending upon the users' requirements, different series-parallel and PV-glass combinations can be made.

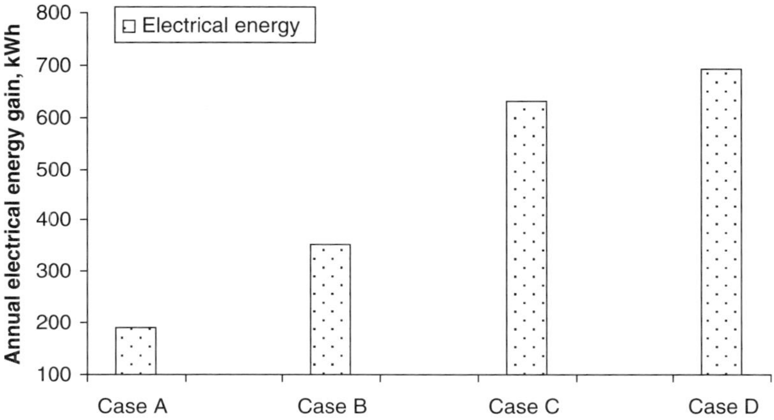

Figure 7.37 Variation of annual electrical energy gain for A, B, C and D cases considering six collectors and $\dot{m} = 0.04\,\mathrm{kg\,s^{-1}}$ for New Delhi conditions.

7.3.5 Comparison of Performance of Liquid and Air Collectors

The comparison of liquid and air flat-plate collectors has been shown in Figure 7.20. The performance of the air collector has been plotted at two different flow rates. It can be seen that the performance of the liquid collector is better in comparison to the air collector. It can also be observed that the flow rate plays an important and significant role in an air collector. Further, one can observe that the difference in performance of both collectors minimizes at higher solar intensities.

7.4 PV/T Solar Distillation System

The shortage of potable water is one of the most important issues in developing countries. In countries like India the availability of drinking water *per capita* is decreasing because of high population growth and this makes it necessary to search for alternative sources of potable water. Different methods have been developed for getting potable water from brackish/saline water and solar distillation is one of the best options to obtain fresh water by utilizing solar energy, which is available in abundance. In the field of distillation many authors reported the performance of different designs of solar still in passive mode and concluded that the passive solar still gives a low yield of around $2.25\,\mathrm{kg\,m^{-2}\,day^{-1}}$, because of low water temperature.[46–49] The yield can be increased further by feeding hot water into the basin by connecting the solar still with a parabolic, flat-plate or evacuated collector. Among these options, the flat-plate collector (FPC) has become more popular because of its easy operation and lower maintenance levels. In the case of an active solar still, the additional thermal energy from the flat-plate collector is fed in to the basin of the solar still, so that the temperature difference between the evaporation and condensing cover increases. The flat-plate collector is integrated to the basin of the solar still. The water in the basin is circulated through the flat-plate collector either in a natural circulation mode or in a forced circulation mode, depending upon the requirement of the user. To reduce/avoid thermal losses from hot water in the pipe to ambient air during hot-water circulation, the connecting pipes are insulated. In an active solar still, the water in the basin is heated directly as well as indirectly through a flat-plate collector. The rise in the temperature of water in the basin mainly depends upon the number of collectors connected in series. The collector should be operated only during sunshine hours. Zaki *et al.*[50] reported the experimental investigation on an active system under the thermosyphon mode of operation where the maximum increase in the yield was up to 33%, when the water in the still was preheated in the collector.

Various authors have studied the heat transfer phenomena inside the still and developed the heat-transfer correlation to study the internal heat-transfer coefficients for different designs of the solar still under different climatic and operational conditions.[51–54] Kumar and Tiwari[55] developed a model to evaluate internal heat-transfer coefficients using regression analysis that does not impose a limitation as in Dunkle's model and gives more realistic values for theoretical prediction. The thermal model to establish the energy balance equations of a passive and an active solar still with different concepts have been developed by previous researchers.[56–59]

7.4.1 Active PV/T Distillation System

A photograph of a self-sustainable hybrid PV/T active solar still is shown in Figure 7.38. The fabricated system consists of a solar still, a PV-integrated flat-plate collector and a DC motor pump. The single slope solar still has an effective basin area of 1 m^2 and is fabricated using glass reinforced plastic (GRP) material. A glass cover with an inclination of 30° to the horizontal is fixed to the top using iron clamps and further sealed with window-putty to prevent vapour leakage to the outside. The inside of the basin is painted black to increase the absorptivity. The orientation of the solar still is kept due south in order to receive maximum solar radiation throughout the year. The still has been mounted on an iron stand and connected to the collector through insulated piping.

Each collector has an effective area of 2 m^2 and is connected in series to feed the water at high temperature in the still basin, to increase the distillate yield. A photovoltaic (PV), glass-to-glass module of area $0.55 \times 1.20\,m^2$ (75 W) has been integrated with one of the collectors at the bottom side. The electrical energy generated by the photovoltaic (PV) module is used to operate the DC water pump, which is used to circulate water under the forced mode of operation during sunshine hours to compensate the pressure drop in the collector and piping arrangement. The radiation that is transmitted through the non-packing area of the PV module is directly absorbed by the blackened surface of the collector; also, the convected thermal energy from the back surface of the PV module to the absorber surface is utilized for water heating.

7.4.1.1 Thermal Modelling of the System

The following assumptions have been made while writing the energy balance equations in the hybrid active solar still:

1) There is no leakage of vapour from the distiller;
2) The heat capacity of the glass cover, insulating material and collector are neglected;
3) The collector, solar still and connecting pipelines are insulated;
4) There is no heat loss from the collector area during off-sunshine hours by reverse convection;
5) The system is in a transient mode during sunshine hours and in steady state during off sunshine hours.

The final expression for the rate of thermal energy available at the end of the second collector (Figure 7.38) is given in eqn (7.36); this heat gain is fed into the solar still. The energy balance equations for solar still are given below.

Energy balance in the solar still
The energy transaction in the solar still among its different components considering area of the basin (A_b) and the glass cover (A_g) is given as[43]

Figure 7.38 Photograph of a PV/T-integrated hybrid active solar still.

Energy balance for inner surface of glass cover:

$$\alpha'_g I_s(t) A_g + h_{lw}(T_w - T_{gi}) A_b = \frac{K_g}{L_g}(T_{gi} - T_{go}) A_g \qquad (7.56a)$$

Energy balance for outer surface of glass cover:

$$\frac{K_g}{L_g}(T_{gi} - T_{go}) = h_{1g}(T_{go} - T_a) \qquad (7.56b)$$

Energy balance for water mass:

$$\dot{Q}_{u(m+C1+C2)} + A_b \alpha'_w I_s(t) + h_{bw}(T_b - T_w) A_b = m_w c_w \frac{dT_w}{dt} + h_{1w}(T_w - T_{gi}) A_b \qquad (7.57)$$

Basin liner:

By neglecting the side heat losses

$$\alpha'_b A_b I_s(t) = h_{bw}(T_b - T_w) A_b + h_{ba}(T_b - T_a) A_b \qquad (7.58)$$

After re-arranging and replacing the various terms from eqns (7.56a, b) and (7.57), eqn (7.58) becomes

$$\dot{Q}_{u(m+C1+C2)} + A_b\left[\alpha'_{eff}\right] I_s(t) = m_w c_w \frac{dT_w}{dt} + U_s(T_w - T_a) \qquad (7.59)$$

The analytical values of different equivalent notations used in the above expression can be obtained as

$$h'_1 = \frac{h_{1w}}{U_{c,ga} A_g + h_{1w} A_b} \qquad U_t = \frac{U_{c,ga} h_{1w}}{U_{c,ga} A_g + h_{1w} A_b}$$

$$h_1 = \frac{h_{bw}}{h_{ba} + h_{bw}} \qquad U_b = \frac{h_{ba} h_{bw}}{h_{ba} + h_{bw}}$$

$$U_{c,ga} = \frac{\frac{K_g}{L_g} h_{1g}}{\frac{K_g}{L_g} + h_{1g}} \qquad U_s = \left[U_t A_g + U_b\right] A_b$$

$$\alpha'_{eff} = \left[\alpha'_w + h'_1 \alpha'_g A_g + h_1 \alpha'_b\right]$$

Equation (7.59) can be written in the following form after replacing the value of $Q_{u(m+C1+C2)}$ from eqn (7.36):

$$\frac{dT_w}{dt} + aT_w = f(t)$$

where

$$a = \frac{U_s}{m_w c_w}$$

$$f(t) = \frac{\left[A_m F_{Rm} h_{p2} (\alpha\tau)_{m,eff} (1 - K_1) + A_{C1} F_{RC1} (\alpha\tau)_{C1,eff} (1 - K_2) + A_{C2} F_{RC2} (\alpha\tau)_{C2,eff}\right] I_c(t)}{m_w c_w}$$

$$- \frac{[A_m F_{Rm} U_{Lm} (1 - K_1) + A_{C1} F_{RC1} U_{LC1} (1 - K_2) + A_{C2} F_{RC2} U_{LC2}](T_{wi1} - T_a)}{m_w c_w}$$

$$+ \frac{A_b \alpha'_{eff} I_s(t) + U_s T_a}{m_w c_w}$$

Theoretical water temperature in basin after time't':

$$T_w = \frac{\bar{f}(t)}{a} \left(1 - e^{-at}\right) + T_{wo} e^{-at} \tag{7.60}$$

where T_{wo} is the temperature of the basin water at time $t = 0$ and $\overline{f(t)}$ is the average value of f_t between two consecutive intervals of time.

Theoretical distillate yield:

The hourly distillate yield per unit area (kg m^{-2} h^{-1}) can be evaluated from known values of T_w and T_{gi} given by

$$\dot{m}_{ew} = \frac{h_{ew} (T_w - T_{gi}) \times 3600}{L} \tag{7.61a}$$

The daily yield from the still is given as

$$m_{ew} = \sum_{i=1}^{i=24} \dot{m}_{ew} \tag{7.61b}$$

Example 7.9

Calculate the hourly output from the still when the water surface is at 20 °C, ambient air is at 8.5 °C and the temperature of the glass = 12 °C. The evaporative heat transfer coefficient = 3.445 W m^{-2} °C^{-1} and $L = 2390 \times 10^3$ J kg^{-1}.

Solution

Given $L = 2390 \times 10^3$ J kg^{-1}, the hourly yield is

$$\dot{m}_{ew} = \frac{h_{ew}(T_w - T_{gi})}{L} \times 3600 \text{ kg m}^{-2} \text{ h}^{-1} \text{ (From Equation (7.61a))}$$

and hence, $\dot{m}_{ew} = \dfrac{3.445 \times (20 - 12)}{2390 \times 10^3} \times 3600 = 0.0415$ kg m^{-2} h^{-1}.

7.4.1.2 Discussion

The monthly outdoor experiments were conducted for 24 hrs on a hybrid active solar still setup for the New Delhi (India) climatic conditions on typical days during the month of April, 2006, to March, 2007. The experiments were conducted by considering different water depths (0.05 m, 0.10 m and 0.15 m) in the solar still. Energy balance equations have been used to predict the hourly water and glass temperature and the hourly yield for a photovoltaic integrated (PV/T) hybrid active solar still by using the design parameters. The monthly variation of measured yield for three different water depths (0.05 m, 0.10 m and 0.15 m) for New Delhi conditions is shown in Figure 7.39. Maximum yield is obtained during the summer period and for lower depth due to the availability of solar radiation.

7.5 PV/T Solar Dryers

The research and development work on PV applications has been increased in recent years in order to conserve the conventional energy sources. The PV applications are many and forced convection crop drying is one of them. A very few researchers have used PV-module powered air circulation for forced convection drying. Saleh and Sarkar[60] studied a PV-operated forced convection solar energy dryer, in which a PV panel of 20 W was installed separately from an air heater collector and drying chamber to drive a 12-volt DC fan. A solar dryer was studied with photovoltaic solar cells, incorporated in the solar air heater section, to drive a DC fan. The dryer dried 90 kg maize grain per batch from an initial moisture content of 33.3 to 20% (dry basis) in just one day. In

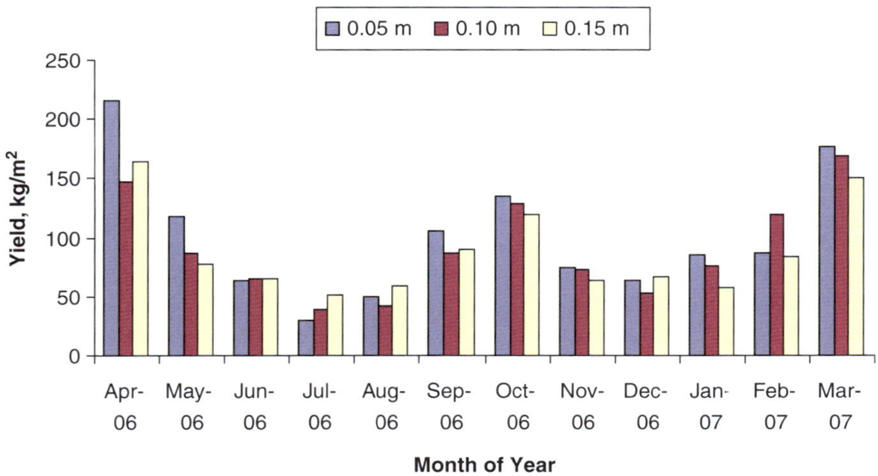

Figure 7.39 Monthly variation of measured yield for three different water depths (0.05 m, 0.10 m and 0.15 m) for New Delhi conditions.

comparison to Sun drying, solar grain drying with a PV-driven DC fan reduces the drying time by over 70%.[61,62] Sopian *et al.*[28] developed and tested a double-pass photovoltaic thermal solar collector suitable for solar drying applications. Farkas *et al.*[63] developed a modular solar dryer in which a PV panel (maximum power: 2 × 20 W), to drive an electrical fan for artificial air circulation, was installed in the front side of the dryer with changeable elevation angle suitable to the different angles of the sunshine in the different periods of the year. Hossain *et al.*[64] optimized a solar tunnel dryer for chilli drying in Bangladesh and reported that the design geometry was not very sensitive to minor material costs, fixed cost and operating cost but more sensitive to costs of major construction materials of the collector, solar radiation and air velocity in the dryer.

The fan or blower, used for forced circulation of heated air from the collector area to the drying beds in active solar energy dryers, can be operated by either grid electricity or DC electricity produced by a photovoltaic (PV) module. The hybrid photovoltaic-solar dryers use DC electricity produced by a PV module to drive the fan or blower for forced circulation of heated air. A schematic view of a conventional hybrid active solar dryer is shown in Figure 7.40.[61,62] The PV module is integrated at the top of the air collector. The electricity produced by the PV module is used to operate a DC fan placed between the air collectors and the drying chamber. The thermal energy available with the PV module is also used for further heating of hot air available from the lower portion of the air collector. Another design of hybrid solar dryer with drying chamber and solar air heater is shown in Figure 7.41.[65] In this case too, the fan is operated by electricity produced by a PV module placed at the top of the collector. In this case, hot air flows over the crop unlike the flow of hot air shown in Figure 7.40. PV-integrated tunnel-type and greenhouse-type hybrid dryers are shown in Figure 7.42[64,66,67] and Figure 7.43.[68] The conventional PV/T mixed mode dryer is shown in Figure 7.44. More details of hybrid photovoltaic-solar dryers are discussed later in this chapter.

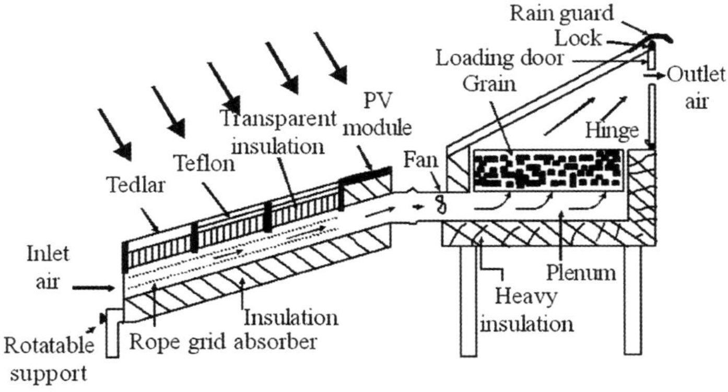

Figure 7.40 Schematic view of a PV-integrated solar dryer.

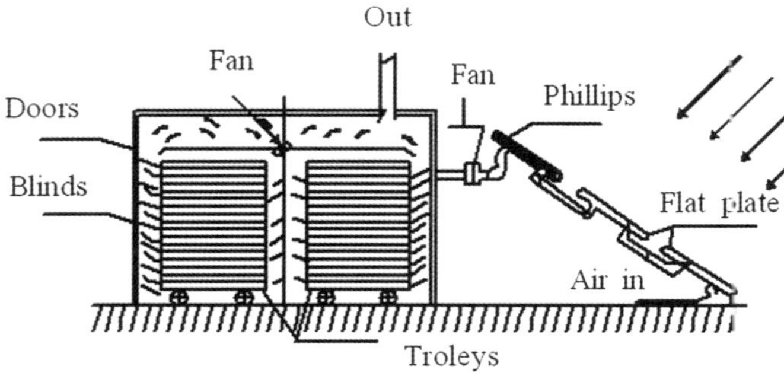

Figure 7.41 Hybrid solar dryer with drying chamber and solar air heater.

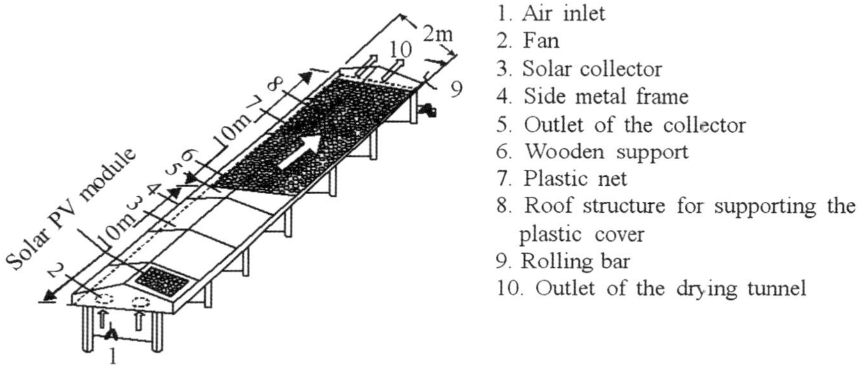

Figure 7.42 Solar tunnel dryer with PV module.

7.5.1 Solar Tunnel Dryer

The solar tunnel dryer mainly consists of a plastic-covered flat-plate solar air heating collector, a drying tunnel, two DC (direct current) fans, a 40-W photovoltaic module, a wooden support, a plastic net, a roof structure for supporting the polyethylene cover and a base structure for supporting the dryer *etc*. (Figure 7.42). The materials used for construction of the collector and dryer are a GI sheet, timber, glass wool, an MS rod, an angle bar, a polyethylene cover, a rubber rope, an aluminium U-channel, a DC fan, a PV module, a GI pipe, a plastic net and miscellaneous materials (screws, rivets, paint, *etc*.). The dryer was 20 m long and 1.80 m wide. The collector and drying chamber units were made of plain metal sheets and wooden frames in a number of small sections and were joined together in series. The collector was painted black to facilitate absorption of solar radiation. Both the collector and the drying units were covered by a 0.2-mm-thick transparent UV-stabilized plastic sheet. The plastic sheet was fixed on the collector side of the dryer to the metal frame

Figure 7.43 Photograph of a hybrid PV/T greenhouse dryer.

using a U-type aluminium channel and a rubber rope. At the drying unit, one end of the plastic sheet was fixed to a metal tube, which allows rolling of the plastic sheet up and down for loading and unloading of the dryer. The drying area of the dryer unit was same as that of the collector. Glass wool was used between the two metal sheets at the bottom of the dryer as an insulation material to reduce the heat loss from the bottom of the dryer. A 40-W solar module was installed at the inlet of the solar collector as a power source to operate the two small fans so that heated air blows over the product in the drying tunnel. The whole system was placed horizontally on tables made of iron angle frame, 0.8 m above the ground floor, for ease of loading and unloading of the products.[64]

The absorber absorbs the solar radiation transmitted through the transparent cover of the collector unit and becomes hot. The air absorbs heat from the hot absorber plate. The heated air from the collector passes over and under the products spread in a single layer in the drying chamber and thus moisture is evaporated and carried away from the products. The crop produce is also heated by the solar radiation transmitted through the transparent cover of the drying unit. Thus there is a further temperature rise in the drying unit and the drying rate increases.

The solar tunnel dryer has been optimized for the drying of chillies (crop produce) in Bangladesh and two optimum designs were obtained. Both the collector and drying units were 14.0 m long and 1.9 m wide for optimum design-I and for optimum design-II they were 13.0 m long and 2.0 m wide. For the basic mode of the dryer, both the collector and drying units were 10.0 m long

Figure 7.44 Conventional PV/T mixed mode dryer.

and 1.8 m wide.[64] The ratios of the length of the collector to that of the drying tunnel of basic mode and optimum mode solar dryers are the same and this ratio was found to be 1 : 0.

7.5.2 Solar Greenhouse Dryer

A hybrid PV/T greenhouse (roof-type even span) dryer (Figure 7.43) has been developed at Solar Energy Park, Indian Institute of Technology (IIT), New Delhi, India. The dryer was constructed using aluminium sections (*e.g.* L angles, Tee-sections, flats, *etc.*), two PV modules (glass-to-glass), a DC fan and a UV-stabilized polyethylene sheet covering *etc.* Aluminium sections were used in construction to avoid rusting/corrosion from the surroundings and thus to extend the life of the dryer.

The dryer consists of two PV modules (glass-to-glass; dimensions: 1.20 m × 0.55 m × 0.01 m; 75 Wp) on the south roof, two openings (dimensions: 1.10 m × 0.55 m) at the north roof symmetrical to the PV modules for natural convection and an aluminium frame door (size: 0.62 m × 0.88 m) on the east side. Arrangement for easy opening/closing of the PV modules (south side) and symmetrical air vent (north side) has been made using hooks *etc*. The dryer has a three-tier drying system which may be used for drying of different crops simultaneously. Each tier consists of two wire mesh trays, having a base area of 0.9 m × 1.30 m, fitted in the centre of the greenhouse. It has a floor area of 2.50 m × 2.60 m with 1.80 m central height and 1.05 m side walls height from the ground. Its roof has a slope of 30°.

At the bottom side, 0.15 m height is open and a further 0.10 m is provided with wire mesh to provide air movement in the greenhouse air heater for drying purposes. The air at the bottom becomes hot and moves from the bottom to the top through a three-tier system of perforated wire mesh trays. Wire mesh trays have been made, which may be easily taken out and kept in the dryer at specific places. A DC fan has been fitted at the upper end of the east side wall for forced-mode operation *i.e.* for rapid removal of humid air and thus to expedite the drying process to the required level.

The specifications of the PV module at 1000 W m^{-2} and 25 °C are given below:

I_{max}	4.4 amp
V_{max}	17 volt
Area of module	0.60534 m^2
Efficiency	12%
Packing factor	83%

The solar radiation incident on the greenhouse may be utilized in the following two ways: (i) on PV modules (glass-to-glass) and (ii) on a UV-stabilized polyethylene sheet.

The solar radiation incident on the PV modules (glass-to-glass) provides thermal heat as well as DC electricity. The thermal heat of the PV module is utilized to heat the air inside the greenhouse and the produced DC electricity is used to operate the DC fan for forced-mode operation of the dryer. The temperature of the PV module will reduce as it transfers heat to greenhouse air, which will help in the drying of crops. It will also help to increase the efficiency of the PV module. This is because the efficiency of the PV module decreases with an increase in temperature.

The incident solar radiation on the UV-stabilized polyethylene sheet is transmitted to the greenhouse to produce the greenhouse effect *i.e.* an increase in greenhouse air temperature. The sheet helps in trapping of infrared radiation and in preventing unnecessary circulation of ambient air, which helps in maintaining the desired temperature inside the greenhouse.

The PV module is considered in analysis as it is glass-to-glass and the area is not negligible in comparison to the total surface area of the dryer.

7.5.2.1 Thermal Modelling (without Load)

The energy and exergy balance were used to obtain the required expression for the thermal modelling of the air heater without load for drying applications.

Assumptions

1. The heat storage capacity of the greenhouse cover and wall material is neglected.
2. There is no radiative heat exchange between the walls and roofs of the greenhouse due to negligible temperature differences.
3. The reflected part of the solar radiation from the floor inside the greenhouse is neglected.
4. There is no stratification in greenhouse air temperature.
5. The absorptivity of the glass of the PV module and the enclosed greenhouse air is neglected.
6. The transmissivity of ethyl vinyl acetate (EVA) is approximately 100%.
7. The temperature variation along the thickness as well as along the width is negligible.
8. The ohmic losses in solar cells are negligible.

Based on the first law of thermodynamics, the energy balance equations for the greenhouse air heater are written to account for energy input, energy output and energy losses.

(i) *For PV module*:

$$\tau_G \alpha_c I_s(t) \beta_c A_{tm} = [U_T(T_c - T_a) + h_{cb}(T_c - T_r) + \eta_c I_s(t) \beta_c \tau_G] A_{tm} \quad (7.62)$$

where $U_T = \left[\frac{l_G}{k_G} + \frac{1}{h_o}\right]^{-1}$; U_T = overall top loss heat transfer coefficient between the solar cell of the PV module and ambient air (W m^{-2} K); h_o = convective heat transfer coefficient between the top surface of the PV module and ambient air (W m^{-2} K); $h_{cb} = \left[\frac{l_G}{k_G} + \frac{1}{h_i}\right]^{-1}$; h_i = convective heat transfer coefficient between the bottom surface of the PV module and greenhouse room air (W m^{-2} K); α_c = absorptivity of the solar cell portion of the PV module; β_c = packing factor of the PV module; η_c = efficiency of the solar cell of the PV module; τ_G = transmissivity of the glass portion of the PV module; I_s(t) = total average solar intensity measured on the south roof of the dryer (W m^{-2}); A_{tm} = total area of all PV modules (m^2); l_G = thickness of glass of the PV module (m); k_G = thermal conductivity of glass (W m K); h_{ct} = heat transfer coefficient between the solar cell of the PV module and the greenhouse room air

($W\,m^{-2}\,K$); T_r = greenhouse room air temperature (°C); T_a = ambient air temperature (°C) and T_c = solar cell temperature (°C).

(ii) *For greenhouse room air*:

$$M_a C_a \frac{dT_r}{dt} = \tau\left(\sum I_i A_i - I_s(t)A_{tm}\right) + \tau_G^2(1-\beta_c)I_s(t)A_{tm} + h_{cb}(T_c - T_r)A_{tm} - \sum U_i A_i (T_r - T_a) - 0.33NV(T_r - T_a) - h_f(T_r - T_f)A_f \tag{7.63}$$

where $\sum U_i A_i =$ overall top heat loss from inside the greenhouse room air to ambient air ($W\,K^{-1}$); $\sum U_i A_i = U_{pra}(\sum A_i - A_{tm}) + U_{mra}A_{tm}$; $U_{pra} = \left(\frac{1}{h_o} + \frac{1}{h_i}\right)^{-1}$; A_f = greenhouse floor area (m^2); $U_{mra} = \left(\frac{1}{U_T} + \frac{L_c}{k_c} + \frac{1}{h_{cb}}\right)^{-1}$; $\sum A_i$ = total outer surface area of greenhouse (m^2); C_a = specific heat capacity of air ($J\,kg^{-1}\,K$); h_f = convective heat transfer coefficient between the greenhouse room air and the greenhouse floor ($W\,m^{-2}\,K$); $\sum I_i A_i$ = total solar radiation received at the outer surface of the greenhouse dryer from all surfaces including the PV module (W); L_c = thickness of the solar cell of the PV module (m); k_c = thermal conductivity of the solar cell ($W\,m^{-1}\,K$); M_a = mass of greenhouse room air (kg); N = number of air changes per hour; t = time (s); T_f = greenhouse floor temperature (°C); U_{mra} = overall top loss heat transfer coefficient between greenhouse room air and ambient air through the PV module ($W\,m^{-2}\,K$); U_{pra} = overall top loss heat transfer coefficient between greenhouse room air and ambient air through greenhouse plastic cover ($W\,m^{-2}\,K$); V = volume of greenhouse (m^3) and τ = transmissivity of the greenhouse plastic cover.

(iii) *For floor*:

$$h_f(T_r - T_f)A_f = h_g(T_f - T_\infty)A_f \tag{7.64a}$$

where $h_g = \frac{L_g}{k_g}$, L_g = thickness/depth of ground (m), h_g = conductive heat transfer coefficient between greenhouse floor and Earth ($W\,m^{-2}\,K$), k_g = thermal conductivity of the ground ($W\,m^{-1}\,K$) and T_∞ = inside Earth temperature (°C).

From eqn (7.62), the expression for the solar cell temperature becomes

$$T_c = \frac{1}{U_T + h_{cb}}\left[(\alpha\tau)_1 I_s(t) + U_T T_a + h_{cb}T_r\right]$$

or

$$h_{cb}(T_c - T_r)A_{tm} = \left(h_{p1}(\alpha\tau)_1 I_s(t) - U_{tra}(T_r - T_a)\right)A_{tm} \tag{7.64b}$$

where $h_{p1} = \frac{h_{cb}}{U_T + h_{cb}}$, $(\alpha\tau)_1 = \tau_G\beta_c(\alpha_c - \eta_c)$ and $U_{tra} = \frac{h_{cb}U_T}{U_T + h_{cb}}$; h_{p1} = penalty factor due to presence of solar cell material and EVA and U_{tra} = overall heat transfer coefficient between greenhouse room air and ambient air (W m^{-2} K).

From eqn (7.64a)the floor temperature becomes

$$T_f = \frac{h_f T_r + h_g T_\infty}{h_f + h_g}$$

or

$$h_f(T_r - T_f)A_f = U_{r\infty}(T_r - T_\infty)A_f \tag{7.64c}$$

where $U_{r\infty} = \frac{h_f h_g}{h_f + h_g}$; $U_{r\infty}$ = overall heat transfer coefficient between greenhouse room air and greenhouse ground depth or inside greenhouse ground (W m^{-2} K).

From eqns (7.64a)(7.64b) and (7.64c), we get

$$\begin{aligned} M_a C_a \frac{dT_r}{dt} &= \tau\left(\sum I_i A_i - I_s(t)A_{tm}\right) + \tau_G^2(1-\beta_c)I_s(t)A_{tm} \\ &+ \left(h_{p1}(\alpha\tau)_1 I_s(t) - U_{tra}(T_r - T_a)\right)A_{tm} \\ &- \sum U_i A_i(T_r - T_a) - 0.33NV(T_r - T_a) - U_{r\infty}(T_r - T_\infty)A_f \end{aligned}$$

or

$$\begin{aligned} M_a C_a \frac{dT_r}{dt} &= (\alpha\tau I)_E ff A_{tm} - \left(U_{tra}A_{tm} + \sum U_i A_i + 0.33NV + U_{r\infty}A_f\right)T_r \\ &+ \left(U_{tra}A_{tm} + \sum U_i A_i + 0.33NV\right)T_a + U_{r\infty}T_\infty A_f \end{aligned} \tag{7.65}$$

where $\alpha\tau I)_{eff}A_{tm} = \left[\tau(\sum I_i A_i - I_s(t)A_{tm}) + \tau_G^2(1-\beta_c)I_s(t)A_{tm} + h_{p1}(\alpha\tau)_1 I_s(t)A_{tm}\right]$

Equation (7.65) may be written as

$$\frac{dT_r}{dt} + aT_r = f(t) \tag{7.66}$$

where $a = \left[\frac{\left(U_{tra}A_{tm} + \sum U_i A_i + 0.33NV + U_{r\infty}A_f\right)}{M_a C_a}\right]$

$$f(t) = \left[\frac{(\alpha\tau I)_E ff A_{tm} + (U_{tra}A_{tm} + \sum U_i A_i + 0.33NV)T_a + U_{r\infty}T_\infty A_f}{M_a C_a}\right]$$

By solving eqn (7.66), the expression of greenhouse air temperature becomes

$$T_r = \frac{\bar{f}(t)}{a}(1 - e^{-at}) + T_{r0}e^{-at} \tag{7.67}$$

where T_{r0} is the greenhouse air temperature at time $t = 0$ and $\bar{f}(t)$ is an average value of $f(t)$ over the time interval between 0 and t.

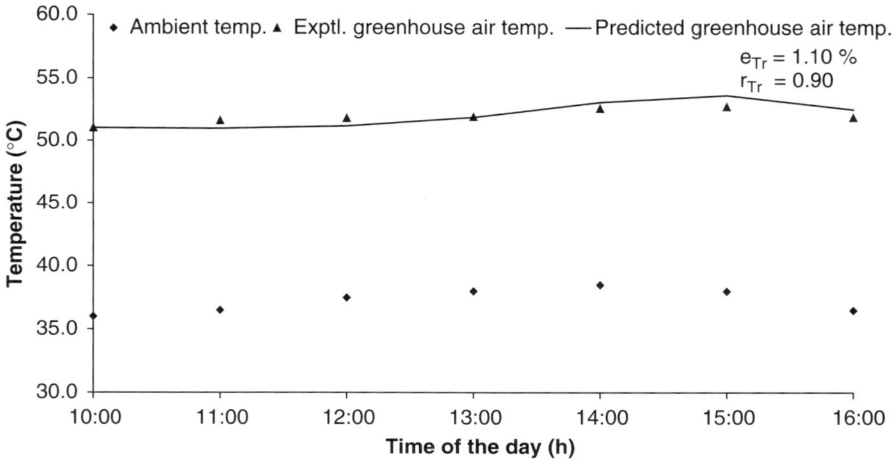

Figure 7.45 Hourly variations of experimental and predicted greenhouse air temperatures for forced mode of operation under no load condition.

Equation (7.67) was used to predict the greenhouse air temperature and there is good agreement of experimental and predicted greenhouse air temperatures for forced mode of operation under no load conditions (Figure 7.45).

7.5.3 Conventional Solar Grain Dryer

The dryer comprises a collector module (air heater and PV section), a drying chamber, a universal joint *etc.* (Figure 7.40). It has a capacity of 90 kg maize grain per batch and it can dry the maize grain from an initial moisture content of 33.3% dry basis to under 20% dry basis in just one day. In this dryer, PV solar cells were incorporated in the solar air heater section to operate a DC fan which provided some passive control over the air flow and hence the drying air temperature.

The collector module comprises the blackened sisal absorber meshes for improved heat transfer, three transparent cover sections to allow any single and double transparent tedlar/Teflon cover combination for high short-wave transmittance and low top heat loss and a PV panel section. The dimensions of the collector module are 2.0 m × 1.1 m. The dimensions of the effective air heater aperture and the PV section are 1.5 m × 1.0 m and 1.0 m × 0.3 m respectively. The path depth of the collector air was 0.05 m, which was filled with an optimized number of sisal grid absorbers and void ratio. To improve heat gain, a transparent insulation material (TIM) was sandwiched between the transparent covers and the absorber grids. The air heater wall insulation, made of plywood pockets filled with dry wood shavings, was 0.8 m thick.

The drying chamber also comprises a 0.08-m-thick insulation wall which was sealed in the plenum to avoid any air leakage. The floor dimensions of the drying chamber were 1.0 m × 0.74 m. For easy rainwater drainage, a slanting roof (30° from the horizontal level) was provided above the drying chamber.

The grain-loading door comprises an air outlet vent, placed above the grain load level. A DC fan (12 volt, 0.42 amp) was placed in suction mode at the drying chamber (plenum) air inlet.

The collector module and the drying chamber were connected through a 0.1-m-diameter flexible insulated duct having a universal joint. The universal joint gave a provision for rotating the collector ±30° from the horizontal to track the Sun for improved collector efficiency. The drying chamber was raised 0.9 m above the ground, while the collector module was tilted 15° from the horizontal level to match the Sun's elevation and to minimize air-flow resistance through the air heater. The hot air from the air heater section was pumped by the DC fan into the drying chamber plenum and then up through the grain bed and exhausted through the loading door air vent. The collector module interior and exterior surfaces and the drying chamber exterior surfaces were painted black in order to ensure maximum heat gain.

7.5.3.1 *Efficiency Parameters*

Three efficiency parameters used in performance evaluation of the dryer are given below:[60]

(i) *Dryer thermal efficiency*:
It can be expressed as

$$\eta_{\mathrm{d,th}} = \frac{\dot{m}_{\mathrm{w}}\lambda}{\dot{m}_{\mathrm{a}}C_{\mathrm{a}}(T_{\mathrm{d}} - T_{\mathrm{fi}})} \tag{7.68}$$

where $\dot{m}_a$ is the air-mass flow rate in the dryer (measured in the connecting duct between the collector and the drying chamber), C_{a} is the air specific heat capacity, T_{d} is the dryer (plenum) air temperature, T_{fi} is the collector inlet air temperature, which can be taken to be equal to ambient air temperature, λ is the latent heat of evaporation of water and $\dot{m}_w$ is the mass of moisture evaporated per unit time.

(ii) *Dryer pick-up efficiency:*
It can be expressed as

$$\eta_{\mathrm{p}} = \frac{\dot{m}_{\mathrm{w}}}{\dot{m}_{\mathrm{a}}\Delta t\left(\omega_{\mathrm{g,e}} - \omega_{\mathrm{g,i}}\right)} = \frac{\dot{m}_{\mathrm{w}}}{\dot{m}_{\mathrm{a}}\left(\omega_{\mathrm{g,e}} - \omega_{\mathrm{g,i}}\right)} \tag{7.69}$$

where m_w is the mass of moisture evaporated in time Δt, $\omega_{\mathrm{g,e}}$ is the grain exit air absolute humidity ($\approx$ adiabatic saturation humidity) and $\omega_{\mathrm{g,i}}$ is the grain inlet (plenum) air absolute humidity.

(iii) *Instantaneous DC fan solar energy utilization efficiency:*
It can be expressed as

$$\eta_{\mathrm{f}} = \frac{I_{\mathrm{f}}V_{\mathrm{f}}}{A_{\mathrm{pv}}\bar{\eta}_{\mathrm{pv}}I(t)} \tag{7.70}$$

where I_{f} is the fan current, V_{f} is the fan voltage, A_{pv} is the photovoltaic solar cell area, $\bar{\eta}_{pv}$ is the solar cell solar energy conversion efficiency (about 8% for amorphous silicon solar cells, used for the dryer work) and $I(t)$ is the incident solar irradiance on the solar cells.

It has been found that the mean thermal, pick-up and solar energy utilization efficiencies of the dryer were 58%, 77% and 33%, respectively, for the drying run.[60]

7.5.3.2 Performance Characteristics

The performance characteristics of the dryer are given below:[61]

i. Without load

The performance characteristics were investigated for the following four management strategies of the collector module and the PV fan:(a)PV fan off without Sun-tracking;(b)PV fan on without Sun-tracking;(c)PV fan off with Sun-tracking;(d)PV fan on with Sun-tracking.In the Sun-tracking mode, the collector module was tilted through the universal joint at the collector head and the best strategy was adopted for the dryer operation.

For the four strategies, temperature profiles along the full length of the dryer are shown in Figure 7.46 and it is clear that the PV fan on with Sun-tracking strategy was the best, giving a uniform air temperature of 60 °C from the collector air outlet to the grain (maize) bin air outlet. From Figure 7.47, it is clear

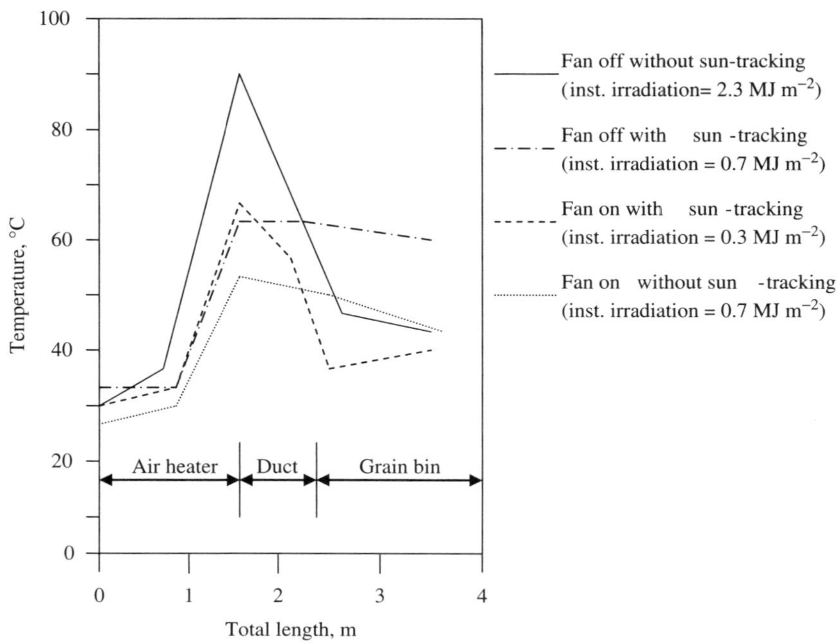

Figure 7.46 Temperature profile along the length of the solar dryer from collector air inlet to grain-bin air outlet.

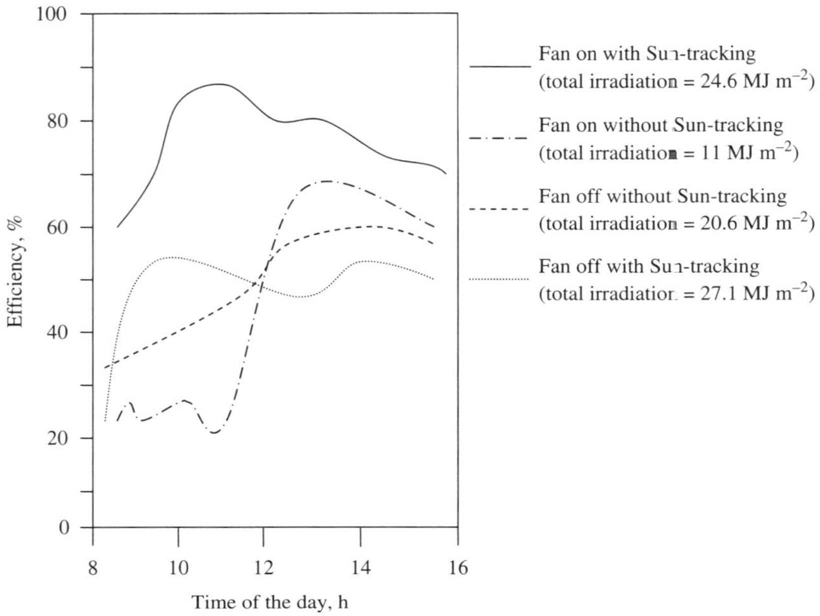

Figure 7.47 Collector efficiency *vs.* time for four collector and DC fan management strategies.

that for the PV fan on with Sun-tracking strategy, the collector thermal efficiency profiles are quite uniform and in the order of 80%.

ii. With Load

The drying curves of the grain (maize) dried through this dryer and the Sun-dried control sample of the same mass are shown in Figure 7.46. For the grain (maize) dried through this dryer and the Sun-dried control, the drying times taken to reach the safe milling moisture content (25% dry basis) were 1.4 h and 2.9 h, respectively, and that to reach the safe storage moisture content (14.3% dry basis) were 7.0 h and 27 h, respectively.

7.5.4 Conventional PV/T Mixed Mode Dryer

A hybrid PV/T conventional mixed mode dryer has been developed at Solar Energy Park, Indian Institute of Technology (IIT), New Delhi, India. The dryer consists of a collector unit, a drying chamber, a DC fan *etc.* (Figure 7.44). The collector unit comprises a PV module (glass-to-glass) and a flat-plate air collector. The PV module (glass-to-glass) was provided at the lower part of the solar collector to operate a DC fan for forced mode of operation. In this case, the solar radiation through the non-packing factor area is also available to the absorber below the PV module for preheating of ambient air. The DC fan is fitted at the junction of the collector module exit and drying chamber inlet to

Table 7.3 Design specifications of PV/T mixed mode dryer.

S. No.	*Details of particulars*	*Specification*
1.	Air duct	2.2 m × 0.65 m × 0.05 m
2.	PV module	0.65 m × 0.55 m; 35 W
3.	Spacing between absorber and glass	0.10 m
4.	DC fan	12 V, 1.3 A
5.	Chimney	0.65 m × 0.26 m × 0.60 m
6.	Number of trays	3
7.	Spacing between two trays	0.15 m
8.	Inclination of absorber (air duct) with horizontal	30°

suck the hot air from the collector module and force it into the drying chamber. The hot air flows from the bottom to the top of the drying chamber through wire mesh trays, takes away moisture from crops placed in the trays and is exhausted to the outside through openings provided at the top of the east and west side walls of the drying chamber. The sides of the dryer are made from plywood/wood for insulation and sealed in to avoid any air leakage. For easy rainwater drainage, a slanting roof was provided above the drying chamber. There are drawers consisting of wire mesh trays, which are placed in the drying chamber from the back portion of the dryer. The design specifications of the PV/T mixed mode dryer are given in Table 7.3.

7.5.4.1 Thermal Modelling of Conventional PV/T Mixed Mode Dryer

In order to write the energy balance equations for a conventional PV/T mixed mode dryer, the following assumptions have been made:

1. The system is in quasi-steady state;
2. Absorptivity of the glass of the PV module and the enclosed air is neglected;
3. The transmissivity of ethyl vinyl acetate (EVA) is approximately 100%;
4. The temperature variation along the thickness as well as along the width is negligible;
5. The ohmic losses in solar cells are negligible.

Based on the first law of thermodynamics, the energy balance equations for the conventional PV/T mixed mode dryer are written to account for energy input, energy output and energy losses.

(i) *For solar cells of PV module (glass-glass):*

$$\tau_G \alpha_c I(t) \beta_c b dx = \{U_T(T_c - T_a) + h_{cair}(T_c - T_{air})\} b dx + \tau_G \eta_c \beta_c I(t) b dx \tag{7.71a}$$

where $U_T = \left[\frac{L_G}{k_G} + \frac{1}{h_o}\right]^{-1}$, U_T = overall top loss heat transfer coefficient between the solar cell of the PV module and ambient air ($W\,m^{-2}\,K^{-1}$), h_o = convective heat transfer coefficient between the top surface of the PV module and ambient air ($W\,m^{-2}\,K^{-1}$), $h_o = 5.7 + 3.8\,V$, V = wind velocity ($m\,s^{-1}$), $h_{cair} = \left[\frac{L_G}{k_G} + \frac{1}{h_i}\right]^{-1}$, h_i = convective heat transfer coefficient between the bottom surface of the PV module and the greenhouse room air ($W\,m^{-2}\,K^{-1}$), $h_i = 2.8 + 3.0\,V$, α_c = absorptivity of the solar cell portion of the PV module, β_c = packing factor of the PV module, η_c = efficiency of the solar cell of the PV module, τ_G = transmissivity of the glass portion of the PV module, I(t) = total average solar intensity measured on the PV module and the collector of the dryer ($W\,m^{-2}$), b = width of the PV module and collector (m), L_G = thickness of the glass of the PV module (m), k_G = thermal conductivity of the glass ($W\,m^{-1}\,K^{-1}$), T_a = ambient air temperature (°C), T_c = solar cell temperature (°C) and T_{air} = temperature of duct air (°C).

From eqn (7.71a), the expression for cell temperature is

$$T_c = \frac{(\alpha\tau)_{1,\text{eff}} I(t) + U_T T_a + h_{cair} T_{air}}{U_T + h_{cair}} \tag{7.71b}$$

where $(\alpha\tau)_{1,\text{eff}} = (\alpha_c - \eta_c)\beta_c\tau_G$.

The temperature-dependent electrical efficiency of a PV module has been expressed by eqn (7.1).

(ii) *For blackened absorber plate*:

$$\left[\alpha_p(1-\beta_c)\tau_G^2 I(t)\right] b dx = \left[h_{pair}\left(T_p - T_{air}\right) + U_{aira}\left(T_{air} - T_a\right)\right] b dx \tag{7.72a}$$

where α_p = absorptivity of the blackened absorber plate, h_{pair} = heat transfer coefficient between the absorber plate and the duct air ($W\,m^{-2}\,K^{-1}$), T_p = absorber plate temperature (°C) and U_{aira} = heat loss between the absorber plate and the ambient air ($W\,m^{-2}\,K^{-1}$).

From eqn (7.80a), the expression for the plate temperature is

$$T_p = \frac{(\alpha\tau)_{2,\text{eff}} I(t) - U_{aira}(T_{air} - T_a)}{h_{pair}} + T_{air} \tag{7.72b}$$

where $(\alpha\tau)_{2,\text{eff}} = \alpha_p(1-\beta_c)\tau_G^2$.

(iii) *For air flowing through the duct:*

The energy balance of flowing air through the duct is given by

$$\dot{m}_a C_a \frac{dT_{air}}{dx} dx = \left[h_{pair}\left(T_p - T_{air}\right) + h_{cair}\left(T_c - T_{air}\right)\right] b dx \tag{7.73}$$

where $\dot{m}$ = mass flow rate of air through the duct ($kg\,s^{-1}$) and C_a = specific heat capacity of air ($J\,kg^{-1}\,K^{-1}$).

The solution of eqn (7.73) with the help of eqns (7.71b) and (7.72b) and initial conditions, namely at $x = 0$, $T_{air} = T_{mairin}$ and at $x = L$, $T_{air} = T_{mairout}$, we get

$$T_{\text{mairout}} = \left[\frac{(\alpha\tau)_{\text{m,eff}} I(t)}{U_{\text{L,m}}} + T_{\text{a}}\right]\left[1 - \exp\left(-\frac{bU_{\text{L,m}}L}{\dot{m}_{\text{a}}C_{\text{a}}}\right)\right] + T_{\text{mairin}}\exp\left(-\frac{bU_{\text{L,m}}L}{\dot{m}_{\text{a}}C_{\text{a}}}\right) \quad (7.74\text{a})$$

where $(\alpha\tau)_{\text{m,eff}} = h_{\text{p1}}(\alpha\tau)_{1,\text{eff}} + (\alpha\tau)_{2,\text{eff}}$

$$h_{\text{p1}} = \frac{h_{\text{cair}}}{U_{\text{T}} + h_{\text{cair}}};\ U_{L1} = \frac{U_{\text{T}}h_{\text{cair}}}{U_{\text{T}} + h_{\text{cair}}};\ U_{\text{L,m}} = U_{L1} + U_{\text{aira}}$$

Here, T_{mairout} is the outlet temperature of the air from the absorber PV module and T_{mairout} becomes the inlet temperature for the remaining part of the collector (T_{cairin}).

The rate of thermal energy available at the end of the absorber PV module (glass-glass) is evaluated as

$$\dot{Q}_{\text{u,m}} = \dot{m}_{\text{a}}C_{\text{a}}(T_{\text{mairout}} - T_{\text{mairin}})$$

After substituting the expression for T_{mairout} from eqn (7.74a), we get

$$\dot{Q}_{\text{u,m}} = A_{\text{m}}F_{\text{Rm}}\left((\alpha\tau)_{\text{m,eff}}I(t) - U_{\text{L,m}}(T_{\text{mairin}} - T_{\text{a}})\right) \quad (7.74\text{b})$$

where A_{m} = area of PV module (m^2) and $F_{\text{Rm}} = \frac{\dot{m}_{\text{a}}C_{\text{a}}}{A_{\text{m}}U_{\text{L,m}}}\left[1 - \exp\left(-\frac{A_{\text{m}}U_{\text{L,m}}}{\dot{m}_{\text{a}}C_{\text{a}}}\right)\right]$.

(iv) *The outlet air temperature at the end of flat-plate collector:*

Following Duffie and Beckman[42] and Tiwari,[43] an expression for the outlet air temperature at the end of the collector (T_{cairout}) will be

$$T_{\text{cairout}} = \left[\frac{(\alpha\tau)_{\text{c,eff}} I(t)}{U_{\text{L,c}}} + T_{\text{a}}\right]\left[1 - \exp\left(-\frac{A_{\text{c}}U_{\text{L,c}}}{\dot{m}_{\text{a}}C_{\text{a}}}\right)\right] + T_{\text{cairin}}\exp\left(-\frac{A_{\text{c}}U_{\text{L,c}}}{\dot{m}_{\text{a}}C_{\text{a}}}\right) \quad (7.75\text{a})$$

where $U_{\text{L,c}}$ = overall heat loss coefficient from absorber to ambient (W m^{-2} K^{-1}), A_c = area of collector (m^2) and $(\alpha\tau)_{\text{c,eff}}$ = product of absorptivity of absorber and transmissivity of glass.

Here, $T_{\text{cairin}} = T_{\text{mairout}}$ can be evaluated from eqn (7.74a).

$$T_{\text{cairin}} = T_{\text{mairout}} = T_{\text{mairin}} + \frac{\dot{Q}_{\text{u,m}}}{\dot{m}_{\text{a}}C_{\text{a}}}$$

The rate of thermal energy available from the first flat-plate collector can be evaluated as

$$\dot{Q}_{\text{u,m}} = \dot{m}_{\text{a}}C_{\text{a}}(T_{\text{mairout}} - T_{\text{mairin}})$$

After substituting the expression for T_{cairout} from eqn (7.75a), we get

$$\dot{Q}_{\mathrm{u,c}} = A_{\mathrm{c}} F_{\mathrm{Rc}} \left[(\alpha\tau)_{\mathrm{c,eff}} I(t) - U_{\mathrm{L,c}} (T_{\mathrm{cairin}} - T_{\mathrm{a}}) \right] \tag{7.75b}$$

where $F_{\mathrm{Rc}} = \frac{\dot{m}_{\mathrm{a}} C_{\mathrm{a}}}{A_{\mathrm{c}} U_{\mathrm{L,c}}} \left[1 - \exp\left(-\frac{A_{\mathrm{c}} U_{\mathrm{L,c}}}{\dot{m}_{\mathrm{a}} C_{\mathrm{a}}} \right) \right]$. Now, $\dot{Q}_{\mathrm{u,(m+c)}} = \dot{m}_{\mathrm{a}} C_{\mathrm{a}} (T_{\mathrm{cairout}} - T_{\mathrm{mairin}})$

$$\begin{aligned} \dot{Q}_{\mathrm{u,(m+c)}} = {} & A_{\mathrm{m}} F_{\mathrm{Rm}} \left[(\alpha\tau)_{\mathrm{m,eff}} I(t) - U_{\mathrm{L,m}} (T_{\mathrm{mairin}} - T_{\mathrm{a}}) \right] \\ & + A_{\mathrm{c}} F_{\mathrm{Rc}} \left[(\alpha\tau)_{\mathrm{c,eff}} I(t) - U_{\mathrm{L,c}} (T_{\mathrm{cairin}} - T_{\mathrm{a}}) \right] \end{aligned}$$

On simplifying the above equation we get

$$\begin{aligned} \dot{Q}_{\mathrm{u,(m+c)}} = {} & \left[A_{\mathrm{m}} F_{\mathrm{Rm}} (\alpha\tau)_{\mathrm{m,eff}} \left(1 - \frac{A_{\mathrm{c}} F_{\mathrm{Rc}} U_{\mathrm{L,c}}}{\dot{m}_{\mathrm{a}} C_{\mathrm{a}}} \right) + A_{\mathrm{c}} F_{\mathrm{Rc}} (\alpha\tau)_{\mathrm{c,eff}} \right] I(t) \\ & - \left[A_{\mathrm{m}} F_{\mathrm{Rm}} U_{\mathrm{Lm}} \left(1 - \frac{A_{\mathrm{c}} F_{\mathrm{Rc}} U_{\mathrm{L,c}}}{\dot{m}_{\mathrm{a}} C_{\mathrm{a}}} \right) + A_{\mathrm{c}} F_{\mathrm{Rc}} U_{\mathrm{Lc}} \right] (T_{\mathrm{mairin}} - T_{\mathrm{a}}) \end{aligned} \tag{7.76}$$

The rate of thermal energy available at the outlet of the air collector ($\dot{Q}_{\mathrm{u},(m+c)}$) is allowed to pass through crops placed in different trays in the vertical direction inside the drying chamber.

(v) *For crop surface in drying chamber*:

$$\dot{Q}_{\mathrm{u},(m+c)} + \tau_{\mathrm{G}} \alpha_{\mathrm{wc}} I_{\mathrm{ch}}(t) A_{Gch} = M_{\mathrm{wc}} C_{\mathrm{wc}} \frac{dT_{\mathrm{wc}}}{dt} + h(T_{\mathrm{wc}} - T_{\mathrm{ch}}) A_{\mathrm{wc}} \tag{7.77}$$

where A_{wc} = surface area of wet crop (m^2), A_{Gch} = glass surface area of drying chamber (m^2), M_{wc} = mass of wet crop (kg), C_{wc} = specific heat capacity of wet crop (J kg^{-1} K^{-1}), T_{wc} = wet crop surface temperature (°C), T_{ch} = drying chamber air temperature (°C), t = time (s), h = total heat transfer coefficient between wet crop and drying air in the drying chamber (W m^{-2} K^{-1}), α_{wc} = absorptivity of wet crop in the drying chamber and I_{ch}(t) = total average solar intensity measured on the glass surface of the drying chamber (W m^{-2}).

(vi) *For drying chamber*:

$$h(T_{\mathrm{wc}} - T_{\mathrm{ch}}) A_{\mathrm{wc}} = 0.33 NV (T_{\mathrm{ch}} - T_{\mathrm{a}}) + U_{\mathrm{ch}} (T_{\mathrm{ch}} - T_{\mathrm{a}}) A_{\mathrm{ch}} \tag{7.78}$$

where N = number of air changes per hour, V = volume of drying chamber (m^3), U_{ch} = overall loss heat transfer coefficient between drying chamber room air and ambient air (W m^{-2} K^{-1}) and A_{ch} = total outer surface area of drying chamber (m^2).

The drying chamber air temperature can be written as

$$T_{\mathrm{ch}} = \frac{h T_{\mathrm{wc}} A_{\mathrm{wc}} + 0.33 NV T_{\mathrm{a}} + U_{\mathrm{ch}} A_{\mathrm{ch}} T_{\mathrm{a}}}{h A_{\mathrm{wc}} + 0.33 NV + U_{\mathrm{ch}} A_{\mathrm{ch}}} \tag{7.79}$$

Now, substituting the value of T_{ch} in eqn (7.77), we get an analytical expression for wet crop temperature (T_{wc}) as a function of time, which can be used for further analysis.

7.6 Statistical Analysis

Statistics is a mathematical science pertaining to the collection, analysis, interpretation or explanation and presentation of data. It also provides tools for prediction and forecasting based on data. It is applicable to a wide variety of academic disciplines. Statistical methods can be used to summarize or describe a collection of data. In addition, patterns in the data may be modelled in a way that accounts for randomness and uncertainty in the observations, and are then used to draw inferences about the process or population being studied.

Arithmetic Mean

The mean is the arithmetic average of a set of values, or distribution.

$$\bar{x} = \frac{1}{N}\sum_{\mathrm{i}=1}^{N} x_{\mathrm{i}} \tag{1.80}$$

where x_{i} = set of values and N = number of values.

Mode

In statistics, the mode is the value that occurs the most frequently in a data set or a probability distribution.

Median

In probability theory and statistics, a median is described as the number separating the higher half of a sample from the lower half. The median of a finite list of numbers can be found by arranging all the observations from the lowest value to the highest value and picking the middle one. If there is an even number of observations, the median is not unique, so one often takes the mean of the two middle values.

Root Mean Square (RMS)

The root mean square is a statistical measure of the magnitude of a varying quantity. It is especially useful when variants are positive and negative, *e.g.* sinusoids.

$$x_{\mathrm{rms}} = \sqrt{\frac{1}{N}\sum_{\mathrm{i}=1}^{N} x_{\mathrm{i}}^2} \tag{7.81}$$

Standard Deviation

The standard deviation is a simple measure of the variability or dispersion of a population, a data set or a probability distribution.

$$\sigma = \sqrt{\frac{1}{N}\sum_{\mathrm{i}=1}^{N} (x_{\mathrm{i}} - \bar{x})^2} \tag{7.82}$$

where $\bar{x}$ = mean of values.

Mean Absolute Error

The mean absolute error is a quantity used to measure how close forecasts or predictions are to the eventual outcomes. The mean absolute error (MAE) is given by

$$MAE = \frac{1}{N}\sum_{i=1}^{N}|X_i - Y_i| \tag{7.83}$$

where X_i = predicted value and Y_i = true value.

Chi-square Distribution

The chi-square distribution (χ^2) is a continuous probability and one of the most widely used theoretical probability distributions in statistical significance tests. The distribution usually arises when a *k*-dimensional vector's orthogonal components are independent and each follow a standard normal distribution. The length of the vector will then have a chi distribution. If X_i are k independent, normally distributed random variables with means μ_i and standard deviations σ_i then the statistic becomes

$$\chi^2 = \sum_{i=1}^{k}\left(\frac{X_i - \mu_i}{\sigma_i}\right) \tag{7.84}$$

Correlation Coefficient and Root Mean Square Percentage Deviation

To compare experimental and theoretical results, the expression for the correlation coefficient (r) and root mean square percent deviation (e) has been given in eqns (7.44a) and (7.44b).

Uncertainty

The uncertainty is a term used in subtly different ways in a number of fields. It applies to predictions of future events, to physical measurements already made or to the unknown. The factors responsible for uncertainty in a model may be:

a) The model structure, *i.e.* how accurately does a mathematical model describe the true system for a real-life situation;
b) The numerical approximation, *i.e.* how appropriately a numerical method is used in approximating the operation of the system;
c) The initial/boundary conditions, *i.e.* how precise are the data/information for initial and/or boundary conditions;
d) The data for input and/or model parameters.
e) The following three types of uncertainties can be identified:
f) Uncertainty due to variability of input and/or model parameters when the characterization of the variability is available;
g) Uncertainty due to variability of input and/or model parameters when the corresponding variability characterization is not available;
h) Uncertainty due to an unknown process or mechanism.

Problems

7.1 Calculate the outlet air temperature for an air duct having cross sectional area 1 m × 0.45 m × 0.04 m. Air is flowing at the rate of $0.5\,\mathrm{m\,s^{-1}}$, $1\,\mathrm{m\,s^{-1}}$ and $2\,\mathrm{m\,s^{-1}}$, the penalty factor is 0.4, gain and loss are 0.8 and $7.2\,\mathrm{W\,m^{-2}\,K^{-1}}$, respectively, $T_a = 25\,°\mathrm{C}$, $T_{airin} = T_a + 2\,°\mathrm{C}$ and $I(t) = 800\,\mathrm{W\,m^{-2}}$. Hint: use eqn (7.9).

7.2 Using the data of Example 7.1, calculate the useful heat gain. Hint: use eqn (7.11).

7.3 Derive an expression for the outlet air temperature when the air collector is covered by a glass-to-glass type PV module.

7.4 Calculate the outlet air temperature for different lengths of collector (2 m–10 m) for the following specifications: $I(t) = 450\,\mathrm{W\,m^{-2}}$, $T_a = 15\,°\mathrm{C}$, $W = 1$ m, $m = 0.02\,\mathrm{kg\,s^{-1}}$, $U_L = 2.81\,\mathrm{W\,m^{-2}\,K^{-1}}$. Also plot the curve between outlet temperature and the length of the collector. Hint: use eqn (7.9).

7.5 Plot the curve of η_i with $\left(\frac{T_{fi} - T_a}{I(t)}\right)$ for a collector using the following specifications: $I(t) = 450$, 600, $750\,\mathrm{W\,m^{-2}}$, $T_{fi} = 25\,°\mathrm{C}$, 35 °C, 50 °C and $T_a = 20\,°\mathrm{C}$. Hint: use eqn (7.9).

7.6 Calculate the variation of instantaneous efficiency for a water collector when a PV module is integrated on the lower and upper portions of the collector. $I(t) = 450$, 600, $750\,\mathrm{W\,m^{-2}}$, $T_{fi} = 35\,°\mathrm{C}$, 45 °C, 70 °C and $T_a = 20\,°\mathrm{C}$. Hint: use eqns (7.26b) and (7.27b).

7.7 Plot the curve between the collector flow factor ($F'' = F_R/F'$) and $m\,C_p/A_c\,U_L\,F$. Hint: use eqn (7.32c) and Example 7.6.

7.8 Derive an expression for the threshold radiation level.

7.9 Derive an expression for a series and parallel combination of collectors fully covered by PV.

7.10 Calculate the hourly output from the still and instantaneous efficiency of a distillation unit when the water surface is at 30 °C, ambient air is at 15.5 °C and the temperature of the glass = 25 °C. Evaporative heat transfer coefficient = $5.6\,\mathrm{W\,m^{-2}\,°C}$ and $L = 2390 \times 10^3\,\mathrm{J\,kg^{-1}}$. Hint: use eqn (7.61a).

7.11 Derive an expression for the outlet water temperature and useful heat gain from a conventional PV/T mixed mode dryer.

References

1. E. C. Kern Jr and M. C. Russell, *in Proc. 13th IEEE Photovoltaic Specialists*, Washington DC, USA, 1978, 1153–1157.
2. S. D. Hendrie, in *Proc. ISES Int. Congress*, Atlanta, USA, 1979, **3**, 1865–1869.
3. L. W. Florschuetz, *Sol. Energ.*, 1979, **22**, 361–366.
4. P. Raghuraman, *Sol. Energ. Eng.*, 1981, **103**, 291–298.
5. B. Lalovic, *Sol. Cell.*, 1986, **19**, 131–138.

6. J. Loferski, J. M. Ahmad and A. Pandey, in *Proc. of the 1988 Annual Meeting, American Solar Energy Society*, Cambridge, Massachusetts, 1998, 427–432.
7. A. K. Bhargava, H. P. Garg and R. K. Agarwal, *Energ. Convers. Manag.*, 1991, **391**(5), 471–479.
8. J. Prakash, *Energ. Convers. Manag.*, 1994, **35**, 967–972.
9. A. D. Jones and C. P. Underwood, *Sol. Energ.*, 2001, **70**(4), 349–359.
10. D. W. Zondag de Vries, W. G. J. van Helden, R. J. C. van Zolengen and A. A. Steenhoven, *Sol. Energ.*, 2003, **74**(3), 253–269.
11. T. T. Chow, *Sol. Energ.*, 2003, **75**, 143–152.
12. A. A. Hegazy, *Energ. Convers. Manag.*, 2000, **41**(8), 861–881.
13. D. Infield, L. Mei and U. Eicker, *Sol. Energ.*, 2004, **76**(1–3), 93–98.
14. Y. Tripanagnostopoulos, T. H. Nousia, M. Souliotis and P. Yianoulis, *Sol. Energ.*, 2002, **72**(3), 217–234.
15. B. P. Cartmell, N. J. Shankland, D. Fiala and V. Hanby, *Sol. Energ.*, 2004, **76**, 45–53.
16. A. S. Joshi and A. Tiwari, *Renew. Energ.*, 2007, **32**(13), 2223–2241.
17. A. Tiwari and M. S. Sodha, *Renew. Energ.*, 2006, **31**(15), 2460–2474.
18. A. Tiwari, M. S. Sodha, A. Chandra and J. C. Joshi, *Sol. Energ. Mater. Sol. Cell.*, 2006, **90**(2), 175–189.
19. A. Guiavarch and B. Peuportier, *Sol. Energ.*, 2006, **80**, 65–77.
20. H. A. Zondag, *Renew. Sustain. Energ. Rev.*, 2008, **12**(4), 891–959.
21. H. A. Zondag, D. W. Vries, W. G. J. van Helden, R. J. C. van Zolengen and A. A. Steenhoven, *Sol. Energ.*, 2002, **72**(2), 113–128.
22. S. A. Kalogirou, *Renew. Energ.*, 2001, **23**, 247–260.
23. H. P. Garg, R. K. Agarwall and J. C. Joshi, *Energ. Convers. Manag.*, 1994, **35**, 621–633.
24. T. T. Chow, W. He and J. Ji, *Sol. Energ.*, 2006, **80**, 298–306.
25. R. Zakharchenko, L. Licea-Jiménez, S. A. Pérez-García, P. Vorobiev, U. Dehesa- Carrasco, J. F. Pérez-Robels, J. González-Hernández and Y. Vorobiev, *Sol. Energ. Mater. Sol. Cell.*, 2004, **82**(1–2), 253–261.
26. B. Sandnes and J. Rekstad, *Sol. Energ.*, 2002, **72**(1), 63–73.
27. A. Tiwari and M. S. Sodha, *Sol. Energ.*, 2006, **80**(7), 751–759.
28. K. Sopian, H. T Liu, S. Kakac and T. N. Veziroglu, *Energ. Convers. Manag.*, 2000, **41**, 353–365.
29. E. Radziemska, *Progr. Energ. Combust. Sci.*, 2003, **29**(5), 407–424.
30. J. S. Coventry, *Sol. Energ.*, 2005, **78**(2), 211–222.
31. J. K. Tonui, Y. Tripanagnostopoulos, *Sol. Energ.*, **81** (4), 498–511.
32. J. K. Tonui and Y. Tripanagnostopoulos, *Renew. Energ.*, 2007, **32**(4), 623–637.
33. M. Y. H. Othman, B. Y. Kamaruzzaman, K. Sopian and M. N. Abu Bakar, *Renew. Energ.*, 2005, **30**(13), 2005–2017.
34. M. Y. H. Othman, B. Yatim, K. Sopian and M. N. Abu Bakar, *Desalination*, 2007, **209**(1–3), 43–49.
35. S. Dubey, G. S. Sandhu and G. N. Tiwari, *Appl. Energ.*, 2009, 86(5), 697–705.

36. S. Dubey, S. C. Solanki and A. Tiwari, *Energ. and Build.*, 2009, **41**, 863–870.
37. F. Kreith, A. Rabl and R. Lof, *Winston Progr. Energ. Combust. Sci.*, 1980, **6**(1), 1–34.
38. A. Tiwari and M. S. Sodha, *Sol. Energ. Mater. Sol. Cell.*, 2007, **91**(1), 17–28.
39. J. K. Tonui and Y. Tripanagnostopoulos, *Sol. Energ.*, 2008, **82**(1), 1–12.
40. B. J. Huang, T. H. Lin, W. C. Hung and F. S. Sun, *Sol. Energ.*, 2001, **70**(5), 443–448.
41. B. Robles-Ocampo, E. Ruíz-Vasquez, H. Canseco-Sánchez, R. C. Cornejo-Meza, G. Trápaga-Martínez, F. J. García-Rodriguez, J. González-Hernándeze and Y. V. Vorobiev, *Sol. Energ. Mater. Sol. Cell.*, 2007, **91**, 1966–1971.
42. J. A. Duffie and W. Beckman, *Solar Engineering of Thermal Processes*, John Wiley and Sons, New York, 1991.
43. G. N. Tiwari, *Solar Energy: Fundamentals, Design, Modeling and Applications*, Narosa Publishing House, New Delhi, 2004.
44. Swapnil Dubey and G. N. Tiwari, *Sol. Energ.*, 2008, **82**, 602–612.
45. Swapnil Dubey and G. N. Tiwari, *Int. J. Energ. Res.*, 2008, **32**, 1362–1372.
46. A. Cipollina, C. Sommariva and M. Giorgio, *Desalination*, 2005, **183**, 127–136.
47. P. I. Cooper, *Sol. Energ.*, 1969, **12**, 313–331.
48. M. A. Hamdan, A. M. Musa and B. A. Jubran, *Energ. Convers. Manag.*, 1999, **40**, 495–503.
49. H. A. Kumze, *Desalination*, 2001, **139**, 35–41.
50. G. M. Zaki, A. Al-Turki and M. Al-Fatani, *Sol. Energ.*, 1992, **11**, 193–199.
51. R. S. Adhikari, A. Kumar and A. Kumar, *J. Energ. Res.*, 1990, **14**, 737–744.
52. R. V. Dunkle, *International Developments in Heat Transfer, A.S.M.E, Proceedings of International Heat Transfer*, part V, University of Colorado, 1961, p. 895.
53. M. A. S. Malik, G. N. Tiwari, A. Kumar and M. S. Sodha, *Solar Distillation*, Pergamon Press, Oxford, UK, 1982, pp. 8–17.
54. A. T. Shawaqfeh and M. M. Farid, *Sol. Energ.*, 1995, **55**, 527–535.
55. S. Kumar and G. N. Tiwari, *Sol. Energ.*, 1996, **57**, 459–464.
56. S. Kumar and S. Sinha, *Energ. Convers. Manag.*, 1996, **37**(5), 629–636.
57. E. Sartori, *Sol. Energ.*, 1996, **56**(2), 199–206.
58. R. Tripathi and G. N. Tiwari, *Sol. Energ.*, 2006, **80**, 956–967.
59. A. K. Tiwari and G. N. Tiwari, *Desalination*, 2007, **207**, 184–204.
60. T. Saleh and M. A. R. Sarkar, in *8th International Symposium for Renewable Energy Education (ISREE-8)*, Orlando, University of Florida, USA, August 4–8, 2002 (http://www.doce-conferences.ufl.edu/isree8/papers.asp & http://www.fsec.ucf.edu/ed/iasee/isree/sarkar-dryer.pdf).

61. J. Mumba, *Renew. Energ.*, 1995, **6**(7), 855–862.
62. J. Mumba, *Energ. Convers. Manage.*, 1996, **37**(5), 615–621
63. I. Farkas, I. Seres and C. S. Meszaros, *Renew. Energ.*, 1999, **16**, 773–778.
64. M. A. Hossain, J. L. Woods and B. K. Bala, *Optimisation Renew. Energ.*, 2005, **30**, 729–742.
65. M. Tsamparlis, *Drying Technology*, 1990, **8**(2), 261–285.
66. M. A. Hossain and B. K. Bala, *Sol. Energ.*, 2007, **81**(1), 85–92.
67. B. K. Bala, M. R. A. Mondol, B. K. Biswas, B. L. D. Chowdury and S. Janjai, *Renew. Energ.*, 2003, **28**(2), 183–190.
68. P. Barnwal and A. Tiwari, *Int. J. Agr. Res.*, 2008, **3**(2), 110–120.

CHAPTER 8

Energy and Exergy Analysis

8.1 Energy Analysis

Energy drives human life and is crucial for continued human development. Energy is inevitable for human life and a secure and accessible supply of energy is crucial for the sustainability of modern societies. In recent years, with the advancement of civilization, energy has become the integral part of the human life for almost every activity *e.g.* domestic, transport, industrial, medical, *etc.* So, there is a need for energy security for sustainability of the growing world population. Continuation of the use of fossil fuels is set to face multiple challenges: depletion of fossil fuel reserves, global warming and other environmental concerns, geopolitical and military conflicts and, of late, continued and significant fuel price rises. These problems will create an unsustainable situation. Renewable energy is the solution to the growing energy challenges. Renewable energy resources such as solar, wind, biomass and wave and tidal energy are abundant, inexhaustible and environmentally friendly. Bentley[1] has overviewed the global oil and gas depletion and reported that conventional energy resources are being exhausted through their uncontrolled harnessing and limited resources. The world relies heavily on fossil fuels to meet its energy requirements – fossil fuels such as oil, gas and coal provide almost 80% of the global energy demands. On the other hand, presently renewable energy and nuclear power are, respectively, only contributing 13.5% and 6.5% of the total energy needs. The enormous amount of energy being consumed across the world is having adverse implications on the ecosystem of the planet.

Fossil fuels are inflicting enormous impacts on the environment. Climatic changes driven by human activities cause the production of greenhouse gas (GHG) emissions in particular, which has a direct impact on the environment. According to the World Health Organization (WHO) as many as 160,000 people die each year from the side-effects of climate change and the numbers could almost double by 2020. These side-effects range from malaria to malnutrition and diarrhoea that follow in the wake of floods, droughts and warmer temperatures.

RSC Energy Series No. 2
Fundamentals of Photovoltaic Modules and Their Applications
By G. N. Tiwari and Swapnil Dubey

Published by the Royal Society of Chemistry, www.rsc.org

With the exception of humans, every organism's total energy demand is its supply of energy in the form of food derived directly or indirectly from the Sun's energy. For humans the energy requirements are not just for heating, cooling, transport and manufacture of goods but also those related to agriculture. Solar energy is a renewable, environmentally friendly, pollution-free and freely available energy source on planet Earth. In this perspective, over the last two decades solar-energy systems have experienced rapid growth in areas receiving a high-level of solar radiation. However, energy analysis can be used to estimate the environmental impact of different activities for producing materials *i.e.* the more energy is requierd, the greater the environmental impact.

The photovoltaic (PV) system converts solar radiation into direct current (DC) electricity, which can be converted to alternating current (AC) electricity by using an inverter. The electrical efficiency of a PV module is reported to be around 10%, which is further reduced due to the involvement of a storage battery, a converter, distribution through wires and efficiency of electrical appliances *etc.*

The photovoltaic (PV) applications of the environmentally friendly solar energy source serve as one of the most promising alternatives to conserve the limiting conventional energy resources. The PV applications of solar energy can provide electricity, thermal energy, day lighting *etc.* depending on the mode of application *e.g.* distillation; air-heating collector; water-heating collector and greenhouse applications *etc.* Development work for PV applications has increased in recent years with the aim to conserve conventional energy sources.

PV systems are the one of the most important, reliable and environmental friendly technologies for energy conversion, with the potential to contribute significantly to a sustainable energy system. They also play an important role in the mitigation of CO_2 emissions. Most of the Indian Territory is blessed with a high potential of solar radiation, which is most suitable for the development of solar photovoltaic (PV) systems for power generation. In view of the above, PV technology has to meet the following two main criteria:

(i) Cost effectiveness;
(ii) The maximum net annual energy yield.

The net maximum annual energy yield for PV/T systems means the sum of annual electrical energy output of a PV system and annual thermal energy.

The total energy requirement for manufacturing a PV system, energy pay back time (EPBT) and also CO_2 emissions has been evaluated.[2] However, the following parameters were not considered:

(i) Support structure;
(ii) Battery replacement intervals;
(iii) Balance-of-system efficiency.

The system efficiency was considered to be 14% uniformly throughout the lifetime of a PV system. Krauter and Ruther[3] have evaluated only the energy requirement for manufacturing the PV system and CO_2 emissions without considering the above-mentioned parameters. Frankl *et al.*[4] considered the support structure for an open field mounted on a rooftop to evaluate the energy requirements for manufacturing PV systems. They considered the same life span for a battery and for a PV system.

The performance of PV/T systems is better under forced mode than in natural mode. For the forced mode of operation a pump/fan is required to circulate the fluid (air/water) in the system. So, PV/T systems cannot be operated in the absence of electricity. Hence, there is a need for electricity to be obtained from a PV, an environmentally friendly source, rather than from a conventional source.

When a PV module is integrated with the solar thermal system, then it can be a sustainable alternative in remote areas of under-developed and developing countries. The system is more cost effective and economical when DC equipment (pump/fan) is used, which eliminates the requirement of an inverter, a battery and a complicated circuit and wirings.

8.2 Energy Matrices

Developments in the design and manufacture of photovoltaic cells have, over the last few years, been very rapid such that they are now predicted to become a major renewable energy source. The embodied energy pay back is important for renewable technologies as their use makes no sense if the energy used in their manufacture is more than they can save in their lifetime. The embodied energy pay back period should always be one of the criteria used for comparing the viability of one renewable technology against another. The energy analysis of a PV module was conducted by Hunt[5] and it was reported that the energy pay back time (EPBT) of a PV module is 12 years. The results reported by Hunt[5] are also in general agreement with those of Kato *et al.*[6] for a crystalline silicon (c-Si) solar cell module. Aulich *et al.*[7] evaluated the EPBT for a crystalline silicon module and it was concluded that the EPBT is 8 years; in this case plastic materials were used for encapsulation for the Siemens C process. The energy pay back time for a crystalline silicon (c-Si) solar cell module under Indian climatic conditions for annual peak load duration is about 4 years.[8] Lewis and Keoleian[9] predicted the energy pay back time (EPBT) for an amorphous silicon (a-Si) solar cell module with efficiency of 5% as 7.4 years for the climatic conditions of Detroit, USA; the EPBT gets reduced to 4.1 years with the increase in the efficiency of the module to 9%. Srinivas *et al.*[10] reported that the energy pay back time for an amorphous silicon (a-Si) solar cell module reduces to 2.6 years after considering the gross energy requirement (GER) and the hidden energy. Battisti and Corrado[11] investigated the energy pay back time for a conventional multi-crystalline building integrated system, retrofitted on a tilted roof, located in Rome, Italy; the yearly global insolation on a horizontal plane was taken as 1530 $kWh\,m^{-2}\,y$. They concluded that the energy pay back time gets reduced from 3.3 years to 2.8 years.

8.2.1 Energy Pay Back Time (EPBT)

The EPBT depends on the energy spent to prepare the materials used for fabrication of the system and its components, *i.e.* embodied energy and the annual energy yield (output) obtained from such a system. To evaluate the embodied energy of various components of the system, the energy densities of different materials are required. The total time period required to recover the total energy spent to prepare the materials (embodied energy) is used for fabrication of the hybrid PV/T systems. It is the ratio of embodied energy and the annual energy output from the system, which can be expressed as

$$EPBT = \frac{Embodied\ Energy(E_{\text{in}})}{Annual\ Energy\ Output(E_{\text{out}})} \tag{8.1}$$

8.2.2 Energy Production Factor (EPF)

The EPF is used to predict the overall performance of the system. It is defined as the ratio of the output energy and the input energy or it can also be expressed as the inverse of EPBT.

$$\chi_{\text{a}} = \frac{E_{\text{out}}}{E_{\text{in}}} \tag{8.2a}$$

or

$$\chi_{\text{a}} = \frac{1}{T_{\text{epb}}} \tag{8.2b}$$

If $\chi_{\text{a}} \rightarrow 1$, for $T_{\text{epb}} = 1$ the system is worthwhile, otherwise it is not worthwhile from an energy point of view.

On whole life time basis, $\frac{\chi_a = T(\text{years})}{T_{\text{epb}}}$.

8.2.3 Life Cycle Conversion Efficiency (LCCE)

LCCE is the net energy productivity of the system with respect to the solar input (radiation) over the lifetime of the system (T years), given by

$$\phi(t) = \frac{E_{\text{out}} \times T - E_{\text{in}}}{E_{\text{sol}} \times T} \tag{8.3}$$

8.3 Embodied Energy

The concept of embodied energy is a relatively new area of environmental assessment that has started to be included in life cycle energy calculations of buildings. Embodied energy is defined as: "*the quantity of energy required by all of the activities associated with a production process, including the relative proportions consumed in all activities upstream to the acquisition of natural resources and the share of energy used in making equipment and in other supporting functions i.e. direct energy plus indirect energy.*"[12] Thus the aim of any embodied energy analysis is to quantify the amount of energy used to manufacture a material, product, component and element. This involves the assessment of the overall expenditure of energy required to extract the raw material, manufacture products and components, and to build and maintain the component element, whichever is being assessed. A secondary aim is to establish the embodied energy required to construct and maintain the item, component or building over the whole life cycle.

Like operational energy, embodied energy is an indicator of the level of energy consumption. Reducing energy consumption through better design has been a goal of designers for many years, but the embodied energy portion of this consumption has largely been ignored. There are several reasons for this omission, including no clear assessment methodology, lack of data, lack of understanding and a common belief that the embodied energy portion of an asset's energy consumption is insignificant. However, over recent years, the methodologies for assessment have improved, data reliability and access have increased and recent reports have indicated that the embodied energy portion may be as high as 20 times the annual operational energy of an office building.[12]

8.3.1 Embodied Energy Analysis

Embodied energy analysis involves identifying energy-consuming processes and calculating their contribution within the total product creation process. This usually involves several individual actions.

To be able to quantify the energy embodied in the construction of an asset, the quantities of materials must first be estimated through a process of desegregation and decomposition to a level of detail which allows for the separation of components into their principal materials. Energy intensities of each material can then be multiplied by the quantities of individual materials and the products aggregated to obtain the total for each material element. In addition to the embodied energy value, other environmental indicators can also be calculated, such as CO_2 emissions. This is the basis of life cycle cost analysis (LCA) work.

8.3.2 Embodied Energy Density

Embodied energy densities (intensities) are derived from energy analysis studies from various national and international sources. Among the difficulties encountered in using a wide variety of sources to verify values is the need to clarify definitions of system boundaries or whether the values are in terms of primary energy or delivered energy. To obtain an accurate and reliable database of embodied energy intensities for all materials used in water assets is an enormous task in itself and is a necessity for detailed comparisons of materials.

The main requirement of embodied energy calculations at the design stage is obtaining accurate and useable material quantities and then combining them with currently available embodied energy intensity values.[13] There are several methods used to carry out an energy analysis including:

- Process analysis – a commonly used procedure which involves identifying a system boundary around a particular process and determining the requirements for direct energy and indirect energy (through the provision of other goods and services crossing the system boundary and capital equipment, including buildings). The critical aspect of a process analysis is the definition of the system boundary. Considerable ranges of results are possible by the selection of different system boundaries. For a particular manufacturing process the system boundary may be the factory fence, or may include the requirements 'upstream' for the provision of natural resources within the system boundary.
- Input-output analysis – developed for economic analysis, used by government economists who have collected data for the compilation of input-output matrices, which trace economic flows of goods and services between sectors of an economy. In Australia, the Australian Bureau of Statistics publishes input-output matrices for the 109 economic sectors every five years. A row in the matrix lists all the sales of a sector and a column lists all the purchases (in dollars of input per \$100 of output). Thus, the energy intensity of a sector, expressed in gigajoules (10^9 joules) of energy per \$100 of sector output (GJ/\$100), can be derived by dividing purchases from individual energy supply sectors by the appropriate tariffs.
- Hybrid analysis – direct energy and quantities of goods and services are obtained for critical aspects of the process under consideration by process analysis. This could, for example, mean that for materials where the manufacture represents the main bulk of the overall environmental impact, the production processes are examined and quantified in detail by the process analysis method. The energy intensities of goods and services further upstream are then obtained using input-output analysis. With this approach the errors associated with input-output analysis are thus removed from a large proportion of the results, but the energy intensities derived only apply to materials and products manufactured by the specific process(es) audited and can not be applied globally.

Traditionally, input-output analyses have been used to derive the embodied energy intensities, as the resultant energy intensities were more complete than those derived from process analysis. Nevertheless, the accuracy of input-output analyses are inherently unreliable, but provide a common basis for comparison purposes. This method greatly reduces the errors associated with input-output analysis and is now considered the preferred method for embodied energy studies.

8.4 Embodied Energy of PV Module (Glass-to-Glass)

The total embodied energy required for making individual components of the PV module is shown in Figure 8.1, with their manufacturing energy needs to be evaluated. The specification and design data of a PV module (glass-to-glass) are given in Table 8.1.

Tiwari and Ghosal[14] reported that 2.4 kg of MG-Si and 2.3 kg of EG-Si are required for 0.729 kg of solar cells. Therefore 0.4032 kg of solar cells requires 1.327 kg of MG-Si and 1.273 kg of EG-Si. Table 8.2 gives the energy requirement in different processes for production of a PV module.

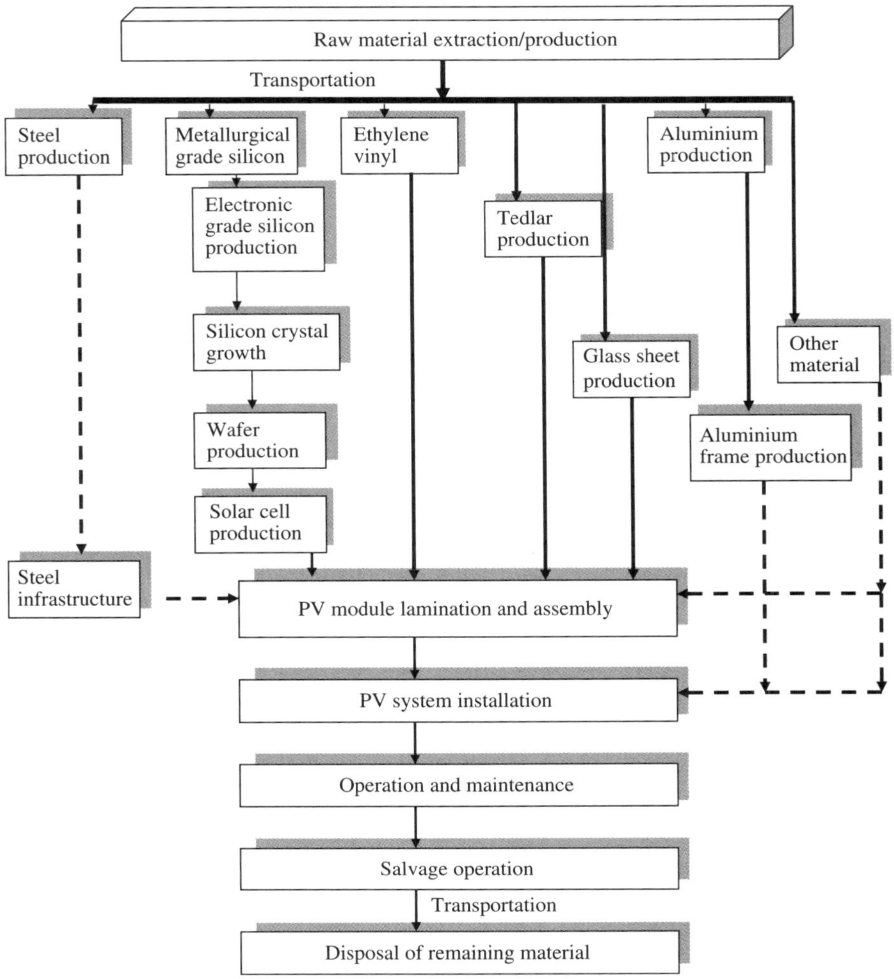

Figure 8.1 Processes to calculate embodied energy of PV module.

Table 8.1 Specification and design data of a PV module (glass-to-glass).

Size of PV module	$1.20 \times 0.55 \times 0.01\,m^3$
Effective area of a PV module	$0.60534\,m^2$
Area of a cell	$0.0139\,m^2$
Thickness of a solar cell	0.00035 m
Density of silicon	$2.3 \times 10^3\,kg\,m^{-3}$
Mass of a single cell	$11.2 \times 10^{-3}\,kg$
Fill factor of solar cell	0.72
Solar cell efficiency	15%
No. of cells in a PV module	36
Total mass of cells	0.4032 kg
Module efficiency	12%
Packing factor of PV module	83%

Table 8.2 Energy requirement in different processes for production of a PV module.

Process	*Energy requirement*	*Reference*
Silicon purification and processing		
(a) Metallurgical grade silicon (MG-Si) production from silicon dioxide (quartz, sand)	20 kWh per kg of MG-Si	Dones and Frischknecht,[48] Blakers and Weber[49] and Kato *et al.*[6]
(b) Electronic grade silicon (EG-Si) production from MG-Si	100 kWh per kg of EG-Si	
(c) Czochralski Silicon (Cz-Si) production from EG-Si	290 kWh per kg of EG-Si	
Solar cell fabrication	$120\,kWh$ per m^2 of silicon cell	Nawaz and Tiwari[50]
PV module assembly	$190\,kWh$ per m^2 of PV module	Tiwari and Ghosal[14]
Roof-top integrated PV system	$200\,kWh$ per m^2 of PV module	Nawaz and Tiwari[50]

The embodied energy required to produce a PV module (glass-to-glass) for different processes is computed as follows:

The embodied energy of a PV module (glass-to-glass) can be derived in the following steps:

(i) Silicon purification and processing
 (a) Production of 1.327 kg of MG-Si $= 1.327 \times 20 = 26.54$ kWh
 (b) Production of 1.273 kg of EG-Si $= 1.273 \times 100 = 127.30$ kWh
 (c) Production of 1.273 kg of EG-Si for Cz-Si $= 1.273 \times 210 = 267.33$ kWh

(ii) Solar cell fabrication = 120 × (0.60534 × 0.83) = 60.29 kWh
(iii) PV Module assembly = 190 × 0.66 = 125.40 kWh
(iv) assuming that the energy required for assembly of a glass-to-tedlar PV module and a glass-to-glass PV module are approximately the same).
(v) Installation/integration = 200 × 0.66 = 132 kWh

Hence, the total embodied energy required for installation/integration of a PV module (glass-to-glass) with PV/T systems = 738.86 ≈739 kWh.

8.5 Balance of System (BOS)

The PV module itself is called the system. Other components are called balance of system (BOS). It comprises wiring, electronic components, foundation, support structure, battery, installation, *etc*. For an open-field installation, the concrete, cement and steel are the main components used for the foundation and frame, which requires maximum energy. The energy requirement for an open-field installation is 500 kWh m^{-2} of panel. For a rooftop-integrated PV system, the energy requirement is reduced from 500 to 200 kWh m^{-2} of panel due to the absence of the foundation and structure for the frame

The requirements for the BOS (that is all components that are a part of the modules) will depend largely on the desired application. Solar PV technology is also used for producing grid quality power. In a grid-connected PV system, we consider here a DC-to-AC converter, cables and some module support materials will be needed. In an autonomous (decentralized) system a battery for energy storage will be required, since solar cells cannot store the energy themselves.

8.6 Analysis of Embodied Energy and EPBT of PV/T Solar Systems

The embodied energy and EPBT for the following PV/T solar systems have been discussed:

(i) Distillation system	(Figure 7.1)
(ii) Air collector	(Figure 7.13)
(iii) Solar water heater (SWH)	(Figure 7.30)
(iv) PV-integrated greenhouse dryer	(Figure 7.39)
(v) Conventional PV/T solar dryer	(Figure 7.39)

8.6.1 Hybrid PV/T Active Distillation System

The different materials used for the construction of an active distillation system (2 m^2 collector area and 1 m^2 still area) are flat-plate collectors, one PV module (glass-to-glass), a DC motor and mild steel stand *etc*. For energy analysis of a hybrid distillation system, the energy required for a flat-plate collector, a PV module, solar still *etc*. are necessary.

The embodied energy of the hybrid PV integrated distillation system is the sum of embodied energies of its different components (Figure 8.2). The list of different materials and embodied energies used are given in Table 8.3.

The total embodied energy used for the hybrid photovoltaic/thermal (PV/T) integrated distillation system = 3868.6 kWh.

The experiments were conducted on clear days during the year 2006–2007 at Solar Energy Park, IIT, Delhi. The total energy output is the sum of the net electrical output and the thermal output from the system.

Net thermal output is defined as

Annual thermal output = Total mass of annual yield (M_{yield}) $\times$ L
M_{yield} = mass of the distilled water, kg (Figure 7.2)
L = latent heat of vaporization, J kg^{-1}

For a water depth of 0.05 m in the still the annual thermal output is calculated as

$$1203.46\,\text{kg} \times 2390 \times 10^3\,\text{J kg}^{-1} = 798.96\,\text{kWh}$$

Net annual electrical output from a PV module

$$= \text{No load output} - \text{On load output}$$
$$= (0.8 \times I_{sc} \times V_{oc} - I_L \times V_L)$$
$$= 83.96\,\text{kWh}$$

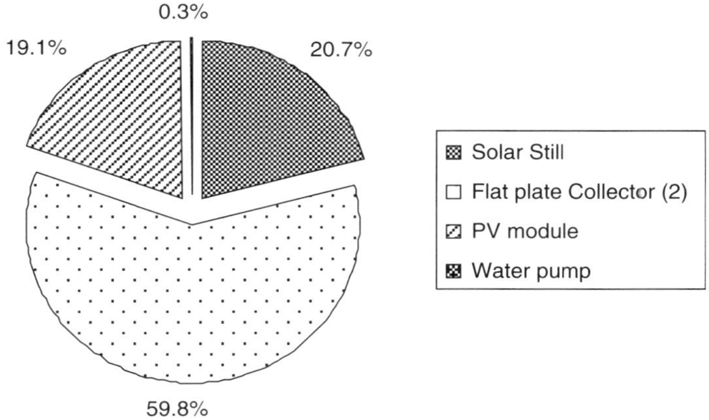

Figure 8.2 Break-up of embodied energy of different components of hybrid PV/T active solar still.

Net annual average equivalent thermal output

$$= \frac{\text{Net annual electrical output}}{0.38} = \frac{83.96}{0.38} = 220.9 \text{ kWh per year}$$

Therefore the total annual energy output

= equivalent thermal of annual net average electrical output + average annual thermal output
= (220.9 + 798.96) kWh
= 1019.91 kWh

Therefore, from eqn (8.1) we get:

$$\text{Energy Pay Back Time, EPBT} = \frac{3868.6}{1019.91} = 3.79 \text{ years (Energy point of view)}$$

$$\text{Energy Pay Back Time, EPBT} = \frac{3868.6}{178.96} = 21.6 \text{ years (Exergy point of view)}$$

8.6.2 PV/T Air Collector

Equations (8.1) have been used to calculate the EPBT for the hybrid PV module under study with and without BOS under standard test (solar intensity, $I(t) = 1000\ \text{W m}^{-2}$, air mass = 1.5 and ambient air temperature $T_a = 25\ ^\circ\text{C}$) and outdoor conditions at Solar Energy park, IIT, Delhi (Figure 8.3) for N = 1. The total embodied energy required is 1667.3 kWh and 1128.5 kWh with and without BOS, respectively. The results of EPBT have been summarized in Table 8.4. It is assumed that the hybrid PV module for standard test conditions for a PV module (without extraction of thermal energy) includes the thermal energy output from the outdoor condition. It is clear that the EPBT under standard test conditions with BOS is about 5.23 years; it gets reduced to 3.65 years in the case of the hybrid system, which allows for extraction of the thermal energy. The EPBT without BOS is reduced by 1.3 years due to the reduced value of the numerator in eqn (8.1). These parameters have a significant effect on the EPBT under outdoor conditions as can be seen from Table 8.4. The EPBT under outdoor conditions is more than the value obtained under standard conditions due to a reduction in the value of the denominator in eqn (8.1). It may be noted that the effect of air velocity in the duct has a marginal effect on EPBT. Hence, one can conclude that air velocity with one fan is near optimum velocity for the present set of experiments.

The EPBT under outdoor conditions (≈150 Wp) obtained in the present study is in very close agreement with the EPBT obtained by Kato *et al.*[6] for 3 kWp residential PV power system. Further EPBT of a hybrid system can be reduced by using the system in high solar intensity condition (irradiation) and by reducing the embodied energy due to development of high technology with improved electrical efficiency as mentioned by Alsema and Nieuwaar[2] in 2000.

Table 8.3 Break-up of embodied energy of different components of hybrid PV/T active solar still.

Components	*Items*	*Quantity*	*Total weight (kg)*	*Embodied energy ($MJ\,kg^{-}$)*	*Total embodied energy*	
					MJ	*kWh*
Solar Still	GRP body	1	21.17	92.3	1954.0	542.8
	Glass cover 4 mm	1	1.16	40060 MJ m^{-3}	185.9	51.6
	M S clamping frame	1	5	34.2	171.1	47.5
	M S clamp	8	2	34.2	68.4	19.0
	Mild steel stand	1	14/20	34.2	478	133
	Inlet/outlet nozzle	2	0.100	44.1	4.4	1.2
	Gaskets 8.9 m	1	2.1	11.83	24.8	6.9
Sub total						**802**
Flat-plate collector Quantity 2	Copper riser 1/2″	20 × 1.8 = 36 m	8.2	81.0	664.2	184.5
	Header 1″	4 × 1.15 = 4.6 m	3.8	81.0	307.8	85.5
	Al box	2	10	199.0	1990.0	552.0
	Cu sheet	2	11	132.7	1460	405.6
	Glass cover toughened 4 mm	2 (3.75 m^2)	0.01464 m^3	66,020 MJ m^{-3}	966.5	268.3
	Glass wool	13 m^2	0.064 m^3	139 MJ m^{-3}	8.89	2.5
	Nuts/bolts/screws	32	1	31.06	31.06	8.6
	Union/elbow	8	1.5	46.8	70.2	19.5
	Nozzle/flange	8	1	62.1	62.1	17.3
	Mild steel stand	1	40	34.2	1368	380

	Paint	1L	1L	90.4	90.4	25.1
	Rubber gasket	18 m	4.2	11.83	49.7	13.8
	G I pipes 1/2″		9.5	44.1	418.9	116.4
	Al frame 1″	12 m	2.5	170	425	118
	Al sheet 24 gauge		2.5	170	425	118
Sub total						**2315.1**
PV module	Glass to glass	1	$0.605\,m^2$	$3612\,m^{-2}$	2185.2	607
	BOS			475.2	475.2	132
Sub total						**739**
Water pump	Copper wire		0.150	110.19	16.5	4.6
	Copper commuter	2	0.04	70.6	2.8	0.78
	Si-steel armature	1	0.05	*	*	*
	Wire insulation	2	0.01	*	*	*
	Motor body (SS)	1	0.100	36.1	3.61	1.0
	Casing (brass)	1	0.300	62.0	18.6	5.2
	Bearings	2	0.030	*	*	*
	Steel shaft	1	0.050	12.5	0.625	0.17
	Impellers (plastic)	1	*	*	*	*
	Nuts/screws/flange		0.100	31.06	3.1	0.86
Sub total						**12.61**
Total embodied energy of hybrid active still						**3868.6**

Figure 8.3 Hybrid PV/T air collectors connected in series.

Table 8.4 The values of the energy pay back time (EPBT) under different conditions. For N = 1.

Condition	*EPBT with standard test conditions*		*EPBT with outdoor condition*	
	Single fan	*Two fan*	*Single Fan*	*Two fan*
With BOS				
1. PV module	5.23	5.23	12.7	12.56
2. Hybrid PV module	3.65	3.53	05.7	05.69
Without BOS				
1. PV module	3.86	3.86	9.37	9.27
2. Hybrid PV module	2.69	2.60	4.21	4.20

8.6.3 Hybrid PV/T Solar Water Heater

The different materials used for the construction of a hybrid water-heating system are flat-plate collectors (2 m^2), a storage tank, one PV module (glass-to-glass), a DC motor and mild steel stand *etc*. For energy analysis of a hybrid water heater, the annual energy output and embodied energy of all the components of the system, flat-plate collector, storage tank, PV module, *etc*. are required (Figure 8.4).

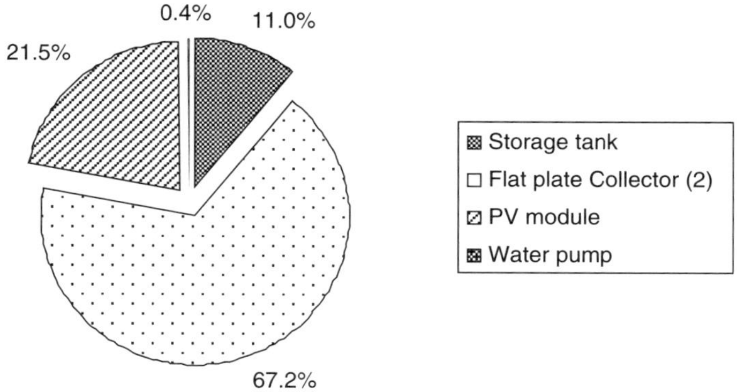

Figure 8.4 Break-up of embodied energy of different components of hybrid PV/T solar water heater.

The total embodied energy used for the hybrid photovoltaic/thermal (PV/T) integrated distillation system = 3443.9 kWh (Table 8.5).[15]

The experiments were conducted on clear days during the year 2007 at Solar Energy Park, IIT, Delhi. The total energy output per year is the sum of the net electrical output and thermal output from the system.

If we consider without withdrawal from the tank then the net thermal output is defined as

$$\text{Thermal output} = M_w C_p (T_w - T_a)$$

M_w = mass of water in tank, kg
C_p = specific heat of water, J kg^{-1} K^{-1}
T_w, T_a = Tank water and ambient temperature, °C

Net annual electrical output from a PV module

= No load output – On load output

Annual overall thermal energy output

= equivalent thermal of annual net average electrical output + average annual thermal output
= 2887.9 kWh

Therefore, from eqn (8.1)we get

$$\text{Energy Pay Back Time, EPBT} = \frac{3443.9}{2877.9} = 1.2 \text{ years (Energy point of view)}$$

$$\text{Energy Pay Back Time, EPBT} = \frac{3443.9}{264.1} = 13 \text{ years (Exergy point of view)}$$

Table 8.5 Break-up of embodied energy of different components of a hybrid PV/T solar water heater.

Components	*Items*	*Quantity*	*Total weight(kg)*	*Embodied energy($MJ\ kg^{-1}$)*	*Total embodied energy*	
					MJ	*kWh*
Storage tank	Storage tank	1	18.0	36.1	649.8	180.5
	Glass wool	23 m^2	0.1 m^3	139 MJ m^{-3}	13.9	3.86
	Al sheet	1	1.2	170	204	56.7
	Mild-steel stand	1	14	34.2	478.8	132.7
	Brass water tap	1	0.2	62.7	12.54	3.48
Sub total						**377.2**
Flat-plate collector		2				2315.1
PV module		1				739
Water pump		1				12.61
Total embodied energy of hybrid solar water heater						**3443.9**

8.6.4 Hybrid PV-integrated Greenhouse Dryer

The different materials used for the construction of a hybrid photovoltaic/thermal (PV/T) integrated greenhouse dryer (2.50 m × 2.60 m floor area; 1.80 m central height and 1.05 m side walls height from ground) are aluminium sections, two PV modules (glass-to-glass), a DC fan and a UV-stabilized polyethylene sheet covering *etc*. For energy analysis of a hybrid photovoltaic/thermal (PV/T) integrated greenhouse dryer, the energy required for the PV module, DC fan, *etc*. is necessary.

The embodied energy of the hybrid PV-integrated greenhouse dryer is the sum of embodied energies of its different components *e.g.* PV module (glass-to-glass), aluminium sections, ultraviolet (UV) plastic sheet, direct current (DC) fan, wire mesh trays and fittings *etc*. The different materials and embodied energy used for a hybrid PV-integrated greenhouse dryer are given in Table 8.6

The total embodied energy used for the hybrid photovoltaic/thermal (PV/T) integrated greenhouse dryer = 5555.13 kWh.

Table 8.6 Embodied energy calculation data for a hybrid PV/T-integrated greenhouse dryer.

S. No.	*Item*			*Weight (kg)*	*Embodied energy (kWh kg^{-1})*	*Embodied energy (kWh)*
1.	Aluminium sections:					
	(i)	$0.0381 \times 0.003\ m^2$ angle		23.338	55.28	1290.12
	(ii)	$0.0254 \times 0.003\ m^2$ angle		25.985	55.28	1436.45
	(iii)	$0.0381 \times 0.0381 \times 0.003\ m^3$ tee		18.179	55.28	1004.94
	(iv)	$0.012 \times 0.012 \times 0.001\ m^3$ channel/U-Clip		0.395	55.28	21.84
	(v)	$0.0254 \times 0.0055\ m^2$ flat		8.060	55.28	445.56
	(vi)	$0.02 \times 0.003\ m^2$ flat		3.203	55.28	177.06
2.	UV plastic/PVC sheet			10.955	25.64	280.89
3.	Wire mesh			11.788	8.89	104.80
4.	PV module (glass to glass)			1 No.	739.000	739.00
5.	DC fan (exhaust fan) (1 No.)					
	(i)	Aluminium		0.390	55.28	21.56
	(ii)	Iron		0.220	8.89	1.96
	(iii)	Plastic		0.120	19.44	2.33
	(iv)	Copper wire		0.050	19.61	0.98
6.	Fittings					
	(i)	Hinges/*kabza*	Aluminium	0.200	55.28	11.06
	(ii)	Kundi (Door lock)		0.025	55.28	1.38
	(iii)	Hooks		0.100	55.28	5.53
	(iv)	Nut/bolt with washer, steel screws and rivets	Galvanized steel	1.000	9.67	9.67
Total embodied energy of hybrid PV/T integrated greenhouse dryer						5555.13

Energy output per year is the sum of the net electrical output and thermal output from the dryer.

Net average electrical output from a PV module

$$
\begin{aligned}
&= \text{No load output} - \text{On load output} \\
&= (0.8 \times I_{sc} \times V_{oc} - I_L \times V_L) \\
&= (42\text{–}8)\,\text{W} \\
&= 34\,\text{W}.
\end{aligned}
$$

Net annual average electrical output

$$
\begin{aligned}
&= \text{Net average electrical output (W)} \times \text{peak sunshine hours per day (h)} \times \\
&\quad \text{number of clear sunny days in a year} \times 10^{-3}\,\text{kWh year}^{-1} \\
&= 34 \times 7 \times 300 \times 10^{-3}\,\text{kWh year}^{-1} \\
&= 71.4\,\text{kWh year}^{-1}.
\end{aligned}
$$

Net annual average equivalent thermal output

$$= \frac{\text{Net annual average electrical output}}{0.38} = \frac{71.4}{0.38} = 187.9\ \text{kWh per year}$$

The dryer is of 100 kg capacity. The experiments were conducted in April, 2007, to dry Thompson seedless grapes (mutant: Sonaka). The grapes were purchased from a local market , manually sorted, washed with fresh groundwater to remove undesired materials *e.g.* dust and foreign materials and the surface water from grapes was removed by using cotton cloths. The drying time was 15 clear sunny days of 7 hours (9:00 to 16:00 hrs) by using a hybrid PV/T-integrated greenhouse dryer.

The total moisture evaporated from Thompson seedless grapes (mutants: Sonaka) by greenhouse drying was 69.2%. The remaining moisture was evaporated during the night when it was kept under a plastic covering.

So, 49.1 kg moisture was removed in 15 days from drying 100 kg grapes. Therefore the average annual thermal output of the dryer

$$
\begin{aligned}
&= \left(\text{moisture evaporated (kg)} \times \text{latent heat of evaporation (J/kg)} \times \frac{300}{15}\right) \\
&\quad \times \left(\frac{1}{3.6 \times 10^6}\right) \\
&= \left(69.2 \times 2.26 \times 10^6 \times \frac{300}{15}\right) \times \left(\frac{1}{3.6 \times 10^6}\right) \text{kWh per year} \\
&= 868.84\,\text{kWh per year.}
\end{aligned}
$$

Therefore the total annual energy output

= equivalent thermal of annual net average electrical output + average annual thermal output
= (187.9 + 868.84)kWh
= 1056.74 kWh.

Therefore, from eqn (8.1) we get

$$\text{Energy Pay Back Time, EPBT} = \frac{5555.13}{1056.74} = 5.26 \text{ years (Energy point of view)}$$

$$\text{Energy Pay Back Time, EPBT} = \frac{5555.13}{162.6} = 34.6 \text{ years (Exergy point of view)}$$

Since the life of a greenhouse dryer made up of aluminium sections along with a PV module can be considered to be more than 30 to 40 years, the EPBT for the present PV/T greenhouse dryer is much less than the expected life of the dryer.

8.6.5 Hybrid Conventional PV/T Solar Dryer

The different materials and embodied energy used for a hybrid conventional PV/T solar dryer are given in Table 8.7.

The total embodied energy used for a hybrid conventional PV/T solar dryer

= 1257.39 kWh.

Table 8.7 Total embodied energy of a hybrid conventional PV/T solar dryer.

S. No.	*Item*	*Weight* (*kg*)	*Energy embodied* (*kWh kg*$^{-1}$)	*Total energy embodied* (*kWh*)
1.	Glass	14.00	7.28	101.92
2.	Steel	10.00	8.89	88.9
3.	Paint	1.00	25.11	25.11
4.	Rubber gasket and polyethylene sheet	1.00	25.64	25.64
5.	Fittings (nut/bolt with washer, steel screws and rivets *etc.*)	1.00	8.89	8.89
6.	Aluminium sheet	10.00	55.28	552.8
7.	Wood material	20.00	2.89	57.8
8.	PV module (glass-to-glass; size: 0.60 × 0.55 × 0.01 m)	1 No.	369.5	369.5
9.	DC fan	1 No.	26.83	26.83
Total embodied energy of hybrid conventional PV/T solar dryer				**1257.39**

The annual useful energy for a hybrid conventional PV/T solar dryer

$= \eta \times \alpha_c \tau \times \bar{I} \times A_c \times N \times n \times 10^{-3}$ kWh.

For dryer thermal efficiency $(\eta) = 0.60$, $\alpha_c \tau = 0.40 \times 0.8 = 0.32$, $A_c = 2.0\,\text{m}^2$, annual average insolation $(\bar{I}) = 500\,\text{W m}^{-2}$, $N = 5$ sunshine hours and $n = 300$ clear days per year.

The annual useful thermal energy for a hybrid conventional PV/T solar dryer

$= 0.6 \times 0.32 \times 500 \times 2.0 \times 5 \times 250 \times 10^{-3}$ kWh
$= 288$ kWh.

Net average electrical output from a PV module

= No load output – On load output
$= (0.8 \times I_{sc} \times V_{oc} – I_L \times V_L)$
$= (29 - 8)$ W
$= 21$ W.

Net annual average electrical output

= Net average electrical output (W) × peak sunshine hours per day (h) × number of clear sunny days in a year × 10^{-3} kWh year^{-1}
$= 21 \times 7 \times 300 \times 10^{-3}$ kWh year
$= 44.1$ kWh year^{-1}.

Net annual average equivalent thermal output

$$= \frac{\text{Net annual average electrical output}}{0.38} = \frac{44.1}{0.38} = 116.1 \text{ kWh per year}$$

So, from eqn (8.1) we get

$$\text{Energy pay back time} = EPBT = \frac{\textit{Embodied Energy}}{\textit{Annual Energy Output}}$$

$$\text{or } EPBT = \frac{1257.39\,\text{kWh}}{(288 + 116.1)\,\text{kWh per year}} = 3.11 \text{ years (Energy point of view)}$$

$$EPBT = \frac{1257.39\,\text{kWh}}{66.7\,\text{kWh}} = 18.1 \text{ years (Exergy point of view)}$$

If the drying chamber efficiency (40%) is considered, then the annual useful energy for a hybrid conventional PV/T solar dryer $= 0.4 \times 288 = 115.2$ kWh.

$$\text{Then } EPBT = \frac{1257.39\,\text{kWh}}{155.2\,\text{kWh per year}} = 10.9 \text{ years}$$

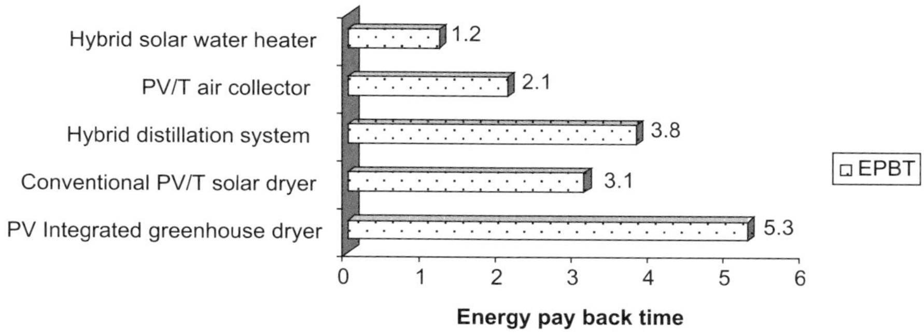

Figure 8.5 Energy pay back time (EPBT) for different PV/T systems.

Since the life of a hybrid conventional PV/T solar dryer can be considered more than 30 years, the EPBT for the present hybrid conventional PV/T solar dryer is much less than the expected life of the dryer.

The comparative figure of Energy Pay Back Time (EPBT) for different PV/T systems is shown in Figure 8.5.

8.7 Energy Pay-back Periods of Roof-mounted Photovoltaic Cells

The energy pay back time of photovoltaic (PV) cells has been a contentious issue for more than a decade. Some studies claim that the joule content of the energy and materials that were put into the process of making the PV cell will be equalled by the joule content of the electrical output of the cell within a few years of operation. Other studies claim that the useful electrical energy output of the PV cell will never exceed the total amount of useful energy contained within all the inputs of the manufacturing, installation and lifetime operating processes of the PV cell. These studies are often loosely referred to as measuring the energy 'pay back' of the PV cell.[16] In order to attempt to draw some conclusions as to the actual energy pay back time of PV cells, several previous studies were reviewed. A summary of their findings is presented in Table 8.8. These studies are all based on different assumptions, and evaluate different types of modules, and therefore cannot be directly compared. The details of abbreviations used in the table are:

sc-Si – Single-crystalline silicon
mc-Si – Multi-crystalline silicon
a-Si – Amorphous silicon

It can be observed that an energy pay back time (EPBT) increases as the design and structure of the system become more complicated. Hence, it is

Table 8.8 Summary of energy pay back periods of roof-mounted photovoltaic cells found by reviewed literature.

Author	*Low estimate (years)*	*Low estimate key assumptions*	*High estimate (years)*	*High estimate key assumptions*
Schaefer and Hagedorn[51]	2.6	25 MWp a-Si module	7.25	2.5 MWp sc-Si module
Lewis and Keoleian[9]	1.4	36.7 kWh yr^{-1} frameless a-Si module located in Boulder, CO	13	22.3 kWh yr^{-1} a-Si module with frame located in Detroit, MI
Kato *et al.*[52]	4	Sc-Si module. Excludes all processes required for micro-electronics industries.	15.5	sc-Si module. Includes all processes required for micro-electronics industries.
Kato *et al.*[6]	1.1	a-Si module. Excludes all processes required for micro-electronics industries.	11.8	sc-Si module. Includes all processes required for micro-electronics industries.
Alsema[53]	2.5	Roof-mounted thin-film module	3.1	Roof-mounted mc-Si module
Alsema and Nieuwlaar[2]	2.6	Thin-film module	3.2	mc-Si module
Kato *et al.*[54]	1.1	100 MW yr^{-1} a-Si, modules including BOS	2.4	10 MW yr^{-1} mc-Si module including BOS
Knapp and Jester[55]	2.2	Production thin-film module	12.1	Pre-pilot thin-film module
Pearce and Lau[56]	1.6	a-Si module	2.8	sc-Si module
Jester[57]	3.2	150 W peak power mc-Si module	5.2	55 W peak power mc-Si module
Meijer *et al.*[58]	3.5	mc-Si module	6.3	Thin-film module
Battisti and Corrado[59]	1.7	Hybrid photovoltaic/thermal module	3.8	Tilted roof, retro-fitted mc-Si module
Jungbluth[60]	4	mc-Si module if emissions are not taken into account	25.5	sc-Si module if emissions are taken into account
Peharz and Dimroth[61]	0.7	FLATCON (Fresnel-lens all-glass tandem-cell concentrator) module – 1900 kWh (m^{-2} yr) insolation	1.3	FLATCON (Fresnel-lens all-glass tandem-cell concentrator) module – 1000 kWh (m¯2 yr) insolation
Raugei *et al.*[62]	1.9	CdTe module including BOS	5.1	mc-Si module including BOS
Tripanagnostopoulos *et al.*[63]	1	Glazed hybrid photovoltaic/thermal	4.1	Unglazed hybrid photovoltaic/thermal

important to note that the following points should be taken into consideration before manufacturing any PV/T system:

(i) Materials with less energy density should be used for construction;
(ii) Materials should have longer life;
(iii) Maintenance should be minimum;
(iv) There should be a maximum use of the system per year.

However, increasing an annual energy saving can further reduce the EPBT of solar systems, which can be increased by increasing insolation, increasing sunshine hours and reducing overall heat loss *etc*.

8.8 Exergy Analysis

A deeper analysis reveals that in real processes energy is not destroyed, but rather transformed into other forms, less suitable for feeding and driving real processes. Hence, besides energy, another physical quantity should be introduced to characterize the quality of the kind of energy under consideration. The ability to perform useful work in a natural environment has been suggested and investigated as a measure of energy quality by many researchers.[17] The term exergy was proposed in the 1950s, and has since been broadly accepted. This marked the beginning of a new branch of thermodynamics, which developed mainly in Europe in the 1950s and later worldwide. The energy analysis provides quantitative study of losses in different sections of the system. It is the general base for comparing the performance of different designs of most of the solar systems in different climatic conditions based on thermal efficiency. However, there are certain limitations on energy-based analysis which are defined as:

- It does not provide a measure of how nearly the system performance approaches ideal
- Energy losses do not represent the true losses that exist to generate the desired product;
- Temperatures of supply, recoverable energy source and surroundings;
- Storage duration.

Exergy analysis is based upon the second law of thermodynamics, which stipulates that all macroscopic processes are irreversible. Every such irreversible process entails a non-recoverable loss of exergy, expressed as the product of the ambient temperature and the entropy generated (the sum of the values of the entropy increase for all the bodies taking part in the process). Some of the components of entropy generation can be negative, but the sum is always positive.[18]

Energy vs. exergy: As water drops over the falls, its potential energy is converted via kinetic energy to thermal energy, but on the whole it is conserved. Still, we can see that something – its ease of use in performing work – is being lost here. This lost quantity is called exergy.

The fact is that quality of energy is more important than quantity. The exergy analysis of a solar thermal system enables us to identify the sources of irreversibility and inefficiencies with the aim of reducing the losses and achieving the maximum resource and capital savings. This can be achieved by careful selection of the technology and optimization of design of the system and components. The alternative means of comparing the thermal system meaningfully is exergy analysis.

"*Exergy is the property of a system which gives the maximum amount of useful work obtained from the system when it comes into equilibrium with a reference to the environment.*"

Every irreversible phenomenon causes exergy losses leading to a reduction of the useful effects of the process or to an increased consumption of energy from whatever source the energy was derived. The chief aim of exergy analysis is to detect and to evaluate quantitatively the causes of the thermodynamic imperfection of the process under consideration. Exergy analysis can, therefore, indicate the possibilities of thermodynamic improvement of the process under consideration, but only an economical analysis can decide the expediency of a possible improvement.[19] According to the second law of thermodynamics, heat cannot be completely converted to work in a cyclic manner and some part of the heat supplied by the system is necessarily rejected to the sink. The maximum part of the input thermal energy which can be converted to work is called the available energy and that rejected to the surroundings is called unavailable energy. Therefore

$$\begin{aligned}\text{Heat supplied (energy)} &= \text{Available energy (exergy)}\\ &\quad + \text{unavailable energy(anergy)}\end{aligned} \tag{8.4}$$

An exergy-based performance analysis of a system is based on the second law of thermodynamics that overcomes the limitations of an energy-based analysis.[20] The exergy transfer can be associated with mass, with work interaction and with heat interaction in renewable energy systems.[21] However, for a system using solar energy, the exergy transfer takes place with mass flow and heat interaction. In recent years, the use of exergy analysis in system design, analyses and optimization of thermal systems has been recognized by many engineers/researchers as a powerful tool for evaluation of the thermodynamic systems.[22] The comparative difference between energy and exergy has been shown in Table 8.9.

Exergetic analysis usually predicts the thermodynamic performance of an energy system and provides a clearer view of energy losses in the system by providing qualitative and quantitative study of different losses. Thus, the exergy analyses can predict whether or not and by how much it is possible to design more efficient thermal systems by reducing the sources of existing inefficiencies. Dincer and Sahin[22] presented a new model for thermodynamic

Table 8.9 Comparative difference between energy and exergy.

Energy	*Exergy*
It is dependent on the parameters of matter or energy flow only, and independent of the environment parameters.	It is dependent both on the parameters of matter or energy flow and on the environment parameters.
It is governed by the first law of thermodynamics for all the processes.	It is governed by the first law of thermodynamics for reversible processes only (in irreversible processes it is destroyed partly or completely).
It is limited by the second law of thermodynamics for all processes (incl. reversible ones).	It is not limited for reversible processes by the second law of thermodynamics.
It is always conserved in a process, so can neither be destroyed nor produced.	It is always conserved in a reversible process, but is always consumed in an irreversible process
It is a measure of quantity only.	It is a measure of quantity and quality due to entropy.
It is dependent on the parameters of matter or energy flow only, and independent of the environment parameters.	It is dependent both on the parameters of matter or energy flow and on the environment parameters.

analysis of a drying process of moist solids subject to air drying in terms of exergy and reported that exergy analysis was useful for thermodynamic assessments of drying processes and also for providing insights into their performances and efficiencies. Rosen[23] carried out the exergy of stratified thermal energy storages using different temperature distribution models and concluded that the use of stratification can therefore increase the exergy storage capacity of thermal storage.

In general, the exergy of any matter is defined as the maximum ability of this matter to carry out work in relation to the given human environment.[24,25] It is generally not conserved as energy but destroyed in the system. The exergy destruction is the measure of irreversibility that is the source of performance loss. Therefore, an exergetic analysis should be carried out for assessing the magnitude of exergy destruction by identifying the location, magnitude and the source of thermodynamic inefficiencies in a thermal system.

8.9 Importance of Exergy

Dincer[26] reported the linkages between energy and exergy, exergy and the environment, energy and sustainable development, energy policy making and exergy in detail and provided the following key points to highlight the importance of exergy and its essential utilization in numerous ways:

(a) It is a primary tool in best addressing the impact of energy resource utilization on the environment;

(b) It is an effective method of using the conservation of mass and conservation of energy principles together with the second law of thermodynamics for the design and analysis of energy systems;
(c) It is a suitable technique for furthering the goal of more efficient energy resource use, for exergy enables the waste and losses in the system to be located and determined;
(d) It is an efficient technique revealing whether or not and by how much it is possible to design more efficient energy systems by reducing the inefficiencies in existing systems;
(e) It is a key component in obtaining sustainable development.

Studies on the exergetic evaluation of various energy systems, namely solar collector, photovoltaic, hybrid, wind, geothermal, biomass, *etc.*, by various authors is available in the literature.

García-Rodríguez and Gómez-Camacho[27] had done exergy analysis of a solar multi-effect distillation system (SOL-14 plant) located in Almeria Solar Research Center in south-eastern Spain. Similarly, Sow *et al.*[28] had done energetic and exergetic analysis of a triple-effect distiller driven by solar energy and obtained exergetic efficiencies between 19 and 26% for the triple-effect system, 17 and 20% for the double-effect system and less than 4% for the single-effect system. This work quantifies power consumption per unit mass of pure water. The exergetic analysis has been widely used in the design, simulation and performance evaluation of energy systems reported by Hepbasli.[29] He has given a key review on exergetic analysis and an assessment of renewable energy resources for sustainable future for the solar collector, solar cooker, solar drying, solar desalination, solar thermal power plants and the hybrid PV/thermal solar collector. Hepbasli and Akdemir[30] have carried out energy and exergy analysis of a ground source (geothermal) heat-pump system. Fujisawa and Tani[31] have carried out an annual exergy-based evaluation of a PV/T hybrid collector and predicted to achieve higher output density than in a unit PV module or liquid FPC. The exergy analyses of the drying of various foods have been reported in the literature.[22,32–35] Akpinar *et al.*[33] studied the thermodynamic (first and second law) analyses of a single-layer drying process of pumpkin slices *via* a cyclone-type dryer and reported that the exergy losses went up with the increase of the energy utilization in both the trays and the drying chamber. Rosen and Dincer[36] developed an original methodology for the analysis of thermal systems and processes that is based on four quantities: exergy, cost, energy and mass. It was referred to as EXCEM analysis. The relations between exergy loss and capital cost and those between exergy and environmental impact were also investigated.

An exergy analysis (or second-law analysis) has proven to be a powerful tool in the simulation of thermodynamic analyses of energy systems. In other words, it has been widely used in the design, simulation and performance evaluation of energy systems. Although numerous studies have been conducted on the performance evaluation of SWH systems by using the energy analysis method in the literature, very few papers have appeared on exergy analysis of these systems.

Earlier studies on domestic-scale solar water heaters were based on the first thermodynamic law. In fact, as we know, it is the quality of energy that is important not the quantity of energy. So, it is necessary to evaluate domestic-scale solar water heaters from the point of view of the second thermodynamic law.

The exergy analysis method is employed to detect and to evaluate quantitatively the causes of the thermodynamic imperfection of the process under consideration. It can, therefore, indicate the possibilities of thermodynamic improvement of the process under consideration.[22,36] Exergy analysis is conducted with the aim of providing some methods to save costs and to keep the efficiency of domestic-scale solar water heaters to the desired extent. The study shows that for an ordinary thermally insulated domestic-scale solar water heater the exergy losses are mainly due to imperfectly thermal insulation in the collector and storage barrel. Exergy losses due to irreversibility in the collector are mainly caused by irreversibility of heat transfer and in the storage barrel is dominated by the mixing of water at different temperatures. Exergy losses due to irreversibility in the collector acts as the driving force for the system while exergy losses due to irreversibility in the storage barrel are of little contribution. Exergy is also a measure of the maximum useful work that can be done by a system interacting with an environment, which is at a constant pressure P_o and a temperature T_o. The simplest case to consider is that of a reservoir with a heat source of infinite capacity and invariable temperature T_o. It has been considered that the maximum efficiency of heat withdrawal from a reservoir that can be converted into work is the Carnot efficiency.[37,38] Xiaowu and Bena[39] performed an exergy analysis of a domestic-scale water heater and investigated the effects of collector design parameters on the collector exergy efficiency. They reported that large exergy losses occurred in the storage barrel and, to improve the exergy efficiency of domestic-scale water heaters, a judicious choice of width of plate and layer number of cover was necessary. Ucar and Inalli[40] studied the exergoeconomic analysis and optimization of a solar-assisted heating system for residential buildings in Elazig, Turkey. They obtained the optimal sizes of the collector area and storage volume in a seasonal storage solar heating system using the exergoeconomic optimization technique. An exergetic performance of SWH based on exergy efficiency correlation has been studied by Gunerhan and Hepbasli[41] and they found that exergy efficiency values range from 2.02% to 3.37% and 3.27% to 4.39% at a dead (reference) state temperature of 32.77 °C, for the solar collector and the entire SWH system, respectively. Hepbasli[42] studied the exergetic performance of solar-assisted domestic hot-water tank integrated ground source heat pump (GSHP) systems for residences in Turkey. He has found that the exergy efficiency values are found to be 72.33% for the GSHP unit, 14.53% for the solar domestic hot-water system and 44.06% for the whole system at dead (reference) state values for 19 °C and 101.325 kPa. The energy and exergy analysis of different configurations of hybrid PV/T water collectors is conducted by Dubey and Tiwari[43] and found that the collectors fully covered by a PV module combine the production of hot water in addition to electricity generation and it is beneficial in terms of exergy, thermal energy and electrical energy gain.

8.10 Exergy of a Process

The maximum work available (W_{max}) from the heat source at T_1 (in K) and sink at (ambient) temperature T_0 (K) is expressed as

$$W_{max} = exergy = \left(1 - \frac{T_0}{T_1}\right) \times Q_1 \qquad (8.5a)$$

where Q_1 is the heat energy supplied at T_1.

For a given ambient temperature T_0, an increase in source temperature T_1 gives more exergy and less anergy for the same heat transfer/energy input. The exergy of a system decreases as the process loses its quality.

$$\text{The unavailable part of the energy} = T_o \Delta s \qquad (8.5b)$$

where, Δs is the change in the entropy of the system during the change in process.

Example 8.1

Calculate the maximum work available (W_{max}) from the heat source at $T_1 = 40\,°C$, $60\,°C$ and $80\,°C$ and ambient temperature $= 20\,°C$ when $Q_1 = 150\,kWh$.

Solution

Using eqn (8.5a) for 40 °C, we have

$$W_{max} = \left(1 - \frac{20 + 273}{40 + 273}\right) \times 150 = 9.58\,kWh$$

Similarly, for 60 °C and 80 °C

$$W_{max} = 18\,kWh \text{ and}$$
$$W_{max} = 25.5\,kWh$$

It is concluded that the maximum work is available at a higher source temperature when the sink temperature is constant.

8.10.1 Solar Radiation Exergy

Exergy is the property of a matter and not of any phenomenon. The matter may be either a substance (which has a rest mass larger than zero) or a field matter, for which the rest mass is zero *e.g.* the matter of the considered heat radiation, a field of surface tension, magnetic field, acoustic field or gravitational field. The terms 'radiation' or 'emission' mean either the radiation phenomenon or the radiation product (the matter of the electromagnetic field). Therefore, 'the exergy of a phenomenon' is an inadequate scientific jargon often used by various researchers. It should be the 'change in exergy of the heat source' instead of the 'exergy of heat'.[24] The term Solar Radiation Exergy is generally referred to as the exergy of the Sun and it is the exergy input from the Sun to any solar system or device.

The conversion of thermal radiation can be through various processes *e.g.* work, heat and other various processes (*e.g.* growth of natural plants or plant vegetation *etc.*). The energetic and exergetic conversion efficiency of thermal radiation into work or heat is given in Table 8.10.

Thus, from Table 8.10, solar radiation exergy (radiation to work conversion) can be expressed as

$$\dot{E}x_{\mathrm{sun}} = b = e \times U_{\mathrm{ee}} \tag{8.6}$$

If I(t) is incident solar radiation (*i.e.* solar intensity/energy from the Sun) on surface area A of the solar device/system at Earth, the energy of thermal radiation (e) can be expressed as $\{I(\mathrm{t}) \times \mathrm{A}\}$ and thus the exergy input *i.e.* radiation exergy (radiation to work conversion) can be written as[19,24]

$$\begin{aligned}\dot{E}x_{\mathrm{sun}} &= \{A \times I(t)\} \times U_{\mathrm{ee}} \\ &= \{A \times I(t)\} \times \left[1 - \frac{4}{3} \times \left(\frac{T_0}{T_{\mathrm{s}}}\right) + \frac{1}{3} \times \left(\frac{T_0}{T_{\mathrm{s}}}\right)^4\right]\end{aligned} \tag{8.7}$$

The exergy input to the greenhouse can be similarly expressed as

$$\dot{E}x_{\mathrm{sun}} = \sum (I_{\mathrm{i}}A_{\mathrm{i}}) \times \left[1 - \frac{4}{3}\frac{T_0}{T_{\mathrm{s}}} + \frac{1}{3}\left(\frac{T_0}{T_{\mathrm{s}}}\right)^4\right] = U_{\mathrm{ee}} \sum (I_{\mathrm{i}}A_{\mathrm{i}}) \tag{8.8}$$

where $I_{\mathrm{i}}A_{\mathrm{i}}$ = Total incident solar energy (W) at the ith surface of the greenhouse

T_0 = Surrounding or environment temperature (K) = T_{a};
T_{s} = Sun surface temperature = T_{Sun} = 6000 K;
T_{a} = Ambient air temperature (K)

Table 8.10 Conversion efficiency of thermal radiation.[24]

S. No.	*Efficiency*	*Radiation to work conversion*	*Radiation to heat conversion*
1.	Energetic, η_{e}	$\eta_{\mathrm{e}} = \frac{W}{e}$; $\eta_{\mathrm{e\,max}} = \frac{b}{e} = U_{\mathrm{ee}}$ *or* ψ	$\eta_{\mathrm{e}} = \frac{e - e_{\mathrm{a}}}{e} = 1 - \left(\frac{T_{\mathrm{a}}}{T}\right)^4$
2.	Exergetic, η_{ex} [a]	$\eta_{\mathrm{ex}} = \frac{W}{b} = \frac{W}{e \times U_{\mathrm{ee}}}$	$\eta_{\mathrm{ex}} = \frac{b_{\mathrm{q}}}{b}$

[a]The exergetic efficiency, η_{ex}, is also denoted by 'ε' by some researchers.
W is the work performed due to utilization of the radiation, b (= $W_{\mathrm{max.}}$) is the exergy of radiation and U_{ee} = unified efficiency expression.

Table 8.11 The input, output and unified efficiency expression (U_{ee}) of utilization of thermal radiation.

S.No.	*Researcher*	*Input*	*Output*	U_{ee}
1.	Petela[64]	Radiation energy	Absolute work	$1-\frac{4}{3}\times\left(\frac{T_0}{T_s}\right)+\frac{1}{3}\times\left(\frac{T_0}{T_s}\right)^4$
2.	Spanner[65]	Radiation energy	Useful work radiation exergy	$1-\frac{4}{3}\times\left(\frac{T_0}{T_s}\right)$
3.	Jeter[66]	Heat	Net work of a heat engine	$1-\left(\frac{T_0}{T_s}\right)$

T_s and T_0 are the surface temperature of the Sun and the environment temperature at Earth, respectively.

The input, output and unified efficiency expression (U_{ee}) of utilization of thermal radiation given by three researchers is shown in Table 8.11.

Example 8.2

Calculate the unified efficiency (U_{ee}) using the expression of Petela model and radiation exergy when surrounding temperature = 20 °C, A = 2m^2 and $I(t) = 750\ \mathrm{W\,m^{-2}}$.

Solution

Using Table 8.11 and eqn (8.7), we have

$$\dot{E}x_{\text{sun}} = 2\times 750\times\left(1-\frac{4}{3}\times\left(\frac{20+273}{6000}\right)+\frac{1}{3}\times\left(\frac{20+273}{6000}\right)^4\right) = 1.4\ \text{kW}$$

8.10.2 Exergy of Stratified Thermal Energy Storages

The energy (E) and exergy (Ex) of a stratified thermal energy storage system can be obtained by integrating over the entire storage–fluid mass, m, within the thermal energy storage and expressed as

$$E = \int_{\mathrm{m}} e\,dm \tag{8.9}$$

and

$$Ex = \int_{\mathrm{m}} \xi\,dm \tag{8.10}$$

where e and ξ denote the specific energy and specific exergy, which are functions of temperature (T) alone for an ideal fluid and can be expressed as

$$e(T) = C(T - T_o) \tag{8.11}$$

$$\xi(T) = e(T) - CT_o \ln\frac{T}{T_o} \tag{8.12}$$

where C and T_0 are the fluid specific heat and reference environment temperature, respectively. The temperature (T) of the fluid in the storage tank is a function of the height ($T(h)$) of fluid in the storage.

8.10.3 Exergy Efficiency

The exergy efficiency is a very useful performance parameter for the evaluation of the thermodynamic systems and is being recognized by various researchers. The thermal efficiency of the system is defined on the basis of the first law of thermodynamics, which includes the energy balance equation for the system to account for energy input, desired energy output and energy losses. The exergy efficiency of the system is based on the second law of thermodynamics, which accounts for total exergy inflow, exergy outflow and exergy destruction for the process.

The general exergy balance for the system can be written as

$$\sum \dot{E}x_{\text{in}} - \sum \dot{E}x_{\text{out}} = \sum \dot{E}x_{\text{dest}} \tag{8.13a}$$

or

$$\left(\sum \dot{E}x_{\text{heat}} + \sum \dot{E}x_{\text{mass,in}}\right) - \left(\sum \dot{E}x_{\text{work}} + \sum \dot{E}x_{\text{mass,out}}\right) = \sum \dot{E}x_{\text{dest}} \tag{8.13b}$$

or

$$\left(\sum \left(1 - \frac{T_0}{T_{\text{k}}}\right)\dot{Q}_{\text{k}} + \sum \dot{m}_{\text{in}}\psi_{\text{in}}\right) - \left(\dot{W} + \sum \dot{m}_{\text{out}}\psi_{\text{out}}\right) = \sum \dot{E}x_{\text{dest}} \tag{8.13c}$$

where $\dot{Q}_k$ is the rate of heat transfer through the boundary at location k at temperature T_{k} (in K).

The Ψ is the specific flow exergy, which is defined as

$$\psi = (h - h_0) - T_0(s - s_0) \tag{8.14}$$

where h and s are the specific enthalpy and entropy, respectively, and subscript 0s refer to these properties at restricted dead state.

Now, the exergy destruction or the irreversibility may be written as

$$\dot{E}x_{\text{dest}} = \dot{I} = T_0 \dot{S}_{\text{gen}} \quad (8.15)$$

where rate of entropy generation $\dot{S}_{\text{gen}} = \sum \dot{m}_{\text{out}} s_{\text{out}} - \sum \dot{m}_{\text{in}} s_{\text{in}} - \sum \frac{\dot{Q}_k}{T_k}$.It is proposed that when exergy destruction or irreversibility $\sum \dot{E}x_{\text{in}} - \sum \dot{E}x_{\text{out}}$ is minimized, there will be maximum improvement in the exergy efficiency for a process or system.[44] Van Gool[44] also suggested that it is useful to employ the concept of an exergetic 'improvement potential' while analysing different processes or sectors of the economy. The rate of 'improvement potential' can be expressed as

$$IP = (1 - \varepsilon) \sum \dot{E}x_{\text{in}} - \sum \dot{E}x_{\text{out}} \quad (8.16)$$

The exergy efficiency or second law efficiency is the ratio of the actual performance of the system to the ideal performance of the system or it is defined as the ratio of exergy output (product exergy) to exergy input and expressed as[34,45]

$$\varepsilon = \frac{\textit{Rate of useful product energy}}{\textit{Rate of exergy input}} = \frac{\dot{E}x_{\text{out}}}{\dot{E}x_{\text{in}}} = 1 - \frac{\dot{E}x_{\text{dest}}}{\dot{E}x_{\text{in}}} \quad (8.17)$$

where $\dot{E}x_{\text{dest}}$ is the rate of exergy destruction.

8.11 Exergetic Analysis of Flat-plate Collector

An exergy analysis of a flat-plate collector (FPC) can be carried out with the aim of providing some ways to save costs and keep the efficiency of the integrated system to the desired extent and at the same time figuring out related exergy losses. The change in kinetic exergy in the utilization procedure is negligible since most domestic-scale solar water heaters are driven by the difference of density of water. Exergetic analysis of the collector-integrated system involves analysis of the collector and analysis of the integrated system.

The following equation can be used to calculate exergy input ($\dot{E}xc$) from the collector to the integrated system.

$$\dot{E}xc = \dot{m}C(T_{\text{fo}} - T_{\text{a}}) - \dot{m}CT_{\text{a}} \ln \frac{T_{\text{fo}}}{T_{\text{a}}} \quad (8.18)$$

where $\dot{E}xc$ = exergy output from collector (W), $\dot{m}$ = mass flow rate of collector fluid (kg s^{-1}), T_{fo} = outlet temperature of fluid from collector (K) and T_{fi} = inlet temperature of fluid from collector (K).

The exergy efficiency of the collector can be expressed as

$$\varepsilon_{\text{c}} = \frac{\dot{E}xc}{\dot{E}x_{\text{sun}}} \quad (8.19)$$

or in other terms

$$\varepsilon_c = \frac{\dot{m}\Delta e}{\dot{E}x_{sun}} = \frac{\eta_c I_c(t) A_c \Delta e}{\dot{E}x_{sun}\Delta h} \tag{8.20}$$

where Δe = exergy increase of collector (kJ kg^{-1} K^{-1}), $I_s(t)$ = incident solar radiation flux (kW m^{-2}), Δh = enthalpy increase of collector (kJ kg^{-1} K^{-1}), η_c = efficiency of solar collector and A_c = area of solar collector (m^2).

The efficiency of the solar collector is given as

$$\eta_c = \frac{\dot{m}\Delta e}{I_c(t) A_c} \tag{8.21}$$

In eqn (8.20)

$$\frac{\dot{E}x_{Sun}}{I_c(t) A_c} \approx 0.933$$

Equation (8.20) will reduce in the following form

$$\varepsilon_c = 0.933\eta_c \left[1 - \frac{T_a}{\Delta T}\right] \ln \frac{T_{fo}}{T_{fi}} \tag{8.22}$$

where ΔT = temperature increase of fluid in the collector (K).

The exergy efficiency of the solar collector will increase with an increase in collector efficiency. The exergy efficiency of the FPC is low as it transfers low-entropy (high-temperature) solar radiations to high-entropy (low-temperature) energy of the working fluid. However, the concentrating collectors have high exergy efficiency because they produce low-entropy (high-temperature) fluids.

Example 8.3

Calculate the exergy output from a collector when $T_{fo} = T_a$ and $T_{fo} - T_a = 35\,°C$. When mass flow rate is 0.06 kg s^{-1}, $C = 4190$ J kg^{-1} K^{-1}.

Solution

Using eqn (8.18), we get

$$\dot{E}x_c = 0.06 \times 4190 \times 35 - 0.06 \times 4190 \times 25 \ln \frac{60}{35} = 5.41\text{kW}$$

8.11.1 The Effects of Collector Design Parameters on the Collector Exergy Efficiency

In a collector, the most important design parameters are the width of plate, corresponding to the thermal transferring performance and the property of the

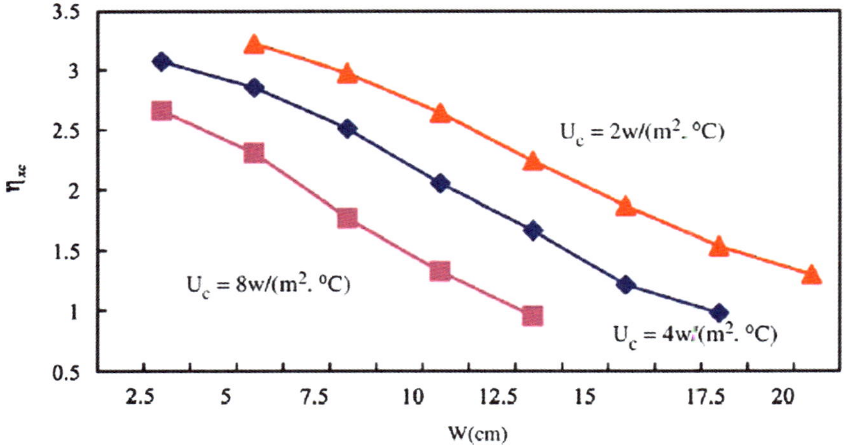

Figure 8.6 Effect of width on collector exergy efficiency for different cover designs.

cover related to the thermal loss. Wang and Ben[46] presented a study for a collector area of $2.59\,\mathrm{m}^2$, solar radiation of $466\,\mathrm{W\,m^{-2}}$, mass flow of $0.015\,\mathrm{kg\,s^{-1}}$, ambient temperature of $25\,^\circ\mathrm{C}$ and heat exchange coefficient between the fluid and the pipe of $300\,\mathrm{W\,m^{-2}\,^\circ C^{-1}}$. Figure 8.6 shows the collector exergy efficiency η_{xc} *versus* the width of plate W under three collector thermal loss coefficients U_c (2, 4 and $8\,\mathrm{W\,m^{-2}\,^\circ C^{-1}}$), which represent three kinds of cover design: one layer, two layer and three layer cover with transparent non-selective coat, respectively. We can observe the trend between η_{xc}, W and U_c. η_{xc} decreases with W and U_c. If we expect a higher η_{xc}, we must design a smaller W and U_c using a larger investment. In fact, almost half of the total investment of a domestic-scale water heater is assigned to the collector. The area of the collector is proportional to the investment. It is impractical to make a collector with three or more layers of cover and a very small width of plate. Wang and Ben[46] recommend that two layers of cover and width of the plate ranging from 5 to 10 cm would be a good choice.

8.12 Exergetic Analysis of PV/T Systems

Exergy analysis is a powerful tool for the evaluation of thermodynamic systems. The energy efficiency of the thermal system is the ratio of energy recovered from the product to the original energy input. The exergetic efficiency can be defined as the ratio of the product exergy to exergy inflow.

Exergy analysis is based on the second law of thermodynamics, which includes accounting for the total exergy inflow, exergy outflow and exergy destructed from the system

$$\sum \dot{E}x_{\mathrm{in}} - \sum \left(\dot{E}x_{\mathrm{thermal}} + \dot{E}x_{\mathrm{electrical}}\right) = \sum \dot{E}x_{\mathrm{dest}} \tag{8.23}$$

where exergy of radiation (Table 8.11)

$$\dot{E}x_{\text{in}} = A_c \times N_c \times I(t) \times \left[1 - \frac{4}{3} \times \left(\frac{T_a}{T_s}\right) + \frac{1}{3} \times \left(\frac{T_a}{T_s}\right)^4\right] \quad (8.24a)$$

$$\text{Thermal exergy} = \dot{E}x_{\text{thermal}} = \dot{Q}_u \left[1 - \frac{T_a + 273}{T_{fo} + 273}\right] \quad (8.24b)$$

$$\text{Electrical exergy} = \dot{E}x_{\text{electrical}} = \eta_c \times A_c \times N_c \times I(t) \quad (8.24c)$$

and

$$\textit{overall exergy} = \dot{E}x_{\text{thermal}} + \dot{E}x_{\text{electrical}} \quad (8.25)$$

where A_c is the area of collector and T_s is the Sun temperature in kelvin.

The exergetic analyses of different PV/T systems are described below.

8.12.1 Active Distillation System

In the case of an active solar still, the exergy input will be the sum of the radiation exergy on the solar still and flat-plate collector (FPC) and given as

$$\dot{E}x_{\text{Sun}} - \left(\dot{E}x_{\text{evap}} + \dot{E}x_{\text{work}}\right) = \dot{E}x_{\text{dest}} \quad (8.26)$$

$$\dot{E}x_{\text{in}} = \dot{E}x_{\text{Sun}}(\textit{solar still}) + \dot{E}x_{\text{Sun}}(\textit{FPC}) \quad (8.27)$$

If the exergy input from the flat-plate collector and radiation exergy input to the solar still is combined, then the exergy input to the active solar still can be expressed as follows:

$$\dot{E}x_{\text{in}}(\textit{FPC}) = \dot{Q}_u\left(1 - \frac{T_a + 273}{T_w + 273}\right) \quad (8.28a)$$

$$\dot{E}x_{\text{in}} = \dot{E}x_{\text{sun}}(\textit{solar still}) + \dot{Q}_u\left(1 - \frac{T_a + 273}{\bar{T}_w + 273}\right) \quad (8.28b)$$

The instant exergy of work rate for a hybrid active solar still (because of the PV module and pump work) is given by

$$\dot{E}x_{work} = W = P_m - P_u = (I_{SC} \times V_{OC}) \times (I_L \times V_L) \quad (8.29)$$

The total exergy output from the hybrid active solar still system is calculated using eqn (8.26) as

$$\dot{E}x_{\text{out}} = \dot{E}x_{\text{evap}} + \dot{E}x_{\text{work}}$$

The thermal efficiency of the hybrid active solar still (system) has been evaluated as

$$\eta_{\text{hybrid, system}} = \frac{\sum_{i=1}^{24} \dot{m}_{\text{ew}} \times L}{\sum_{i=1}^{t} [I_s(t) \times A_{\text{g}} + I_{\text{c}}(t) \times (A_{\text{c}} + A_{\text{m}})] \times 3600} \times 100 \quad (8.30\text{a})$$

The thermal efficiency for the solar still alone, connected to a hybrid FPC (based on energy input from collector to solar still) is given by

$$\eta_{\text{hybrid, still}} = \frac{\sum_{i=1}^{24} \dot{m}_{\text{ew}} \times L}{\sum_{i=1}^{t} [I_s(t) \times A_{\text{g}} + \dot{Q}_{\text{u}}] \times 3600} \times 100 \quad (8.30\text{b})$$

The term overall thermal efficiency has been widely used in the performance evaluation of hybrid PV/T systems.[47] For a hybrid active solar still system it can be expressed as

$$\eta_{\text{oth}} = \eta_{\text{th(hybrid system)}} + \eta_{\text{eth}} \quad (8.31)$$

The equivalent thermal efficiency of a PV/T module is given by

$$\eta_{\text{eth}} = \frac{\eta_{\text{e}}}{\eta_{\text{power}}} \quad (8.32)$$

where η_{power} is the electrical power generation efficiency of a conventional thermal power plant and often assigned the value of 38%.

The hourly variation of energy and exergy for different water depths (0.05 m, 0.1 m and 0.15 m) is shown in Figures 8.7 and 8.8. Experiments are conducted during typical days of April, 2006, for New Delhi climatic conditions. It is also observed that during the early hours of the day (7a.m.–8a.m.), the exergy of water fed from the FPC is negative (–0.003 kWh). This is because the exergy is a composite property depending on the state of both the system and the environment. The higher ambient temperature ($T_{\text{a}} > T_{\text{w}}$) is recorded rather than water temperature, fed from the collector to the solar still at the time of drawn. This negative exergy implies that work is to be supplied to operate the system and is worthless for the present system. A similar effect is also observed at low sunshine hours (3p.m.–5p.m.). The negative values of energy and exergy imply that there is destruction of both energy and exergy of water when it flows through the FPC tube during this period. The energy and exergy are transported out from the water circulating through the collector tube ($T_{\text{wi1}} > T_{\text{p}} > T_{\text{wo3}}$) at evening time. The higher exergy destruction is obtained at low water depth, because of the higher water inlet temperature (higher thermal losses in the FPC). It is to be noted that with an increase in water depth (0.15

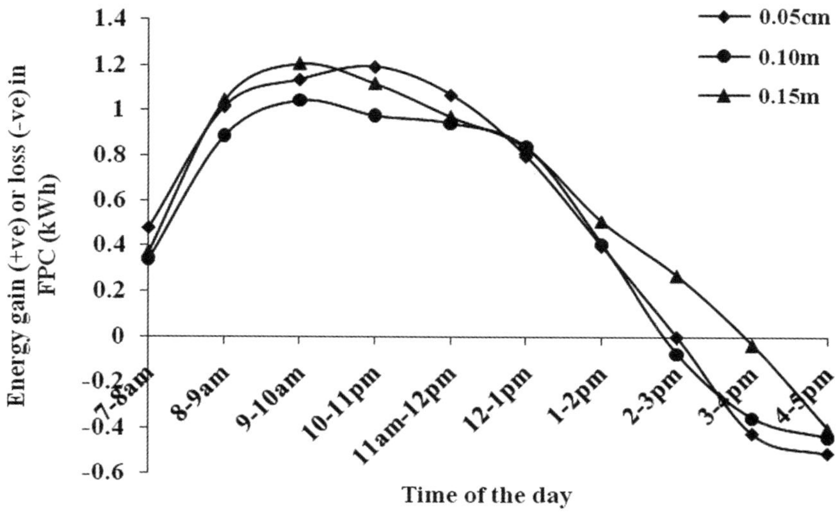

Figure 8.7 Hourly variation of energy for different water depths (0.05 m, 0.1 m and 0.15 m).

m), the destruction period of energy and exergy also reduces due to the lower inlet water temperature in the FPC. The analysis shows that to improve the exergy and energy efficiency of a hybrid active solar still, the water circulation period through the integrated FPC during evening hours (low sunshine) needs adjustment to avoid destruction of both energy and exergy. The daily exergy efficiency of the PV-integrated collector for the present configuration and obtained from experimental data is in the range of 3.3%–4.4% for different water depths.

8.12.2 PV/T Water Heater

The energy analysis is based on the first law of thermodynamics, and the expression for total thermal gain can be defined as (see Section 7.4.3.5)

$$\sum \dot{Q}_{\text{u,total}} = \sum \dot{Q}_{\text{u,thermal}} + \frac{\sum \dot{Q}_{\text{u,electrical}}}{0.38} \tag{8.33a}$$

In the case of withdrawal from the tank the thermal energy output from the tank can be calculated as

$$\dot{Q}_{\text{u,thermal}} = \dot{M}_{\text{w}} C_{\text{w}} (T_{\text{w}} - T_{\text{a}}) \tag{8.33b}$$

M_{w} = mass of water in tank, kg, C_{w} = specific heat of water, J kg^{-1} K^{-1}, and T_{w}, T_{a} = Tank water and ambient temperature, °C.

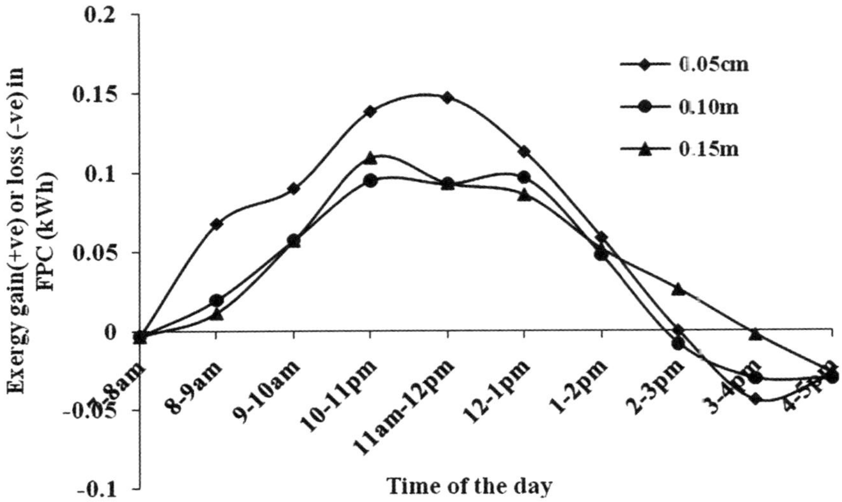

Figure 8.8 Hourly variation of exergy for different water depths (0.05 m, 0.1 m and 0.15 m).

Net annual electrical output from a PV module
= No load output – On load output

$$\dot{E}x_{\text{electrical}} = \eta_c \times A_m \times I(t) - (V_L \times I_L) \tag{8.34}$$

The annual energy and exergy gain have been evaluated for a hybrid PV/T solar water heating system by using the radiation data obtained from the Indian Metrological Department (IMD), Pune, for New Delhi climatic conditions and considering with and without withdrawal from the tank, namely

Case (i) without withdrawal from the tank
Case (ii) continuous withdrawal at the rate of 50 litres h^{-1}
Case (iii) two times in a day at the rate of 100 litres h^{-1}
Case (iv) two hours in the evening at the rate of 100 litres h^{-1} and
Case (v) two hours in the next day morning at the rate of 100 litres h^{-1}.

The hourly variation of tank-water temperature, considering with and without withdrawal from the tank for a typical day in a summer month for Case (i) to Case (v), is shown in Figure 8.9. The monthly variation of energy and exergy gain for Case (iii) is shown in Figures 8.10 and 8.11. The annual energy and exergy gain is 2720.1 kWh and 263.3 kWh, respectively.

Equations (8.25) and (8.33a) have been used for evaluating the annual overall energy and exergy gain for all the five cases. The comparison of annual overall energy and exergy gain for all the five cases for New Delhi conditions is

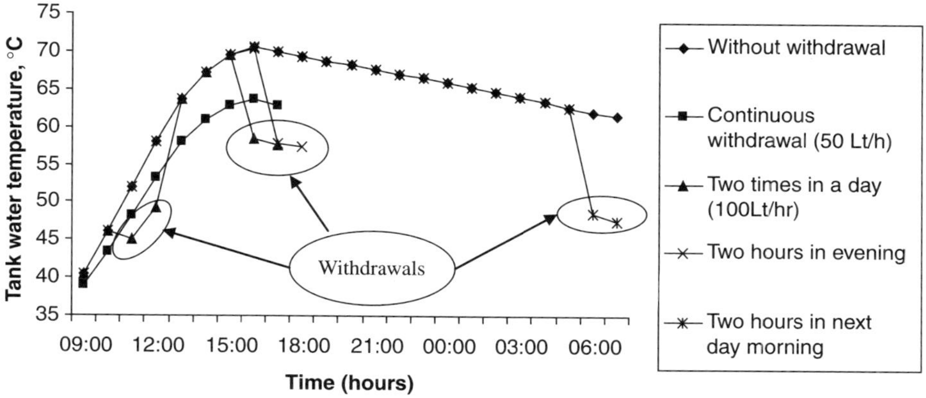

Figure 8.9 Variation of tank water temperature, considering with and without withdrawal from tank for a typical day in a summer month for Case (i) to Case (v).

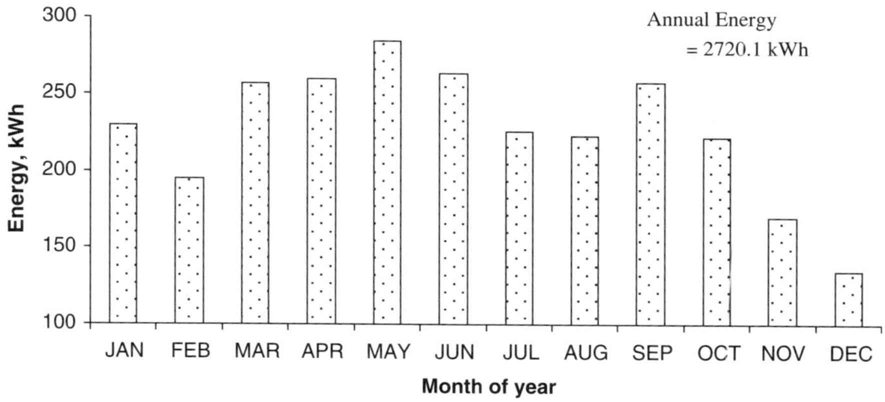

Figure 8.10 Monthly variation of overall energy gain in kWh in Case (iii).

shown in Figures 8.12 and 8.13. Maximum energy and exergy gain are obtained in Case (ii) and minimum in Case (v).

8.12.3 PV/T Solar Dryers

The exergy balance for the drying process in a solar dryer can be expressed as

$$\dot{E}x_{\text{Sun}} - (\dot{E}x_{\text{evap}} + \dot{E}x_{\text{work}}) = \dot{E}x_{\text{dest}} \tag{8.35}$$

In the case of exergy analysis of an active solar dryer, the fan may be driven by either grid electricity or with DC electricity generated by a PV module.

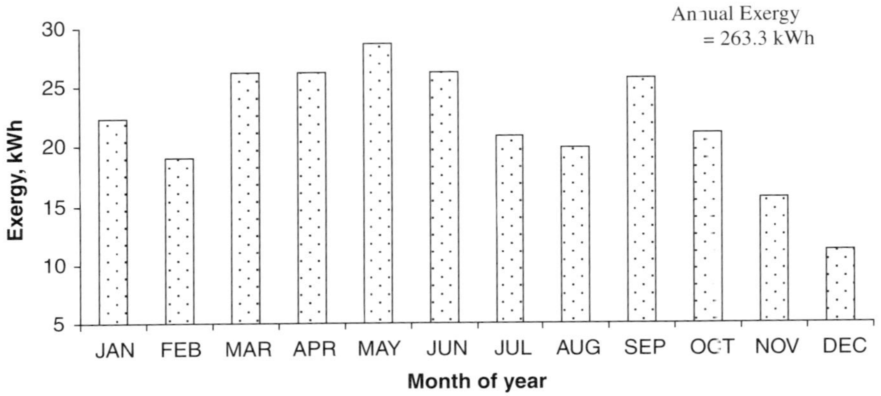

Figure 8.11 Monthly variation of overall energy gain in kWh in Case (iii).

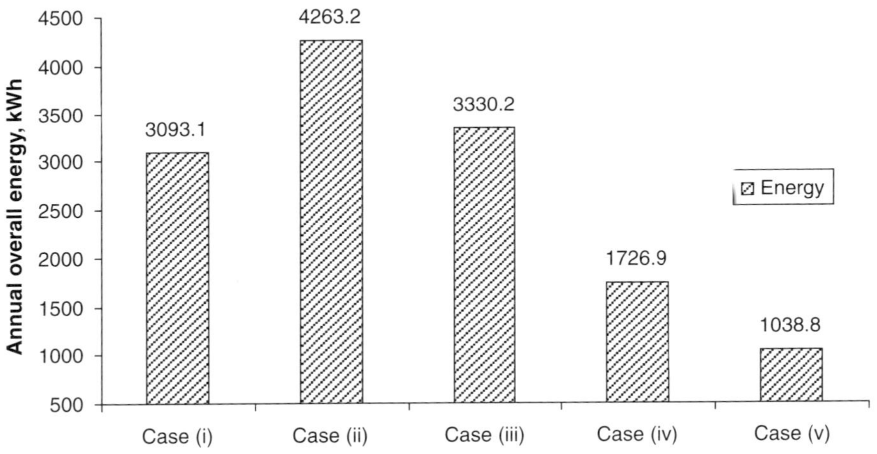

Figure 8.12 Variation of annual overall energy gain for all the five cases for New Delhi conditions.

In this case, the exergy input will be expressed as

$$\dot{E}x_{\text{in}} = U_{\text{ee}} \sum (I_{\text{i}}A_{\text{i}}) + \dot{E}x_{\text{PV module}} \tag{8.36}$$

The exergy of a PV module can be given as

$$\dot{E}x_{\text{PV module}} = \eta_{\text{em}} \times (I_i A_i)_{\text{PV module}} \tag{8.37}$$

where η_{em} is the module efficiency (eqn. (7.17)).

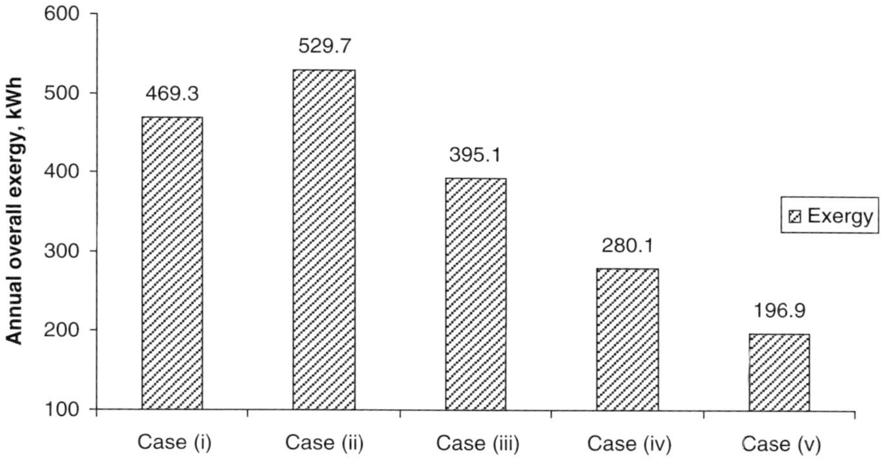

Figure 8.13 Variation of annual overall exergy gain for all the five cases for New Delhi conditions.

If exergy input from a flat-plate collector and radiation exergy input $\dot{E}x_{in}$ to a drying chamber are combined then the exergy input, in this case, can be expressed as follows

$$\dot{E}x_{Sun}(\text{drying chamber}) + \dot{Q}_u\left(1 - \frac{T_a + 273}{T_{ch} + 273}\right) + \dot{E}x_{PV\,module} \tag{8.38}$$

where the temperatures are in kelvin and $\dot{Q}_u$ is the useful thermal energy supplied to the drying chamber from the flat-plate collector. The term $\dot{E}x_{PV\ module}$ is zero if the PV module is integrated in the collector (Figure 7.39 (b and c)).

The exergy of the work rate for a hybrid active solar dryer (because of PV module and fan work) is given by

$$\dot{E}x_{work} = W = P_m - P_u = (I_{SC} \times V_{OC}) - (I_L \times V_L) \tag{8.39}$$

where P_m = power output from PV module (W), P_u = power used to drive the fan (W), I_{sc} = short circuit current (A), V_{oc} = open circuit voltage (V), I_L = load current (A) and V_L = load voltage (V).

The total exergy output from a hybrid active solar dryer can be calculated as

$$\dot{E}x_{out} = \dot{E}x_{evap} + \dot{E}x_{work} \tag{8.40}$$

Thus, exergetic efficiency can be expressed as

$$\varepsilon = \frac{\dot{E}x_{out}}{\dot{E}x_{in}} = \frac{\dot{E}x_{evap} + \dot{E}x_{work}}{\dot{E}x_{in}} \tag{8.41}$$

where $\dot{E}x_{in}$ will be given by eqn (8.7) or (8.8).

Problems

8.1 Draw a pie-chart of mass and embodied energy of a greenhouse crop dryer of 0.96 m^2 effective area and calculate the percentage of mass and embodied energy used for different components. Hint: see Sections 8.3.1 and 8.3.2.

8.2 Repeat Problem 8.1 for a reverse absorber cabinet dryer (RACD). Hint: see Sections 8.3.1 and 8.3.3.

8.3 Repeat Problem 8.1 for an active (conventional) solar dryer (ASD). Hint: see Sections 8.3.1 and 8.3.4.

8.4 Repeat Problem 8.1 for a hybrid photovoltaic/thermal (PV/T) integrated greenhouse dryer. Hint: see Sections 8.3.1 and 8.3.5.

8.5 Repeat Problem 8.1 for a hybrid photovoltaic/thermal (PV/T) solar dryer. Hint: see Sections 8.3.1 and 8.3.6.

8.6 Calculate the maximum work available (W_{max}) from the heat source at $T_1 = 20\,°C$, $50\,°C$ and $90\,°C$ and ambient temperature $= 30\,°C$ when $Q_1 = 500$ kWh. Hint: use eqn (8.5a).

8.7 Calculate the unified efficiency (U_{ee}) using the expression given by different researchers and radiation exergy when surrounding temperature $= 25\,°C$, $A = 4 m^2$ and $I(t) = 850$ W m^{-2}. Hint: use Table 8.11 and eqn (8.7).

8.8 Calculate the exergy output from a collector when $T_{fo} = T_a$ and $T_{fo} - T_a = 45\,°C$, mass flow rate is 0.08 kg $s^{-1}$and $C = 4190$ J kg^{-1} K. Hint: use eqn (8.18).

8.9 Calculate the exergy of the work rate for a hybrid active solar dryer when $I_{sc} = 3.2$ A, $V_{oc} = 16$ V, $I_L = 0.5$ A and $V_L = 14$ V. Hint: use eqn (8.39).

8.10 Calculate the exergetic efficiency of a hybrid active solar dryer when $\dot{E}_{evap} = 12$ W. Hint: use eqn (8.41).

References

1. R. W. Bentley, *Energ. Pol.*, 2002, **30**(3), 189–205.
2. E. A. Alsema and E. Niluwlaar, *Energ. Pol.*, 2000, **28**, 999–1010.
3. S. Krauter and R. Ruther, *Renew. Energ.*, 2004, **29**, 345–355.
4. P. Frankl, A. Masini, M. Gamberale and D. Toccaceli, *Progress in Photovoltaics: Research and Applications*, 1998, **6**(2), 137–146.
5. L. P. Hunt, *IEEE PV Specialists Conference*, Piscataway, NJ, 1986, pp. 347–352.
6. K. Kato, A. Murata and K. Sakuta, *Progress in Photovoltaics: Research and Applications*, 1998, **6**(2), 105–115.
7. H.A. Aulich, F.W. Schulz and B. Strake, *IEEE PV Specialist Conference*, Piscataway, NJ, 1986, pp. 1213–1218. .
8. R. Prakash and N. K. Bansal, *Energ. Sourc.*, 1995, **17**, 605–613.

9. G. Lewis and G. Keoleian, *National Pollution Prevention Center*, School of Natural Resources and Environment, University of Michigan, 1996.
10. K. S. Srinivas, M. Vuknic, A. V. Shah and R. Tscharner, *6th International Photovoltaic Science and Engineering Conference (PVSEC-6)*, New Delhi, India, 1992, pp. 403–413.
11. R. Battisti and A. Corrado, *Energy*, 2005, **30**, 952–967.
12. G. J. Treloar, *Master of Architecture Thesis*, Deakin University, Geelong, 1994.
13. I. Boustead and G. F. Hancock, Handbook of Industrial Energy Analysis, Ellis Horwood Publishers, West Sussex, England, 1979, pp. 309–410.
14. G. N. Tiwari and M. K. Ghosal, *Renewable Energy Resources: Basic Principles and Applications*, Narosa Publishing House, New Delhi, India, 2005.
15. S. Dubey and G. N. Tiwari, *Open Environ. J.*, 2008, **2**, 15–25.
16. C. Bankier and S. Gale, Energy Payback of Roof Mounted Photovoltaic Cells, http://www.energybulletin.net/node/17219, accessed 5 November 2007.
17. J. Szargut, A. Ziebik and W. Stanek, *Energ. Convers. Manag.*, 2002, **42**, 1149–1163.
18. J. Szargut, *Bull. Pol. Acad. Sci.*, 1997, **45**, 241–250.
19. J. Szargut, *Energy*, 2003, **28**(11), 1047–1054.
20. M. J. Ebadi and M. Gorji-Bandpy, *International Journal of Exergy*, 2005, **2**(1), 31–39.
21. R. Cornelisse, The method of exergy analysis, University of Twente, http://www.thw.wb.utwente.nl/topics/exergy/theoryFe.htm, accessed 21 July 2007.
22. I. Dincer and A. Z. Sahin, *Int. J. Heat Mass Tran.*, 2004, **47**, 645–652.
23. M. A. Rosen, *Sol. Energ.*, 2001, **71**(3), 173–185.
24. R. Petela, *Sol. Energ.*, 2003, **74**(6), 469–488.
25. J. Szargut, D. R. Morris and F. R. Steward, *Exergy Analysis of Thermal, Chemical, and Metallurgical Processes*, Hemisphere Publishing, New York, 1988.
26. I. Dincer, *Energ. Pol.*, 2002, **30**(2), 137–149.
27. L. García-Rodríguez and C. Gómez-Camacho, *Desalination*, 2001, **137**(1–3), 251–258.
28. O. Sow, M. Siroux and B. Desmet, *Desalination*, 2005, **174**(3), 277–286.
29. A. Hepbasli, *Renew. Sustain. Energ. Rev.*, 2008, **12**(3), 593–661.
30. A. Hepbasli and O. Akdemir, *Energ. Convers. Manag.*, 2004, **45**(5), 737–753.
31. T. Fujisawa and T. Tani, *Sol. Energ. Mater. Sol. Cell.*, 1997, **47**(1–4), 135–148.
32. E. K. Akpinar, *Int. Comm. Heat Mass Tran.*, 2004, **31**(4), 585–595.
33. E. K. Akpinar, A. Midilli and Y. Bicer, *J. Food Eng.*, 2006, **72**, 320–331.
34. A. Midilli and H. Kucuk, *Energy*, 2003, **28**, 539–556.
35. S. Syahrul, I. Dincer and F. Hamdullahpur, *Int. J. Therm. Sci.*, 2003, **42**, 691–701.
36. M. A. Rosen and I. Dincer, *Energ. Convers. Manag.*, 2003, **44**, 1633–1651.

37. M. A. Rosen and I. Dincer, *Int. J. Energ. Res.*, 2003, **27**(4), 415–430.
38. M. A. Rosen, M. N. Le and I. Dincer, *Appl. Therm. Eng.*, 2005, **25**(1), 147–159.
39. W. Xiaowu and H. Bena, *Renew. Sustain. Energ. Rev.*, 2005, **9**, 638–645.
40. A. Ucar and M. Inalli, *Build. Environ.*, 2006, **41**(11), 1551–1556.
41. H. Gunerhan and A. Hepbasli, *Energ. Build.*, 2007, **39**, 509–516.
42. A. Hepbasli, *Energ. Build.*, 2007, **39**, 1211–1217.
43. S. Dubey and G. N. Tiwari, *Int. J. Energ. Res.*, 2008, DOI: 10.1002/er.1422.
44. W. Van Gool, in *Innovation and Technology – Strategies and Policies*, ed. O. D. D. Soares, A. M. da Cruz, G. C. Pereira, I. M. R. T. Soares and A. J. P. S. Reis, Kluwer, Dordrecht, 1997, pp. 93–105.
45. E. H. Kuzgunkaya and A. Hepbasli, *Int. J. Energ. Res.*, 2007, **31**(3), 245–258.
46. X. Wang and H. Ben, *Renew. Sustain. Energ. Rev.*, 2005, **9**(6), 638–645.
47. T. Bergene and O. M. Lovvik, *Sol. Energ.*, 1995, **55**, 453–462.
48. R. Dones and R. Frischknecht, *Progress in Photovoltaics: Research and Applications*, 1998, **6**(2), 117–125.
49. A. Blakers and K. Weber, The energy intensity of photovoltaic (PV) systems, http://www/ecotopia.com/opollo2/pvepbtoz.htm, accessed 27 July 2007.
50. I. Nawaz and G. N. Tiwari, *Energ. Pol.*, 2006, **34**(17), 3144–3152.
51. H. Schaefer and G. Hagedorn, *Renew. Energ.*, 1992, **2**, 15–166.
52. K. Kato, A. Murata and K. Sakuta, *Sol. Energ. Mater. Sol. Cell.*, 1997, **47**, 95–100.
53. E. Alsema, *Progress in Photovoltaics: Research and Applications*, 2000, **8**, 17–25.
54. K. Kato, T. Hibino, K. Komoto, S. Ihara, S. Yamamoto and H. Fujihara, *Sol. Energ. Mater. Sol. Cell.*, 2001, **67**, 279–287.
55. K. Knapp and T. Jester, *Sol. Energ.*, 2001, **71**, 165–172.
56. J. Pearce and A. Lau, in *Proceedings of American Society of Mechanical Engineers Solar 2002: Sunrise on the Reliable Energy Economy*, ed. R. Cambell-Howe, New York, NY, 2002.
57. T. Jester, *Progress in Photovoltaics: Research and Applications*, 2002, **10**, 99–106.
58. A. Meijer, M. Huijbregts, J. Schermer and L. Reijnders, *Progress in Photovoltaics: Research and Applications*, 2003, **11**, 275–287.
59. R. Battisti and A. Corrado, *Energy*, 2005, **30**, 952–967.
60. N. Jungbluth, *Progress in Photovoltaics: Research and Applications*, 2005, **13**, 429–446.
61. G. Peharz and F. Dimroth, *Progress in Photovoltaics: Research and Applications*, 2005, **13**, 627–634.
62. M. Raugei, S. Bargigli and S. Ulgiati, Energy and Environment Research Unit, Department of Chemistry, University of Siena, Italy, http://www.nrel.gov/pv/thin_film/docs/20theuropvscbarcelona4cv114_raugei.pdf, accessed 3 August 2007.

63. Y. Tripanagnostopoulos, M. Souliotis, R. Battisti and A. Corrado, *Progress in Photovoltaics: Research and Applications*, 2005, **13**, 235–250.
64. R. Petela, *Trans. ASME, J. Heat Tran.*, 1964, **2**, 187–192.
65. D. Spanner, *Introduction to Thermodynamics*, Academic Press, London, 1964.
66. S. M. Jeter, *Sol. Energ.*, 1981, **26**(3), 231–236.

CHAPTER 9

CO_2 *Mitigation and Carbon Trading*

9.1 Introduction

Energy consumption of a country is one of the indicators of its socioeconomic development. *Per capita* electricity consumption of India is one of the lowest in the world. Table 9.1 gives the *per capita* electricity consumption of a few countries in the world. *Per capita* energy consumption in India is also one of the lowest in the world. It is about 30% of that in China, about 22% of that in Brazil and about 3.18% of that in the USA. With development, the *per capita* energy consumption is likely to increase. In order to achieve a *per capita* energy consumption equal to that of Brazil (which, like India, is still a developing country) India's energy production and consumption must be quadrupled and to achieve the European average (about 6500 kWh capita^{-1}), the energy production and consumption must be increased by 15.5 times. At present India's annual economic growth rate is 8–10% per annum. To sustain this growth rate we desperately need additional secured and reliable energy sources. For energy, India depends on oil and gas imports, which account for over 65% of its consumption; it is likely to increase further considering the economic development, improvement in living conditions of people and rising prices. Coal, which currently accounts for over 60% of India's electricity production, is the major source of emission of greenhouse gases and of acid rain. India will become the third largest polluter in the world after the USA and China if the country continues to depend on coal as the main source of electricity in the years to come. In the business-as-usual scenario, India will exhaust its oil reserves in 22 years, its gas reserves in 30 years and its coal reserves in 80 years.[1] Even more alarmingly, coal reserves might disappear in fewer than 40 years if India continues to grow at 8% a year.[1]

The present energy scenario in India is alarming. There are serious shortcomings in access to electricity for the rural and urban poor, in meeting the peak demand and in the reliability of the power supply. More than 50% of India's population does not have access to electricity. If the population that, at

RSC Energy Series No. 2
Fundamentals of Photovoltaic Modules and Their Applications
By G. N. Tiwari and Swapnil Dubey

Published by the Royal Society of Chemistry, www.rsc.org

Table 9.1 Per capita electricity consumption of a few countries in the world. (Source: Wikipedia 2009).

Rank	*Country*	*Electricity consumption (MWh/year)*	*Year of Data*	*Average power per capita (Watts)*
—	World	16,830,000,000	2005	297
1	United States	3,816,000,000	2005	1,460
2	People's Republic of China	2,859,000,000	2006	248
—	European Union[4]	2,820,000,000	2004	700
3	Russia	985,200,000	2007	785
4	Japan	974,200,000	2005	868
5	Germany	545,500,000	2005	753
6	Canada	540,200,000	2005	1,910
7	India	488,500,000	2005	50.5
8	France	451,500,000	2005	851
9	South Korea	368,600,000	2007	879
10	Brazil	368,500,000	2005	226
11	United Kingdom	348,700,000	2005	667
12	Italy	307,100,000	2005	603
13	Spain	243,000,000	2005	644
14	South Africa	241,400,000	2007	581
15	Taiwan (Republic of China)	221,000,000	2006	1,101
16	Australia	219,800,000	2005	1,244
17	Mexico	183,300,000	2005	195
18	Ukraine	181,900,000	2006	446
19	Saudi Arabia	146,900,000	2005	682
20	Iran	136,200,000	2005	224
21	Sweden	134,100,000	2005	1,692
22	Turkey	129,000,000	2005	201
23	Poland	120,400,000	2005	356
24	Thailand	117,700,000	2005	209
25	Norway	113,900,000	2005	2,812
26	Netherlands	108,200,000	2005	757
27	Indonesia	108,000,000	2006	55.3
28	Argentina	88,980,000	2005	262
29	Finland	88,270,000	2007	1,918
30	Egypt	84,490,000	2005	130
31	Belgium	82,990,000	2005	909
32	Malaysia	78,720,000	2005	354
33	Kazakhstan	76,430,000	2007	588
34	Venezuela	73,360,000	2005	313
35	Pakistan	67,060,000	2005	48.4
36	Austria	60,250,000	2005	839

Table 9.1 (*Continued*).

Rank	*Country*	*Electricity consumption (MWh/year)*	*Year of Data*	*Average power per capita (Watts)*
37	Czech Republic	59,720,000	2005	667
38	Romania	58,490,000	2007	307
39	Switzerland	58,260,000	2005	916
40	Greece	54,310,000	2005	557
41	United Arab Emirates	52,620,000	2005	1,335
42	Vietnam	51,350,000	2007	69.5
43	Portugal	48,550,000	2006	528
44	Chile	48,310,000	2005	338
45	Uzbekistan	47,000,000	2006	202
46	Philippines	46,860,000	2005	64.4
47	Israel	43,280,000	2005	734
—	Hong Kong	40,300,000	2006	653
48	Colombia	38,910,000	2005	97.3
49	Bulgaria	37,400,000	2006	552
50	New Zealand	37,390,000	2006	1,059

present, does not have access to electricity starts consuming electricity at the current national average (421 kWh per annum) electricity production will have to be more than doubled. Globally, the renewable energy industry is no longer in a state of infancy, with global investments in 2004 totalling $28 billion as compared to $6 billion in 1995. Total installed capacity based on renewable energy was 155,000 MW in 2004.[2]

Solar PV Systems are one of the most promising future sources of energy. They have an advantage over the traditional energy sources like coal-, oil- and gas-fired power plants, as well as nuclear and hydro power plants. Amongst the alternative solar electricity sources, solar PV systems are the most promising. They have a very small gestation period and do not have any moving parts; as a result they are nearly maintenance free. The only disadvantage they have, at present, is the high capital cost. This, too, is likely to go down substantially because of new techniques which have been developed and which are being developed for the manufacture of solar cells. Solar cells are generally made from single crystalline silicon cells which show high efficiency and long-term stability. Solar cells convert sunlight into direct current (DC) electricity. A group of electrically connected PV cells, packed with ethyl vinyl acetate, is known as a PV module. The top surface of the cells is coated with an anti-reflective transparent coating. Solar PV modules connected in series and parallel are known as PV arrays.

In India, the cost of electricity generated by solar PV cells amounts to €0.122 KWh^{-1}, which is equivalent to Rs 7.93 kWh^{-1} (where €1 = Rs 56,

August, 2008). Globally the capital cost of installing a solar PV system comes to €4500–6500 kWh^{-1}.[3] Prakash *et al.*[4] have estimated the capital cost of installing a solar PV system at €6336.2 kWp^{-1} (Pathak, 2007). The solar photovoltaic power cost is expected to reduce by 50% in the next 15–20 years.[4] Both the capital cost and the cost of electricity generated are likely to reduce substantially if the following are taken into account:

(a) economy of scale;
(b) advancement in technology;
(c) carbon credits likely to be earned by such plants are as per the Kyoto Protocol.

Secondly stand-alone PV systems are better suited for Indian conditions. These systems do not require sophisticated grid synchronization equipment and systems. The electricity generated is directly used in running the electrical loads and balance electricity is stored in battery banks. These batteries along with the inverter (a device to convert direct current into AC) are used to run the electrical loads during the night and off-sunshine periods. In India most of the village houses are single storied and can easily support the stand-alone photovoltaic system (SAPV) system either at the roof top or on the open land adjacent to the house.

Carbon Credit Trading (Emission Trading) is an administrative approach used to control pollution by providing economic incentives for achieving reductions in the emission of pollutants. The development of a carbon project that provides a reduction in greenhouse gas emissions is a way by which participating entities may generate tradable carbon credits. Carbon credits are a tradable permit scheme. A credit gives the owner the right to emit one ton of carbon dioxide. International treaties such as the Kyoto Protocol set quotas on the amount of greenhouse gases countries can produce. Countries, in turn, set quotas on the emissions of businesses. Businesses that are over their quotas must buy carbon credits for their excess emissions, while businesses that are below their quotas can sell their remaining credits. By allowing credits to be bought and sold, a business for which reducing its emissions would be expensive or prohibitive can pay another business to make the reduction for it. This minimizes the quota's impact on the business, while still reaching the quota. Credits can be exchanged between businesses or bought and sold in international markets at the prevailing market price. There are currently two exchanges for carbon credits: the Chicago Climate Exchange and the European Climate Exchange. In 2005, 375 million tons of carbon dioxide equivalents (tCO_2e) were transacted at a value of US$2.7 billion with an average price of US$7.23. In the first three months of 2006, the average reported price of carbon dioxide equivalents was US$11.45 per ton. European and Japanese companies were the major buyers and China was the major seller of the carbon credits in 2005–2006. Demand for carbon credits continued to soar in 2006–2007 resulting in an increase in the traded rate of carbon credits. In early May 2006, EU 2008 futures were being quoted at around €20–24 (State and trends of carbon market, 2006). The present market rate is fluctuating at €20–22 in the European Climate Exchange (www.europeanclimateexchange.com).

9.2 CO_2 Emissions

Greenhouse gases are the gases present in the Earth's atmosphere which reduce the loss of heat into space and therefore contribute to global temperatures through the greenhouse effect. Greenhouse gases are essential for maintaining the temperature of the Earth; without them the planet would be so cold as to be uninhabitable. However, an excess of greenhouse gases can raise the temperature of a planet to lethal levels, as on Venus where the 90-bar partial pressure of carbon dioxide (CO_2) contributes to a surface temperatures of about 467 °C (872 °F). Greenhouse gases are produced by many natural and industrial processes, which currently result in CO_2 levels of 380 ppmv in the atmosphere. Based on ice-core samples and records, current levels of CO_2 are approximately 100 ppmv higher than during immediately pre-industrial times, when direct human influence was negligible. Carbon emissions from various global regions during the period 1800–2000 AD are shown in Figure 9.1.

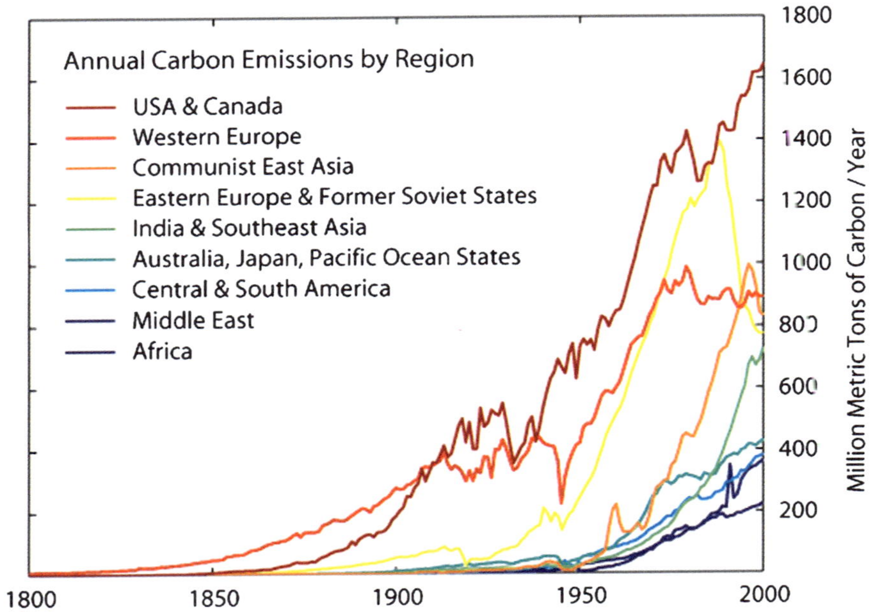

Figure 9.1 Carbon emissions from various global regions during the period 1800–2000 AD. (source: Wikipedia 2009).

The average carbon dioxide (CO_2) equivalent intensity for electricity generation from coal is approximately 0.98 kg of $CO_2\,kWh^{-1}$.[5] If the PV system has a lifetime of 35 years, the CO_2 emissions per year by each component can be calculated as

$$CO_2 \text{ emissions per year} = \frac{\textit{Embodied energy} \times 0.98}{\textit{Lifetime}} \tag{9.1}$$

The CO_2 emissions per year for a PV module (glass-to-glass) (effective area = 0.60534 m^2 and size = 1.20 m × 0.55 m × 0.01 m) in present conditions are given in Table 9.2. The CO_2 emissions for different PV/T systems are shown in Figure 9.2.

Example 9.1

Calculate the carbon dioxide emissions per year from a solar water heater in a lifetime of 10, 20 and 30 years, when the total embodied energy required for manufacturing the system is 3550 kWh.

Table 9.2 CO_2 emissions per year from a PV module (glass-to-glass) (effective area = 0.60534 m^2).

Sl. No.	*Components*	*Embodied energy (kWh)*	*CO_2 emissions (kg)*
1	MG-Si	26.54	0.74
2	EG-Si	127.30	3.56
3	Cz-Si	267.33	7.49
4	Solar cell fabrication	60.29	1.69
5	PV Module assembly	125.40	3.51
Total		**606.86**	**16.99**

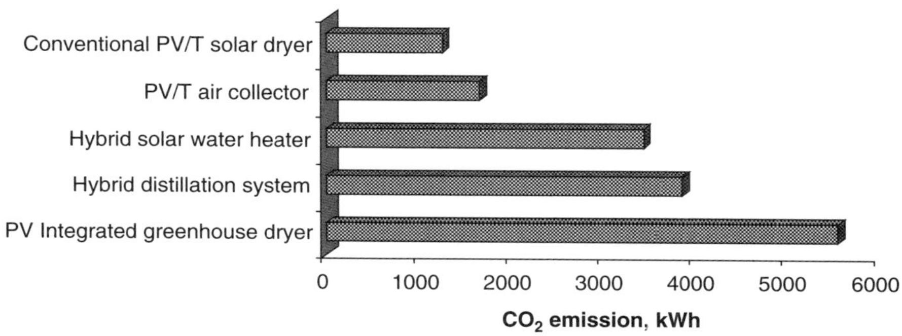

Figure 9.2 CO_2 emissions for different PV/T systems.

Solution

Using eqn (9.1), we have
For lifetime = 10 years

$$CO_2 \text{ emissions per year} = \frac{3550 \times 0.98}{10} = 347.9 \text{ kg of } CO_2$$

Similarly, for lifetime = 20 and 30 years CO_2 emissions per year are 173.9 and 115.9 kg of CO_2 respectively.

9.3 The Kyoto Protocol

The Kyoto Protocol is a protocol to the international Framework Convention on Climate Change with the objective of reducing greenhouse gases that cause climate change.

It was adopted on 11 December, 1997, by the 3rd Conference of the Parties, which was a meeting in Kyoto, and it entered into force on 16 February, 2005. As of May 2008, 182 parties have ratified the protocol. Of these, 36 developed countries (plus the EU as a party in its own right) are required to reduce greenhouse gas emissions to the levels specified for each of them in the treaty (representing over 61.6% of emissions from Annex I countries), with three more countries intending to participate. 137 developing countries have ratified the protocol, including Brazil, China and India, but have no obligation beyond monitoring and reporting emissions. The United States has not ratified the treaty. Among various experts, scientists and critics, there is debate about the usefulness of the protocol, and there have been cost-benefit studies performed on its usefulness.

The Kyoto Protocol is an agreement made under the United Nations Framework Convention on Climate Change (UNFCCC). Countries that ratify this protocol commit to reducing their emissions of carbon dioxide and five other greenhouse gases (GHGs), or engaging in emissions trading if they maintain or increase emissions of these greenhouse gases. The objective is to achieve "stabilization of greenhouse gas concentrations in the atmosphere at a level that would prevent dangerous anthropogenic interference with the climate system". The Intergovernmental Panel on Climate Change (IPCC) has predicted an average global rise in temperature of 1.4 °C (2.5 °F) to 5.8 °C (10.4 °F) between 1990 and 2100.[6]

"*The Kyoto Protocol is an agreement under which industrialized countries will reduce their collective emissions of greenhouse gases by 5.2% compared to the year 1990 (but note that, compared to the emissions levels that would be expected by 2010 without the Protocol, this limitation represents a 29% cut). The goal is to lower overall emissions of six greenhouse gases – carbon dioxide, methane, nitrous oxide, sulfur hexafluoride, hydro fluorocarbons, and per fluorocarbons – averaged over the period of 2008–2012. National limitations range from 8% reductions for the European Union and some others to 7% for the US, 6% for Japan, 0% for Russia, and permitted increases of 8% for Australia and 10% for Iceland.*"

The Kyoto Protocol now covers 181 countries globally but only 60% of countries in terms of global greenhouse gas emissions. As of December, 2007, the US and Kazakhstan are the only signatory nations not to have ratified the act. The first commitment period of the Kyoto Protocol ends in 2012, and international talks began in May 2007 on a subsequent commitment period.

The Kyoto Protocol establishes the following principles:

- Kyoto is underwritten by governments and is governed by global legislation enacted under the UN's aegis;
- Governments are separated into two general categories: developed countries, referred to as Annex I countries (who have accepted greenhouse gas emission reduction obligations and must submit an annual greenhouse gas inventory), and developing countries, referred to as Non-Annex I countries (who have no greenhouse gas emission reduction obligations but may participate in the Clean Development Mechanism);
- Any Annex I country that fails to meet its Kyoto obligation will be penalized by having to submit 1.3 emission allowances in a second commitment period for every ton of greenhouse gas emissions by which they exceed their cap in the first commitment period (*i.e.* 2008–2012);
- As of January, 2008, and running through 2012, Annex I countries have to reduce their greenhouse gas emissions by a collective average of 5% below their 1990 levels (for many countries, such as the EU member states, this corresponds to some 15% below their expected greenhouse gas emissions in 2008). While the average emissions reduction is 5%, national limitations range from an 8% average reduction across the European Union to a 10% emissions increase for Iceland; but, since the EU's member states each have individual obligations, much larger increases (up to 27%) are allowed for some of the less developed EU countries. Reduction limitations expire in 2013;
- Kyoto includes 'flexible mechanisms', which allow Annex I economies to meet their greenhouse gas emission limitation by purchasing GHG emission reductions from elsewhere. These can be bought either from financial exchanges, from projects which reduce emissions in non-Annex I economies under the Clean Development Mechanism (CDM), from other Annex 1 countries under the JI or from Annex I countries with excess allowances. Only CDM Executive Board-accredited Certified Emission Reductions (CER) can be bought and sold in this manner. Under the aegis of the UN, Kyoto established this Bonn-based Clean Development Mechanism Executive Board to assess and approve projects ('CDM Projects') in Non-Annex I economies prior to awarding CERs. (A similar scheme called 'Joint Implementation' or 'JI' applies in transitional economies mainly covering the former Soviet Union and Eastern Europe.)

9.3.1 Kyoto's Flexible Mechanisms

A credit can be an emissions allowance which was originally allocated or auctioned by the national administrators of a cap-and-trade program, or it can be an offset of emissions. Such offsetting and mitigating activities can occur in any developing country which has ratified the Kyoto Protocol, and has a national agreement in place to validate its carbon project through one of the UNFCCC's approved mechanisms. Once approved, these units are termed Certified Emission Reductions, or CERs. The Protocol allows these projects to be constructed and credited in advance of the Kyoto trading period.

The Kyoto Protocol provides for three mechanisms that enable countries or operators in developed countries to acquire greenhouse gas reduction credits:[7]

- Under *Joint Implementation (JI)*, a developed country with relatively high costs of domestic greenhouse reduction would set up a project in another developed country.
- Under the *Clean Development Mechanism (CDM)* a developed country can 'sponsor' a greenhouse gas reduction project in a developing country where the cost of greenhouse gas reduction project activities is usually much lower, but the atmospheric effect is globally equivalent. The developed country would be given credits for meeting its emission reduction targets, while the developing country would receive the capital investment and clean technology or beneficial change in land use.
- Under *International Emissions Trading (IET)*, countries can trade in the international carbon credit market to cover their shortfall in allowances. Countries with surplus credits can sell them to countries with capped emission commitments under the Kyoto Protocol.

9.3.2 Emission Allowances

The Protocol agreed 'caps' or quotas on the maximum amount of greenhouse gases for developed and developing countries, listed in its Annex I. In turn these countries set quotas on the emissions of installations run by local businesses and other organizations, generically termed 'operators'. Countries manage this through their own national 'registries', which are required to be validated and monitored for compliance by the UNFCCC. Each operator has an allowance of credits, where each unit gives the owner the right to emit one metric tonne of carbon dioxide or other equivalent greenhouse gas. Operators that have not used up their quotas can sell their unused allowances as carbon credits, while businesses that are about to exceed their quotas can buy the extra allowances as credits, privately or on the open market.[8] As demand for energy grows over time, the total emissions must still stay within the cap, but it allows industry some flexibility and predictability in its planning to accommodate this.

By permitting allowances to be bought and sold, an operator can seek out the most cost-effective way of reducing its emissions, either by investing in 'cleaner' machinery and practices or by purchasing emissions from another operator who already has excess 'capacity'.

Since 2005, the Kyoto mechanism has been adopted for CO_2 trading by all the countries within the European Union under its European Trading Scheme (EU ETS) with the European Commission as its validating authority. From 2008, EU participants must link with the other developed countries that ratified Annex I of the protocol, and trade the six most significant anthropogenic greenhouse gases.

9.3.3 Additionality and Its Importance

It is also important for any carbon credit to prove a concept called additionality. Additionality is a term used by Kyoto's Clean Development Mechanism (CDM) to describe the fact that a carbon dioxide reduction project (carbon project) would not have occurred had it not been for concern for the mitigation of climate change. More succinctly, a project that has proven additionality is a beyond-business-as-usual project.[9]

It is generally agreed that voluntary carbon offset projects must also prove additionality in order to ensure the legitimacy of the environmental stewardship claims resulting from the retirement of the carbon credit (offset). According to the World Resources Institute/World Business Council for Sustainable Development (WRI/WBCSD): 'GHG emission trading programs operate by capping the emissions of a fixed number of individual facilities or sources. Under these programs, tradable "offset credits" are issued for project-based GHG reductions that occur at sources not covered by the program. Each offset credit allows facilities whose emissions are capped to emit more, in direct proportion to the GHG reductions represented by the credit. The idea is to achieve a zero net increase in GHG emissions, because each tonne of increased emissions is "offset" by project-based GHG reductions. The difficulty is that many projects that reduce GHG emissions would happen regardless of the existence of a GHG program and without any concern for climate change mitigation. If a project "would have happened anyway", then issuing offset credits for its GHG reductions will actually allow a positive net increase in GHG emissions, undermining the emissions target of the GHG program. Additionality is thus critical to the success and integrity of GHG programs that recognize project-based GHG reductions.'

9.4 Emission Trading

Kyoto is a 'cap-and-trade' system that imposes national caps on the emissions of Annex I countries. On average, this cap requires countries to reduce their emissions 5.2% below their 1990 baseline over the 2008 to 2012 period. Although these caps are national-level commitments, in practice most countries will devolve their emissions targets to individual industrial entities, such as a power plant or paper factory. One example of a 'cap-and-trade' system is the 'EU ETS'. Other schemes may follow suit in time. This means that the ultimate buyers of credits are often individual companies that expect their emissions to exceed their quota (their Assigned Allocation Units, AAUs or 'allowances'). Typically, they will purchase credits directly from another party with excess allowances, from a broker, from a JI/CDM developer or on an exchange.

National governments, some of whom may not have devolved responsibility for meeting Kyoto obligations to industry, and that have a net deficit of allowances, will buy credits for their own account, mainly from JI/CDM developers. These deals are occasionally done directly through a national fund or agency. Since allowances and carbon credits are tradable instruments with a transparent price, financial investors can buy them on the spot market for speculation purposes, or link them to futures contracts. A high volume of trading in this secondary market helps price discovery and liquidity, and in this way helps to keep down costs and set a clear price signal in CO_2, which helps businesses to plan investments. This market has grown substantially, with banks, brokers, funds, arbitrageurs and private traders now participating in a market valued at about $60 billion in 2007.[10]

Although Kyoto created a framework and a set of rules for a global carbon market, there are in practice several distinct schemes or markets in operation today, with varying degrees of linkages among them.

Kyoto enables a group of several Annex I countries to join together to create a market-within-a-market. The EU elected to be treated as such a group, and created the EU Emissions Trading Scheme (ETS). The EU ETS uses EAUs (EU Allowance Units), each equivalent to a Kyoto AAU. The scheme went into operation on 1 January, 2005, although a forward market has existed since 2003. The UK established its own learning-by-doing voluntary scheme, the UK ETS, which ran from 2002 through 2006. This market existed alongside the EU's scheme, and participants in the UK scheme.

The Clean Development Mechanism (CDM) allows the creation of new carbon credits by developing emission reduction projects in Non-Annex I countries, while JI allows project-specific credits to be converted from existing credits within Annex I countries. CDM projects produce Certified Emission Reductions (CERs), and JI projects produce Emission Reduction Units (ERUs), each equivalent to one AAU. Kyoto CERs are also accepted for meeting EU ETS obligations and ERUs will become similarly valid from 2008 for meeting ETS obligations (although individual countries may choose to limit the number and source of CER/JIs they will allow for compliance purposes starting from 2008). CERs/ERUs are overwhelmingly bought from project developers by funds or individual entities, rather than being exchange-traded like allowances.

Several non-Kyoto carbon markets are in existence or being planned, and these are likely to grow in importance and numbers in the coming years. These include the New South Wales Greenhouse Gas Abatement Scheme, the Regional Greenhouse Gas Initiative and Western Climate Initiative in the

United States, the Chicago Climate Exchange and the State of California's recent initiative to reduce emissions. These initiatives taken together may create a series of partly linked markets, rather than a single carbon market. The common theme across most of them is the adoption of market-based mechanisms centred on carbon credits that represent a reduction of CO_2 emissions. The fact that some of these initiatives have similar approaches to certifying their credit makes it conceivable that carbon credits in one market may in the long run be tradable in other schemes. This would broaden the current carbon market far more than the current focus on the CDM/JI and EU ETS domains. An obvious precondition, however, is a realignment of penalties and fines to similar levels, since these create an effective ceiling for each market.

9.5 Clean Development Mechanism (CDM)

The Clean Development Mechanism (CDM) is an arrangement under the Kyoto Protocol allowing industrialized countries with a greenhouse gas reduction commitment (called Annex 1 countries) to invest in projects that reduce emissions in developing countries as an alternative to more expensive emission reductions in their own countries. A crucial feature of an approved CDM carbon project is that it has established that the planned reductions would not occur without the additional incentive provided by emission reductions credits, a concept known as 'additionality'. The CDM allows net global greenhouse gas emissions to be reduced at a much lower global cost by financing emissions reduction projects in developing countries where costs are lower than in industrialized countries. The distribution of CDM emission reductions, by country, is shown in Figure 9.3.

9.5.1 CDM Projects

An industrialized country that wishes to get credits from a CDM project must obtain the consent of the developing country hosting the project that it will

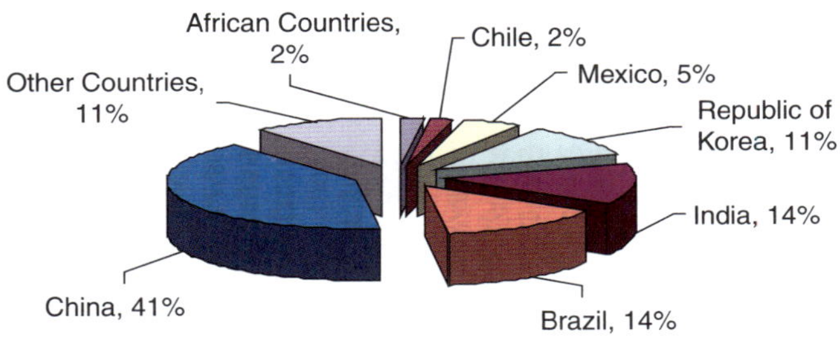

Figure 9.3 Distribution of CDM emission reductions, by country.

contribute to sustainable development. Then, using methodologies approved by the CDM Executive Board (EB), the applicant (the industrialized country) must make the case that the carbon project would not have happened anyway (establishing additionality), and must establish a baseline estimating the future emissions in the absence of the registered project. The case is then validated by a third party agency, called a Designated Operational Entity (DOE), to ensure the project results in real, measurable and long-term emission reductions. The EB then decides whether or not to register (approve) the project. If a project is registered and implemented, the EB issues credits, called Certified Emission Reductions (CERs, commonly known as carbon credits, where each unit is equivalent to the reduction of one metric tonne of CO_2e, *e.g.* CO_2 or its equivalent), to project participants based on the monitored difference between the baseline and the actual emissions, verified by the DOE.[11]

Small-scale renewable energy projects are helping to alleviate poverty and foster sustainable development. However, the low emission reductions per installation are making it difficult for such projects to derive value from participating in the CDM. Negotiators of the Marrakech Accords of November, 2001 (UNFCCC, 2002), as well as the CDM Executive Board, recognized this problem and adopted simplified CDM modalities and procedures for qualifying small-scale projects defined as (a) renewable energy project activities with a maximum output capacity equivalent of up to 15 MW, (b) energy efficiency improvement project activities which reduce energy consumption by an amount equivalent to 60 GW h per year and (c) other project activities whose emission reductions are less than 60 kt CO_2 per year.[12]

9.5.1.1 Baseline

The quantification of climate benefits of a project – *i.e.* the mitigation of GHG emissions – is done by means of a 'baseline'. The amount of emission reduction obviously depends on the emissions that would have occurred without the project minus the emissions of the project. The construction of such a hypothetical scenario is known as the baseline of the project. The baseline may be estimated through reference to emissions from similar activities and technologies in the same country or other countries, or to actual emissions prior to project implementation. A baseline describes the (theoretical) emissions that would have occurred in case the CDM project was not implemented. The amounts of CERs that can be earned by the project are then calculated as the difference of baseline emissions and project emissions. It allows that, for renewable energy technologies that displace technologies using fossil fuels, the simplified baseline is the fuel consumption of the technologies that would have been used in the absence of the project activity times an emission coefficient for the fossil fuel displaced. IPCC default values for emission coefficients may be used. For renewable energy technologies that displace electricity the simplified baseline is the electricity consumption times the relevant grid emission factor.

9.5.1.2 Additionality

To maintain the environmental integrity of the Kyoto Protocol, CDM credits are given only for activities that would otherwise not be expected to occur. Even in the hypothetical case of an off-grid situation where lifecycle costs of the solar water heating system would be cheaper than all other alternatives, the high up-front investment cost to a user in acquiring a solar water heating system would still be a high barrier to widespread market penetration. Most of the SWHs so far disseminated in India are sold with a subsidy.

9.5.1.3 Monitoring

Monitoring under small-scale rules consists in an annual check of all systems or a sample thereof to ensure that they are still operating. Since the installations of photovoltaic/thermal (PV/T) systems are often widely dispersed, monitoring costs could make CDM participation prohibitive if each user with a system is visited. Simple and efficient sampling procedures are therefore required. There are two variables that need to be monitored and verified in order to correctly establish emission reductions from PV/T systems according to small-scale methodology: (i) number of systems operating (evidence of continuing operation, such as on-going rental/lease payments could be a substitute); and (ii) annual hours of operation of an average system, if necessary estimated using survey methods. Annual hours of operation can be estimated from total output and output per hour if an accurate value of output per hour is available.

9.5.2 CDM as an Instrument of Technology Transfer

After a slow start, the CDM market has grown enormously due to greater political certainty after implementation of the Kyoto Protocol and because of the market's increasing experience with the process. As a result, there has been a steep increase from 64 projects and about 100 kt in expected certified emissions reductions (CERs) by 2012 in January, 2005, to 2647 projects and about 2.3 Gt in expected CERs in November, 2007.[13] As current estimates for the compliance shortfall of countries with reduction obligations under the Kyoto Protocol are around 3.3 Gt CO_2, the CDM can, contrary to initial doubts about its potential, contribute significantly to meeting Kyoto's reduction goals.[14] The market has attracted and created many different players from both the public and the private sector, whose objectives have included increasing awareness about the CDM.

The CDM's current contribution to technology transfer can be estimated by assessing empirical work based on Project Design Document (PDD) evaluations.[15–19] Seres[19] uses the most recent data and finds that 64% of expected CERs originate from projects involving technology transfer. Combining the expected 2.3 Gt in CERs with the average price calculated from primary CDM transactions in 2005 and 2006[14] suggests an investment flow of around 9 billion Euro into projects containing technology transfer. This exceeds the investment generated by the Global Environment Facility (GEF), a fund deliberately set up to promote technology transfer,[20] making the CDM the largest technology-transfer mechanism under the United Nations Framework Convention on Climate Change (UNFCCC).

9.6 Carbon Credit Analysis

Carbon credits are a key component of national and international emissions trading schemes that have been implemented to mitigate global warming. They provide a way to reduce greenhouse effect emissions on an industrial scale by capping total annual emissions and letting the market assign a monetary value to any shortfall through trading. Credits can be exchanged between businesses or bought and sold in international markets at the prevailing market price. Credits can be used to finance carbon reduction schemes between trading partners and around the world. *Per capita* greenhouse gas emissions on a world map are shown in Figure 9.4.[21] There are also many companies that sell carbon credits to commercial and individual customers who are interested in lowering their carbon footprint on a voluntary basis. These carbon offsetters purchase the credits from an investment fund or a carbon development company that has aggregated the credits from individual projects. The quality of the credits is based in part on the validation process and sophistication of the fund or development company that acted as the sponsor to the carbon project.

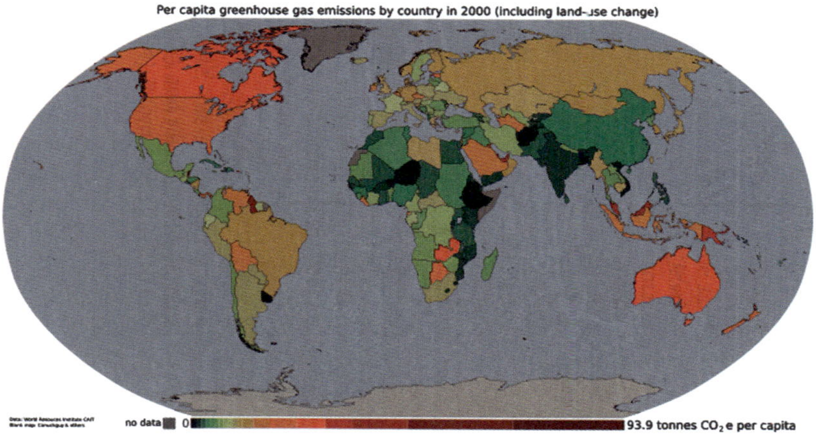

Figure 9.4 *Per capita* greenhouse gas emissions on the world map. (Source: Wikipedia 2009).

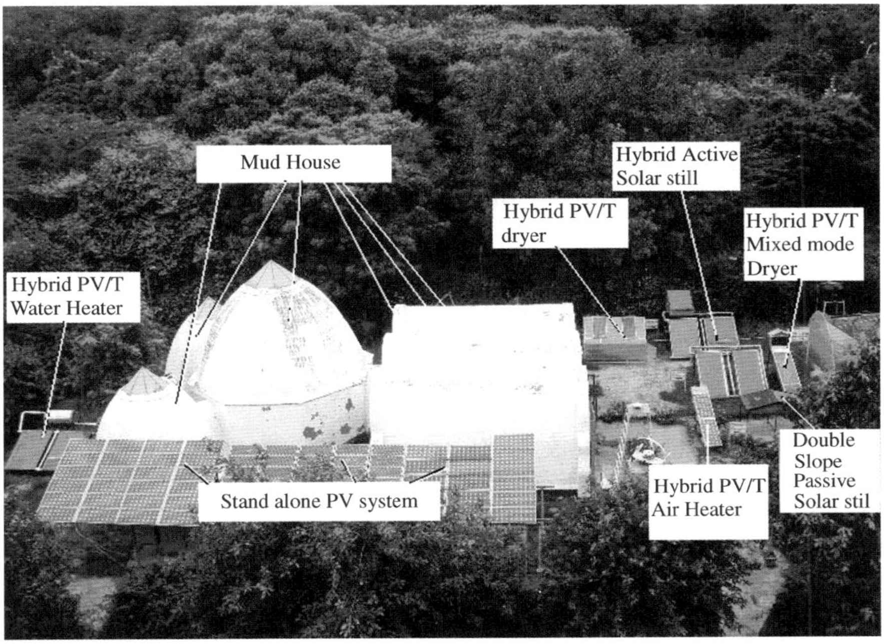

Figure 9.5 Aerial view of solar systems at Solar Energy Park, IIT, Delhi.

9.6.1 Solar Energy Park (SEP)

SEP is located in the campus of IIT, New Delhi (29° 35′ N, 77° 12′ E). It is spread over an area of 23 m × 42 m. It has built-up area (mud house) of 11 m × 13 m. The aerial view of Solar Energy Park at IIT Delhi has been shown in Figure 9.5. The stand-alone PV system and various PV/T solar systems have also been marked in the same figure. Brief descriptions of various PV/T systems are given in Sections 7.2.1, 7.3.2, 7.4.3, 7.5.3 and 7.5.4. The other PV/T systems of SEP are described below.

9.6.1.1 Mud House

The ‘mud house’ is a six-room building, having been built with traditional building material, generally used to build houses in Indian villages. It is a natural conditioned vaulted or curved roof structure for a composite climate, made of a three-layered 23-cm-thick roof. The inside layer is 7-cm-thick brick, the middle layer is mud and is 12-cm thick and the outer layer consists of 4-cm-thick brick tiles. The walls have two layers. The outer layer is 12-cm thick and is made of mud. The inner layer is of brick having 7-cm thickness. Mud forms 70% of the building material, hence the name ‘mud house’. There is provision of day-lighting in the central hall of the mud house. The variation in room temperature is attenuated as compared to ambient air temperature fluctuations because of the thick walls and roof of the mud house and the high thermal heat capacity and low heat conductivity of mud. The maximum monthly heating and cooling potential are 550 megajoules (MJ) in February and 400 megajoules (MJ) in June, respectively, for New Delhi climatic conditions. The mud house is integrated with an Earth air-heat exchanger. The mud house can maintain a constant 14 °C–16 °C inside room air temperature during the winter season. Under natural circulation mode, during summer conditions, the inside room air temperature can be maintained between 32 °C and 35 °C when the outside ambient air temperature reaches 45 °C. Using an Earth air-heat exchanger, the room air temperature can be maintained from 28 °C to 30 °C in the summer season.

9.6.1.2 Photovoltaic Systems

The stand-alone photovoltaic system (SAPV) consists of two arrays of PV modules. The first array is made by CEL and has 32 modules and the second array is made by SIEMENS and has 34 modules. These are integrated with the electrical load of the Solar Energy Park (SEP), including that of the 'mud house'. The direct current produced by the SAPV power system is converted into standard 220 V AC supply by an inverter; the DC produced is also used to charge the battery bank of the backup power system. When there is no solar intensity (cloudy sky or during the night), the backup power system takes over the electrical load. The SAPV system supplies uninterrupted power to all the electrical appliances fitted in the Solar Energy Park (SEP), including that in the mud house. The first array has 1.12 kilowatt peak power output rating. The second array has 2.4 kilowatt peak power output rating. The battery bank has specification 48 volt/360 ampere-hour (Ah). Tubular type, 6-volt/180 ampere-hour batteries have been used in the battery bank. The inverter is of 3 KVA rating. The inverter has efficiency in the range of 90%.

9.6.2 Solar PV/T Systems

The overall annual thermal energy (eqn (8.31a)) and exergy (eqn (8.32)) of each PV/T system on the basis of experimental and theoretical results were obtained for New Delhi climatic conditions. On the basis of validation, we have observed that there are about 5–15% estimated errors between the experimental and theoretical results for each system. The results of an overall thermal energy and exergy evaluation for each system have been shown in Figures 9.6 and 9.7. It can be observed that the maximum overall annual thermal energy and exergy are obtained for the mud house and greenhouse, respectively. The total overall annual thermal and exergy energy for Solar Energy Park are 106,556.00 kWh (106.5 MWh) and 2692 kWh (2.692 MWh).

Carbon dioxide reduction by solar photovoltaic power plants installed all over the world has been compiled by Denis Lenordic. Data for carbon

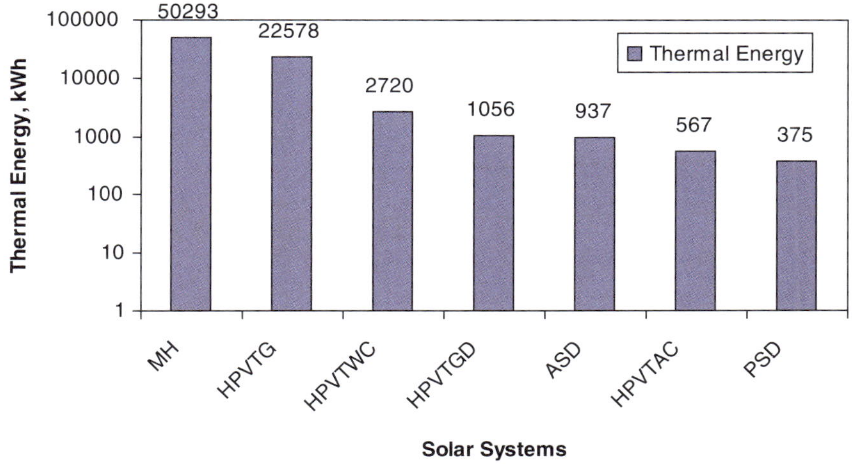

Figure 9.6 Annual thermal energy gain by various solar systems in Solar Energy Park. (where MH = mud house, HPVTG = hybrid greenhouse dryer, HPVTWC = hybrid PV/T water heater, HPVTGD = hybrid greenhouse dryer for cultivation, ASD = active solar still, HPVTAC = hybrid photovoltaic/thermal air heater and PSD = passive solar still).

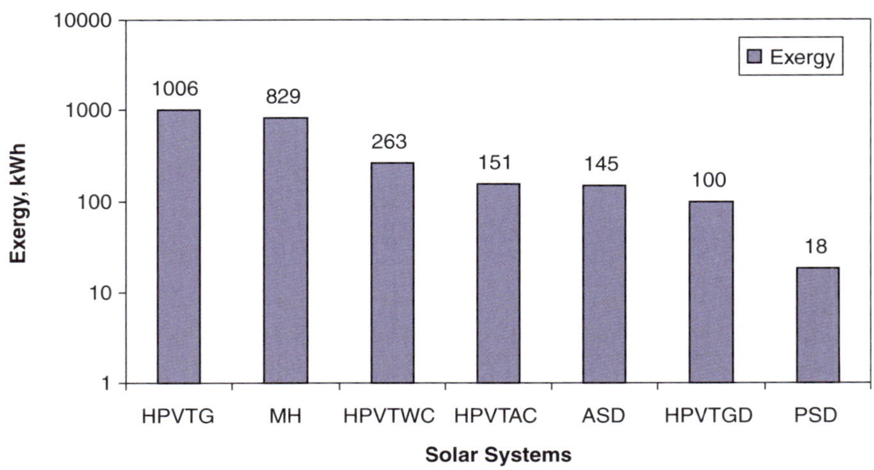

Figure 9.7 Annual thermal exergy gain by various solar systems in Solar Energy Park.

dioxide emission reduction by the top 200 solar photovoltaic power plants are available.[22] The data available include power produced per annum in MWh, annual carbon emission reduction. The average annual carbon emission reduction per MWh of electricity produced, for the top 100 solar voltaic power plants, for which data of electricity produced in MWh and emission reduction per annum are available, comes to 0.982 tons of carbon dioxide emission reduction per MWh of electricity produced.[5] However, 40% is transmission and distribution losses and 20% is due to the inefficient electric equipment used for Indian conditions *i.e.* if 1 MWh of electricity is required at the consumption point then 1.6 MWh of electricity should be produced at the generation point.

$$\begin{aligned}\text{Then, the total } CO_2 \text{ mitigation} &= 1.6 \times 0.982 \\ &\approx 1.58 \text{ tons of } CO_2.\end{aligned}$$

The mitigation of CO_2 per year on the basis of thermal energy and exergy are 99.3 tons and 2.5 tons, respectively.

If carbon dioxide emission reduction is at present being traded at €20 ton^{-1}, then the carbon emission reduction by various solar systems (Figures 9.6 and 9.7) in SEP will be evaluated as follows.

CO_2 credit earned by annual saving = €99.3 × 20 = €1986.00 on the basis of energy. And similarly = €2.5 × 20 = €50.00 on the basis of exergy.

9.6.3 Carbon Credits Earned by Stand Alone Photovoltaic (SAPV) System[23]

The total energy consumption in the Solar Energy Park by instruments/electrical gadgets in various system mentioned in Section 9.6.1 is evaluated as 23.52 kWh per day.

The total installed capacity of the SAPV system is as follows:

(a) Five frames with 32 CEL modules (each of 35 Wp), PV Power = 1120 Wp and
(b) Nine Frames with 34 SIEMEN modules (each 75 Wp) PV power = 2250 Wp.

Total power produced = (1120 + 2250) Wp = 3670 Wp = 3.670 kWp.

Assuming an average 12 hours of sunshine per day, which is generally true for European conditions all over the country,

Total power produced per day = 3670 × 12 = 44,040 Wh = 44.04 kWh.

On average there are 300 days of clear sky per annum, then

$$\begin{aligned}\text{Total power produced per annum} &= 44.04 \times 300\,\text{kWh} \\ &= 13,212\,\text{kWh} = 13.212\,\text{MWh}.\end{aligned}$$

If the unit cost of electricity is Rs 5.5 (€0.1), then

$$\begin{aligned}&\text{Cost of energy produced per annum} \\ &\qquad = \text{Rs}\,13,212 \times 5.5 = \text{Rs}\,72,666(€1327.33).\end{aligned}$$

If the sunshine hours for Indian conditions is considered as six hours per day then

$$\text{Cost of energy produced per annum} = \text{Rs}\frac{13,212}{2} \times 5.5$$
$$= \text{Rs}\,36,333(€663.62)$$

where the conversion of unit cost is €1 = Rs 55 and Rs 40 = 1 USD at the level of 2007.

Taking the value of 0.932 tons of carbon dioxide emission reduction per MWh of electricity for the SAPV plant installed in the solar energy park

$$CO_2\ \text{(carbon)emission reduction} = 13.212 \times 0.932 = 12.310545 = 12.31\ \text{tons.}$$

As was pointed out earlier, if carbon dioxide emission reduction is at present being traded at €20 ton^{-1}, then

$$CO_2\ \text{emission reduction by SAPV plant per annum}$$
$$= €12.31 \times 20 = €246.2(\text{Rs}\,13479.5)$$

For twelve sunshine hours

$$CO_2\ \text{emission reduction by SAPV plan} = \frac{246.2}{2} = €123.1\ \text{per annum}$$

For six sunshine hours

The total carbon credit earned by Solar Energy Park (SEP) will be the sum of the carbon credits earned by the solar PV/T system and SAPV system in SEP.

$$CO_2\ \text{credit earned by Solar Energy Park (SEP)} = €(50 + 246.2) = €296.2$$

For twelve sunshine hours

$$= €(50 + 123.1) = €173.1$$

For six sunshine hours

9.6.4 Carbon Credit on National Level by SAPV System[23]

There are 602 districts in India based on 2005 statistics (Table 9.3) and as per the 2001 census there are approximately 639,000 villages. The census of India regards most settlements of fewer than 5000 as a village. These settlements range from tiny hamlets of thatched huts to larger settlements of tile-roofed stone and brick houses. Most Indian villages are small; nearly 80% have fewer than 1000 inhabitants, according to the 1991 census. Most are nucleated settlements, while others are more dispersed. It is in villages that India's most basic business-agriculture takes place. This means there are

Table 9.3 Statewise distribution of districts in India.[24]

S. No.	*Name of state*	*Number of districts*
1	Andhra Pradesh	23
2	Arunachal Pradesh	15
3	Assam	23
4	Bihar	37
5	Chhattisgarh	16
6	Goa	2
7	Gujarat	25
8	Haryana	19
9	Himachal Pradesh	12
10	Jammu and Kashmir	14
11	Jharkhand	22
12	Karnataka	27
13	Kerala	14
14	Madhya Pradesh	48
15	Maharashtra	35
16	Manipur	9
17	Meghalaya	7
18	Mizoram	8
19	Nagaland	8
20	Orissa	30
21	Punjab	19
22	Rajasthan	32
23	Sikkim	4
24	Tamil Nadu	30
25	Tripura	4
26	Uttrakhand	13
27	Uttar Pradesh	70
28	West Bengal	18
29	Andaman and Nicobar Islands	2
30	Chandigarh	1
31	Dadra and Nagar Haveli	1
32	Daman and Diu	1
33	Lakshadweep	1
34	Pondicherry	1
35	Delhi	9
	Total	**602**

at least 127,800 villages in India each having a population of more than 1000. These villages should be the first to have a mud house. Presuming one SAPV system is built to be the nucleus of the village activities in each of such villages, the total number of SAPV systems required will be about 127,800.

For 12 sunshine hours, the carbon credit earned from all these villages per annum will be

$$\begin{aligned}\text{Annual carbon credits} &= 127,800 \times \text{Rs}\,13,479.5 = \text{Rs}\ 1,722,680,000\\ &= \text{Rs}\,1722.68\,\text{million}(€\,3.15\ \text{million}).\end{aligned}$$

If the cost of the SAPV system installed in the solar park is €19,936.38, then

$$\begin{aligned}\text{Capital cost of installing 127,800 SAPV systems} &= 127,800 \times 19,936.38 \times 54.75\\ &= \text{Rs } 13.9495 \text{ million}\\ &= € \, 2.54786 \times 10^9.\\ \text{Total power produced} &= 13.212 \times 127,800 \text{ MWh}\\ &= 1,688,493.6 \text{ MWh}.\end{aligned}$$

This is equivalent to Rs 9286.71 million (€169 million) worth of electricity.

If in the first stage SAPV systems are installed on each district headquarters of the country

$$\begin{aligned}\text{the capital cost of SAPV system} &= 602 \times 19,936.38\\ &= €11.55 \text{ million} (\text{Rs } 657.093 \text{ million})\end{aligned}$$

Total carbon credits earned by such systems = Rs 13,479.5 × 602 = Rs 8,114,659.00

$$\begin{aligned}\text{Value of total electricity generated by such systems} &= \text{Rs } 13,212 \times 602 \times 5.5\\ &= \text{Rs } 43.75 \text{ million}\\ &= € \, 0.8 \text{ million}.\end{aligned}$$

9.6.5 Effect of Solar Intensity and Number of Clear Days

Power produced by the SAPV system is proportional to solar intensity and to the number of clear days in a year. Power produced, carbon credits earned and return on capital have been calculated and are given in Table 9.4, assuming the

Table 9.4 Variation in power produced, carbon credits earned and return on capital with variation in solar intensity and number of clear days in a year.

Sl. No.	*Solar intensity W m*$^{-2}$	*No. of clear days in a year*	*Power produced MWh annum*$^{-1}$	*Carbon credits earned annum*$^{-1}$ *(€)*
1	700	300	8.397	156.52
2	700	250	6.998	130.442
3	700	200	5.598	104.347
4	500	300	5.998	111.803
5	500	250	4.998	93.163
6	500	200	3.998	74.523
7	350	300	4.199	78.269
8	350	250	3.499	65.220
9	350	200	2.799	52.173

solar intensity to be 700 W m^{-2}, 500 W m^{-2} and 350 W m^{-2} and the number of clear days in a year to be 300, 250 and 200. The power output of the SAPV system varies with variation in solar intensity; the efficiency of the solar PV modules on a typical day (September 11, 2006) used in the Solar Energy Park at 6.08% for the CEL module and 12.52% for the SIEMENS module has been computed.[4] The efficiency of the cells will vary with variation in ambient temperature and inclination of the PV modules.

9.7 Energy Pricing

Government energy pricing policies in most developing countries have multiple implicit or explicit objectives. These include economic efficiency, government revenues, equity and incidence (maintenance or improvement of income distribution, or promotion or protection of particular sectors or groups), demand management, domestic energy resource development and security of supply. Energy pricing is a particularly important issue in developing countries because energy forms a large part of their economies' costs and is often a major source of government revenue through either taxation or domestic resource development.[25] Pricing of energy fuels has been critical in determining the pattern of development of the energy sector.

Carbon dioxide (CO_2) emissions from energy utilization are a major factor contributing to the greenhouse effect. Removing existing price distortions and imposing privately efficient energy pricing makes a substantial impact on energy demand. A goal of pricing policy analysis should be to consider alternative policies in a manner such that the interrelationships and trade-offs between the multiple objectives can be estimated. These estimates can be expressed as impacts on particular economic measures such as government revenues, balance of payments, household cost of living, sectorial output, prices and profitabilities and welfare or efficiency losses. Other measures such as refinery imbalances (difference between refined products produced and those consumed) or employment level changes might also be pertinent. Such an analysis can be undertaken either in a static framework for a particular base year or, preferably, a more dynamic framework in which the effect of alternative prices are examined in the context of important changes in national and international economic conditions. The major elements of this study are:[25]

1. energy pricing policy analysis;
2. international and domestic economic analyses;
3. energy use analyses;
4. impact analyses.

The pricing analysis includes examination of existing prices and policies and generation of alternative policies and the corresponding product-specific prices. The economic analyses include international energy prices (current and forecast), domestic macroeconomic forecasts of GDP, sectoral growth, inflation, foreign exchange rates and estimate of shadow prices and discount rates. The energy use analyses are the largest components and include characterization of current energy use on a detailed sectoral basis (quantities, fuel mixes and energy intensities); energy demand forecasts including conservation and substitution issues, own- and cross-price elasticities of individual energy demands, and if possible energy-capital-labour factor input relationships. These analyses pertain primarily to definition of a baseline and of how energy use varies in response to price, technical and structural changes. A second major aspect of the energy analyses uses this information and the alternative energy pricing scenarios to estimate scenario-specific effects on energy demands and sector costs/prices. Input-output models are useful for representing sectoral energy use and estimating the inter-sectoral interactions needed for impact analyses. The impact analyses estimate the changes in various measures (revenues, consumer price index and efficiency losses) due to the effects of different energy price scenarios. The most important inputs to these analyses are scenario-specific changes in energy demands and in sectoral costs/prices. Pricing policy decisions can then be based on scenario impacts and their trade-offs and government priorities.

Problems

9.1 Explain the factors which reduce the capital cost and cost of electricity generated by solar photovoltaic systems.
9.2 What is carbon credit trading? Explain in detail with examples.
9.3 Calculate the carbon dioxide emission per year from a PV-integrated greenhouse dryer in a lifetime of 20, 30 and 40 years, when the total embodied energy required for manufacturing the system is 2650 kWh. Hint: see eqn (9.1) and Example 9.1.
9.4 Describe the principles of the Kyoto Protocol and its mechanisms.
9.5 What is emission trading?
9.6 Explain the instruments required for implementing/establishing a CDM project.

References

1. R. Kalshian, Energy versus emissions: The big challenge of the new millennium, By Info Change News & Features, www.infochangeindia.org/agenda5_01.jsp, accessed 21 March 2008.
2. International Energy Agency, http://www.iea.org/Textbase/stats/index.asp, accessed 8 August 2008.
3. Photovoltaic system economics, Economics and Environmental Impacts, http:/www.pvresources.com/en/ economics.php, accessed 10 June 2008.

4. O. Prakash, A. Chel and G.N. Tiwari, in *3rd International Conference on Solar Radiation and Day Lighting (SOLARIS 2007)*, New Delhi, India, 2008, Vol. II, pp. 87–101.
5. M. Watt, A. Johnson, M. Ellis and N. Quthred, *Progress In Photovoltaics: Research and Applications*, 1998, **6**(2), 127–136.
6. H. Lund, *Energy*, 2006, **31**, 2325–2332.
7. G. Klepper and S. Peterson, *Energ. J.*, 2006, **27**(2), 1–26.
8. C. Böhringer and T. F. Rutherford, *Environ. Resource Econ.*, 2002, **22**(3), 391–417.
9. E. Johnsona and R. Heinen, *Environ. Int.*, 2004, **30**, 279–288.
10. N. Anger, *Energ. Econ.*, 2008, **30**, 2028–2049.
11. H. de Coninck, C. Fischer, R. G. Newell and T. Ueno, *Energ. Pol.*, 2008, **36**, 335–356.
12. P. Purohit and T. C. Kandpal, *International Journal of Ambient Energy*, 2005, **26**, 135–146.
13. UNEP Risoe, CDM/JI Pipeline Analysis and Database, http://www.cdmpipeline.org/cers.htm, accessed 3 September 2008.
14. K. Capoor and P. Ambrosi, State and Trends of the Carbon Market 2007, www.ieta.org/ieta/www/pages/download.php?docID = 1567-4k, accessed 2 March 2008.
15. H. de Coninck, F. Haake and N. van der Linden, *Clim. Pol.*, 2007, **7**, 444–456.
16. A. Dechezleprêtre, M. Glachant and Y. Ménièrea, *Energ. Pol.*, 2008, **36**, 1273–1283.
17. E. Haites, M. S. Duan and S. Seres, *Clim. Pol.*, 2006, **6**, 327–344.
18. A. P. Velasco, *Joint Implementation Quarterly*, 2007, **13**, 5–6.
19. S. Seres, Analysis of technology transfer in CDM Projects, http://cdm.unfccc.int/Reference/Reports/TTreport/report1207.pdf, accessed 10 March 2008.
20. C. Egenhofer, L. Milford, N. Fujiwara, T. L. Brewer and M. Alessi, *European Climate Platform*, 2007, **4**, 1–32.
21. Per capita greenhouse gas emissions on world map, http://en.wikipedia.org, accessed 20 July 2008.
22. European Climate Exchange, http://www.europeanclimateexchange.com, accessed 6 July 2008.
23. Prabhakant and G. N. Tiwari, *The Open Energy and Fuels Journal*, 2008, **1**, 57–66.
24. Censes of India, http://www.censusindia.gov.in, accessed 12 August 2008.
25. R. J. de Lucia and M. C. Lesser, *Energ. Pol.*, 1985, **13**(4), 345–349.

CHAPTER 10

Economic Analysis

10.1 Introduction

Interest in the development of and dissemination of renewable energy technologies has again reignited in the view of increasing global climate change concerns. In addition to the development of new and appropriate technology, issues related to their financial and economic viability and financing of renewable energy systems are being given considerable importance. Techno-economic analysis is the area of engineering where engineering judgment and experience are utilized. Analysis is used for project cost control, profitability analysis, planning, scheduling and optimization of operational research *etc*. In the case of PV/T systems, it is necessary to work out its economic viability so that the users of the technology may know its importance and can utilize the area under their command to their best advantage.

An effective economic analysis can be made by the knowledge of cost analysis, using cash flow diagrams and some other methods.

Techno-economic analysis of PV/T systems mainly depends on the following factors:

- Initial investment for construction of system;
- Initial cost of additional heating, if any;
- Operating cost;
- Annual maintenance cost;
- Life of the system and its salvage value.

In addition to the above points, it is also necessary to mention the impact on the environment due to CO_2 emissions by embodied energy (one time) of solar systems. The energy used to operate it (annually) and pretreatments *etc*. should be taken into account.

For effective economic analysis of PV/T systems, the subsequent sections deal with the knowledge of cost analysis, cash-flow diagrams, pay back time and benefit-cost analysis *etc*.

RSC Energy Series No. 2
Fundamentals of Photovoltaic Modules and Their Applications
By G. N. Tiwari and Swapnil Dubey

Published by the Royal Society of Chemistry, www.rsc.org

10.2 Cost Analysis

Financial evaluation of PV/T technologies necessitates that various energy resource technology combinations for a given end use are compared with each other. For such comparisons it is necessary that monetary values at different points in time be reduced to an equivalent basis.

10.2.1 Capital Recovery Factor

Let P be the present amount invested at zero ($n=0$) time at the interest rate of i per year and if S_n is its future value at the end of n years, then the cash flow can be diagrammatically shown as follows:

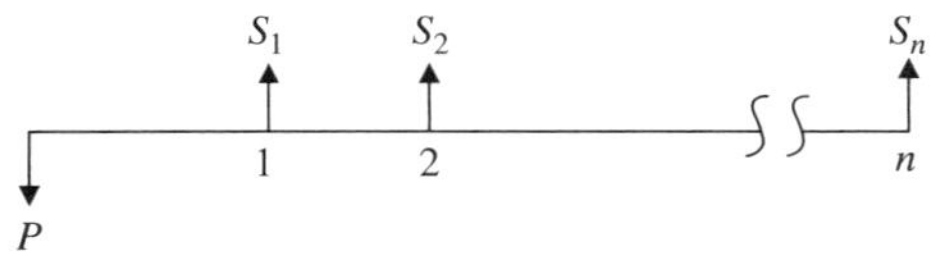

At the end of one year, the time value of investment P is given by

$$S_1 = P + iP = P(1+i)$$

and, at the end of second year the value becomes

$$S_2 = S_1 + iS_1 = P(1+i) + iP(1+i) = P(1+i)(1+i) = P(1+i)^2.$$

Similarly, at the end of the third and the nth years, respectively, the value becomes

$$S_3 = P(1+i)^3 \text{ and } S_n = P(1+i)^n.$$

For simplicity, assuming S_n to be S, the above equation can be written as

$$S = P(1+i)^n \tag{10.1a}$$

Here, $S > P$ for $i > 0$, considering compound interest law. Further, the above equation can be simplified as

$$S = P\,F_{PS} \tag{10.1b}$$

i.e. Future value = (Present value) × (Future value factor)

F_{PS} is more completely designated as $F_{PS,i,n}$, where i is the rate of interest, n is the number of years under consideration, and

$$F_{PS.i,n} = (1+i)^n \tag{10.1c}$$

where $F_{PS,i,n}$ is known as the *compound interest factor* or *future value factor*, which evaluates the future amount if the present amount is known, *i.e.* conversion of P into S. Thus, the compound interest factor when multiplied with the present value gives the future value.

If one year is divided into p equal units of period, then n becomes np and i becomes i/p, which is the rate of return per unit period.

Substitution of these values in eqns (10.1b) and (10.1c) gives

$$S = P\left(1 + \frac{i}{p}\right)^{np}$$

This can be written as

$$S = P\left[\left(1 + \frac{i}{p}\right)^{p}\right]^{n}$$

where the expression $(1 + i/p)^p$ can be expressed as follows:

$$\left(1 + \frac{i}{p}\right)^{p} = 1 + \text{effective rate of return}$$

$$\text{or effective rate of return} = \left(1 + \frac{i}{p}\right)^{p} - 1 = \text{i for p} = 1$$

$$> i \quad \text{for } p > 1 \qquad (10.2)$$

For simple interest

$$S = \mathrm{P}(1 + \mathrm{ni}) = P + (iP)\mathrm{n} \qquad (10.3)$$

Equation (10.1a)can be rewritten as

$$\begin{aligned} P &= S/(1 + i)^{\mathrm{n}} \\ \textit{i.e.} \qquad P &= S(1 + i)^{-\mathrm{n}} \end{aligned} \qquad (10.4a)$$

This shows that the future amount (at the nth year) is reduced when converted against the calendar to the present value (at zeroth time), assuming i to be positive.

$$P = S\,F_{\mathrm{SP}} \qquad (10.4b)$$

or, *Present value = future value × (present value factor).*

The numerical value of F_{SP} will always be less than unity. For this reason, present-worth calculations are generally referred to as *discounted cash flow (DCF)* methods. Other terms generally used in reference to present-worth (PW) calculations are *present value (PV)* and *net present value (NPV).*

From eqns (10.1b) and (10.4b), F_{PS} and F_{SP} can be related as

$$\begin{aligned} F_{\mathrm{PS}} &= \frac{1}{F_{\mathrm{SP}}} \\ \text{or} \qquad F_{\mathrm{PS}} \cdot F_{\mathrm{SP}} &= 1 \end{aligned} \qquad (10.5)$$

Example 10.1

A low interest loan of USD 2000 is provided for the purchase of a low-capacity hybrid solar dryer for a period of 18 months at a simple interest rate of 5%. What is the future amount due at the end of the loan period?

Solution

$$\begin{aligned} \text{Simple interest}(I_s) &= P\,n\,i \\ &= 2000 \times 18/12 \times 5/100 (18\,\text{months} = 12/18\,\text{years}, \\ 5\% &= 5/100) \\ \text{or,} \quad I_s &= \text{USD}\,150. \end{aligned}$$

Thus the total amount due at the end of the loan period

$$= 2000 + 150 = \text{USD}\,2150.$$

Example 10.2

If USD 20,000 compounds to USD 28,240 in 4 years of a given solar system, what will be the rate of return?

Solution

Using eqn (10.1a), $S = P\,(1 + i)^n$ and substituting $S =$ USD 28,240, $P =$ USD 20,000 and $n = 4$, we get

$$28,240 = 20,000(1 + i)^4 \text{ or } (1 + i)^4 = 1.412.$$

Solving the above equation, we get

$$i = 0.09 \text{ or } 9\% \text{ per year.}$$

Example 10.3

How long will it take for money to double if compounded annually at 10% per year?

Solution

Let us assume that the money doubles in n years. Then $S = 2P$.

Using eqn (10.1a) and substituting $S = 2P$, we get

$$\begin{aligned} 2P &= P(1 + 0.10)^n \\ 2 &= (1 + 0.10)^n. \end{aligned}$$

Solving the above equation, we get

$$\log 2 = n \log 1.1, \quad i.e.\ n = 7.3\,\text{yrs.}$$

The money will be doubled in 7.3 years.

Example 10.4

Calculate the effective rate of return for 10% interest for $p = 5$ and $p = 12$.

Solution

From eqn (10.3), we have

$$\textit{Effective rate of return} = \left(1 + \frac{i}{p}\right)^p - 1$$

For $p = 5$; the *Effective rate of return* $= \left(1 + \frac{0.10}{5}\right)^5 - 1 = (1.02)^5 - 1 = 0.104$
For $p = 12$; the *Effective rate of return* $= \left(1 + \frac{0.10}{12}\right)^{12} - 1 = 0.1047.$

Example 10.5

It is estimated that about 120 million households in the country can benefit from the use of improved PV/T drying techniques. What is the required growth rate to achieve the potential in the next 20 years if the number of improved drying techniques disseminated so far is 30 million?

Solution

$F = 120$ million
$P = 30$ million
$n = 20$ years

$$\log(1 + i) = 1/n \log(F/P)$$

$$\text{or} \quad \log(1 + i) = 1/20 \log(120/30)$$

$$\text{or} \quad \log(1 + i) = 0.05 \log 4 = 0.030103$$

which may be solved to give

$$i \approx 0.07177 (\text{or} \approx 7.18\%).$$

Thus, a compound rate growth rate of more than 7 per cent could be required to achieve the estimated potential of improved drying techniques utilization in the country in the next 20 years.

Example 10.6

A farmer borrows USD 2000 to buy a PV/T hybrid solar dryer and returns USD 2100 at the end of six months. What was the rate of interest paid by the farmer?

Solution

We have $S = 2100$, $P = 2000$ and $n = 6/12$. Thus, using eqn (10.3) we can write:

$$2100 = 2000\left(1 + \frac{6}{12}i\right)$$

Simplifying, we can write

$$1.05 = 1 + 0.5\,i,$$

$$\text{or,} \qquad i = 0.10 \text{ or } 10\%.$$

Example 10.7

The owner of a small restaurant borrows USD 10,000 for a hybrid PV/T solar water heater at 10% for 4 yrs and 4 months. Considering compound interest, calculate the money paid.

Solution

Using eqn (10.3), we have

$$S = 10,000\left[1 + \frac{4}{12} \times 0.1\right] = Rs.\ 10,333$$

The future amount after 4 months is evaluated as USD 10,333, which becomes P for another 4 years. For compound interest, using eqns (10.1b) and (10.1c), we have:

$$S = PF_{PS,10\%,4} = 10,333(1 + 0.1)^4$$

Thus, substituting the numerical values in above equation

$$S = 10,333(1.4641) = \text{USD } 15,129.$$

10.2.2 Unacost

In solving engineering economic problems it is convenient to diagram expenditures (debits) and receipts (credits) as vertical lines positioned along a horizontal line representing time. Expenditures and receipts can point in opposite directions. By using this concept, a uniform annual amount will be discussed.

The smallest unit of time normally considered is a year. Consider a uniform end-of-year annual amount R (*unacost*) for every year for a period of n years. The diagram for this is as shown below:

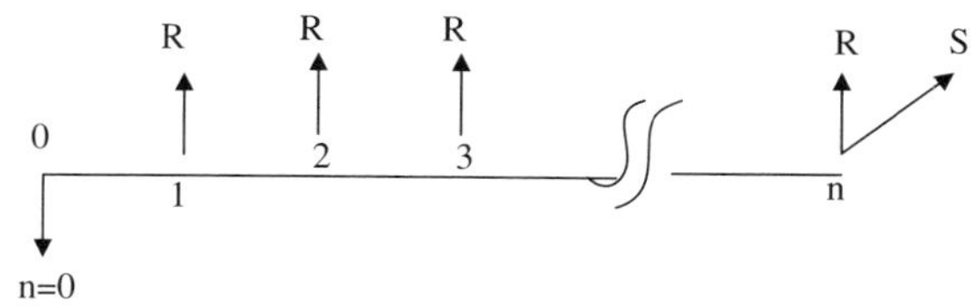

Let P represent single present value at initial time (*i.e.* at $n=0$), then by eqn (10.4a), we get

$$P = R\left[\frac{1}{1+i} + \frac{1}{(1+i)^2} + \ldots\ldots + \frac{1}{(1+i)^n}\right] \quad (10.6a)$$

This can be written as

$$P = R\sum_{1}^{n}\frac{1}{(1+i)^n}$$

Present worth factor $= \frac{1}{(1+i)^n}$

Equation (10.6a) is a geometric series, which has $1/(1+i)$ as the first term and $1/(1+i)$ as the ratio of n successive terms. The term summation of geometric series in eqn (10.6a) can be further evaluated as

$$\sum_{1}^{n}\frac{1}{(1+i)^n} = \frac{1}{(1+i)}\left[\frac{1-\left\{\frac{1}{(1+i)}\right\}^n}{1-\frac{1}{(1+i)}}\right] = \frac{(1+i)^n - 1}{i(1+i)^n}$$

Substituting in eqn (10.6a), we get

$$P = R\left[\frac{(1+i)^n - 1}{i(1+i)^n}\right] = R\, F_{\mathrm{RP,i,n}} \quad (10.6b)$$

or, Present value = (Unacost) × (Unacost present value factor)

$$\text{where} \quad F_{\mathrm{RP,i,n}} = \left[\frac{(1+i)^n - 1}{i(1+i)^n}\right] \quad (10.6c)$$

$F_{\mathrm{RP,i,n}}$ is the *equal-payment series present value factor* or *annuity present value factor*.

Equation (10.6b) can also be rewritten as

$$R = P\left[\frac{i(1+i)^n}{(1+i)^n - 1}\right] = P\, F_{\mathrm{PR,i,n}} \quad (10.7a)$$

or, *Unacost = (Present value) × (Capital recovery factor)*

where, $F_{PR,i,n}$

$$F_{PR,i,n} = \left[\frac{i(1+i)^n}{(1+i)^n - 1}\right] = CRF \tag{10.7b}$$

This is also known as the *capital recovery factor* (CRF). The relation between *equal-payment series present value factor* and *capital recovery factor* can be obtained by eqns (10.6c) and (10.7b) as

$$F_{RP,i,n} = \frac{1}{F_{PR,i,n}} \tag{10.7c}$$

Example 10.8

A large-capacity water heating system is expected to save USD 4,000 every year in terms of fuel savings. If the effect of escalation in the price of fuel saved is neglected, what is the present worth of fuel saving in the 5th, 10th, 15th, 20th, 25th and 30th years for a discount rate of 12%?

Solution

Given that the amount of fuel saving is USD 4000 per year and $i = 0.12$, the values of the present worth factors and the corresponding present worth of annual fuel saving for the desired years are tabulated below.

Year (n)	*Present worth factor*	*Present worth of fuel savings*
	$PWF = \left[\frac{1}{(1+i)^n}\right]$	(4000 × PWF)
5	$1/(1.2)^5 = 0.5670$	2269.7
10	$1/(1.2)^{10} = 0.3220$	1287.8
15	$1/(1.2)^{15} = 0.1827$	730.7
20	$1/(1.2)^{20} = 0.1037$	414.6
25	$1/(1.2)^{25} = 0.0588$	235.2
30	$1/(1.2)^{30} = 0.0334$	133.5

It may be noted that the present worth of fuel savings in later years of the useful life of the domestic solar water heating system is rather small. Thus, the present value analysis of a renewable energy system with longer useful life may not be representative of its actual usefulness to the user.

10.2.3 Sinking Fund Factor

The future value S at the end of n years can be distributed into an equal uniform end-of-year annual amount R as discussed above. It will also be known as a *uniform end-of-year annual amount* but corresponding to the future value S.

Equation (10.7a) can be expressed in terms of S by using eqn (10.4a) as

$$R = S(1+i)^{-n}\cdot\left[\frac{i(1+i)^n}{(1+i)^n-1}\right] = S\cdot\left[\frac{i}{(1+i)^n-1}\right] = S\,F_{\mathrm{SR,i,n}} \tag{10.8a}$$

or, *Unacost* = *(Future amount)* × *(Sinking fund factor)*

$$\text{where,} \qquad F_{\mathrm{SR,i,n}} = \left[\frac{i}{(1+i)^n-1}\right] = \mathrm{SFF} \tag{10.8b}$$

This is referred to as the sinking fund factor (SFF). This is mostly used to calculate the *uniform end-of-year annual amount* corresponding to the salvage value of any system in future after completion of the system life. Equation (10.8a) can be rewritten as

$$S = R\left[\frac{(1+i)^n-1}{i}\right] = R\,F_{\mathrm{RS,i,n}} \tag{10.9a}$$

or, *Future amount* = *(Unacost)* × *(Equal payment series future value factor)*

$$\text{where,} \qquad F_{\mathrm{RS,i,n}} = \left[\frac{(1+i)^n-1}{i}\right] \tag{10.9b}$$

This is known as the *equal payment series future value factor*. The reciprocal relation between the *sinking fund factor* and the *equal payment series future value factor* can be obtained by eqns (11.8b) and (11.9b) as

$$F_{\mathrm{SR,i,n}} = \frac{1}{F_{\mathrm{RS,i,n}}} \tag{10.9c}$$

A uniform beginning of year annual amount, say R_b, can be derived in terms of P and S as R and R_b have the following relationship

$$R = R_b(1+i) \tag{10.9d}$$

The values of various conversion factors with the number of years for a given rate of interest have been given in Table 10.1.

Example 10.9

Derive an expression for R_b in terms of P and S.

Solution

Substitute the expression of R from eqn (10.6b) into eqn (10.9a)

$$R_b(1+i) = PF_{\mathrm{PR,i,n}}$$

$$\text{so,} \qquad R_b = \frac{P}{(1+i)}.F_{\mathrm{PR,i,n}}$$

Table 10.1 The values of conversion factors.[5]

n	F_{PS}	F_{SP}	F_{RP}	F_{PR}	F_{RS}	F_{SR}	F_{PK}
				$i=0.04$			
1	1.04	0.962	0.962	1.04	1	1	26
2	1.082	0.925	1.886	0.53	2.04	0.49	13.255
3	1.125	0.889	2.775	0.36	3.122	0.32	9.009
4	1.17	0.855	3.63	0.275	4.246	0.235	6.887
5	1.217	0.822	4.452	0.225	5.416	0.185	5.616
6	1.265	0.79	5.242	0.191	6.633	0.151	4.769
7	1.316	0.76	6.002	0.167	7.898	0.127	4.165
8	1.369	0.731	6.733	0.149	9.214	0.109	3.713
9	1.423	0.703	7.435	0.134	10.583	0.094	3.362
10	1.48	0.676	8.111	0.123	12.006	0.083	3.082
11	1.539	0.65	8.76	0.114	13.486	0.074	2.854
12	1.601	0.625	9.385	0.107	15.026	0.067	2.664
13	1.665	0.601	9.986	0.1	16.627	0.06	2.504
14	1.732	0.577	10.563	0.095	18.292	0.055	2.367
15	1.801	0.555	11.118	0.09	20.024	0.05	2.249
16	1.873	0.534	11.652	0.086	21.825	0.046	2.146
17	1.948	0.513	12.166	0.082	23.697	0.042	2.055
18	2.026	0.494	12.659	0.079	25.645	0.039	1.975
19	2.107	0.475	13.134	0.076	27.671	0.036	1.903
20	2.191	0.456	13.59	0.074	29.778	0.034	1.84

n	F_{PS}	F_{SP}	F_{RP}	F_{PR}	F_{RS}	F_{SR}	F_{PK}
				$i=0.06$			
1	1.06	0.943	0.943	1.06	1	1	17.667
2	1.124	0.89	1.833	0.545	2.06	0.485	9.091
3	1.191	0.84	2.673	0.374	3.184	0.314	6.235
4	1.262	0.792	3.465	0.289	4.375	0.229	4.81
5	1.338	0.747	4.212	0.237	5.637	0.177	3.957
6	1.419	0.705	4.917	0.203	6.975	0.143	3.389
7	1.504	0.665	5.582	0.179	8.394	0.119	2.986
8	1.594	0.627	6.21	0.161	9.897	0.101	2.684
9	1.689	0.592	6.802	0.147	11.491	0.087	2.45
10	1.791	0.558	7.36	0.136	13.181	0.076	2.264
11	1.898	0.527	7.887	0.127	14.972	0.067	2.113
12	2.012	0.497	8.384	0.119	16.87	0.059	1.988
13	2.133	0.469	8.853	0.113	18.882	0.053	1.883
14	2.261	0.442	9.295	0.108	21.015	0.048	1.793
15	2.397	0.417	9.712	0.103	23.276	0.043	1.716
16	2.54	0.394	10.106	0.099	25.672	0.039	1.649
17	2.693	0.371	10.477	0.095	28.213	0.035	1.591
18	2.854	0.35	10.828	0.092	30.906	0.032	1.539
19	3.026	0.331	11.158	0.09	33.76	0.03	1.494
20	3.207	0.312	11.47	0.087	36.786	0.027	1.453

n	F_{PS}	F_{SP}	F_{RP}	F_{PR}	F_{RS}	F_{SR}	F_{PK}
				$i=0.08$			
1	1.08	0.926	0.926	1.08	1	1	13.5
2	1.166	0.857	1.783	0.561	2.08	0.481	7.01
3	1.26	0.794	2.577	0.388	3.246	0.308	4.85
4	1.36	0.735	3.312	0.302	4.506	0.222	3.774

Table 10.1 (*Continued*).

5	1.469	0.681	3.993	0.25	5.867	0.17	3.131
6	1.587	0.63	4.623	0.216	7.336	0.136	2.704
7	1.714	0.583	5.206	0.192	8.923	0.112	2.401
8	1.851	0.54	5.747	0.174	10.637	0.094	2.175
9	1.999	0.5	6.247	0.16	12.488	0.08	2.001
10	2.159	0.463	6.71	0.149	14.487	0.069	1.863
11	2.332	0.429	7.139	0.14	16.646	0.06	1.751
12	2.518	0.397	7.536	0.133	18.977	0.053	1.659
13	2.72	0.368	7.904	0.127	21.495	0.047	1.582
14	2.937	0.34	8.244	0.121	24.215	0.041	1.516
15	3.172	0.315	8.559	0.117	27.152	0.037	1.46
16	3.426	0.292	8.851	0.113	30.324	0.033	1.412
17	3.7	0.27	9.122	0.11	33.75	0.03	1.37
18	3.996	0.25	9.372	0.107	37.45	0.027	1.334
19	4.316	0.232	9.604	0.104	41.446	0.024	1.302
20	4.661	0.215	9.818	0.102	45.762	0.022	1.273

$i = 0.10$

n	F_{PS}	F_{SP}	F_{RP}	F_{PR}	F_{RS}	F_{SR}	F_{PK}
1	1.1	0.909	0.909	1.1	1	1	11
2	1.21	0.826	1.736	0.576	2.1	0.476	5.762
3	1.331	0.751	2.487	0.402	3.31	0.302	4.021
4	1.464	0.683	3.17	0.315	4.641	0.215	3.155
5	1.611	0.621	3.791	0.264	6.105	0.164	2.638
6	1.772	0.564	4.355	0.23	7.716	0.13	2.296
7	1.949	0.513	4.868	0.205	9.487	0.105	2.054
8	2.144	0.467	5.335	0.187	11.436	0.087	1.874
9	2.358	0.424	5.759	0.174	13.579	0.074	1.736
10	2.594	0.386	6.145	0.163	15.937	0.063	1.627
11	2.853	0.35	6.495	0.154	18.531	0.054	1.54
12	3.138	0.319	6.814	0.147	21.384	0.047	1.468
13	3.452	0.29	7.103	0.141	24.523	0.041	1.408
14	3.797	0.263	7.367	0.136	27.975	0.036	1.357
15	4.177	0.239	7.606	0.131	31.772	0.031	1.315
16	4.595	0.218	7.824	0.128	35.95	0.028	1.278
17	5.054	0.198	8.022	0.125	40.545	0.025	1.247
18	5.56	0.18	8.201	0.122	45.599	0.022	1.219
19	6.116	0.164	8.365	0.12	51.159	0.02	1.195
20	6.728	0.149	8.514	0.117	57.275	0.017	1.175

$i = 0.12$

n	F_{PS}	F_{SP}	F_{RP}	F_{PR}	F_{RS}	F_{SR}	F_{PK}
1	1.12	0.893	0.893	1.12	1	1	9.333
2	1.254	0.797	1.69	0.592	2.12	0.472	4.931
3	1.405	0.712	2.402	0.416	3.374	0.296	3.47
4	1.574	0.636	3.037	0.329	4.779	0.209	2.744
5	1.762	0.567	3.605	0.277	6.353	0.157	2.312
6	1.974	0.507	4.111	0.243	8.115	0.123	2.027
7	2.211	0.452	4.564	0.219	10.089	0.099	1.826
8	2.476	0.404	4.968	0.201	12.3	0.081	1.678
9	2.773	0.361	5.328	0.188	14.776	0.068	1.564
10	3.106	0.322	5.65	0.177	17.549	0.057	1.475

Table 10.1 (*Continued*).

11	3.479	0.287	5.938	0.168	20.655	0.048	1.403
12	3.896	0.257	6.194	0.161	24.133	0.041	1.345
13	4.363	0.229	6.424	0.156	28.029	0.036	1.297
14	4.887	0.205	6.628	0.151	32.393	0.031	1.257
15	5.474	0.183	6.811	0.147	37.28	0.027	1.224
16	6.13	0.163	6.974	0.143	42.753	0.023	1.195
17	6.866	0.146	7.12	0.14	48.884	0.02	1.17
18	7.69	0.13	7.25	0.138	55.75	0.018	1.149
19	8.613	0.116	7.366	0.136	63.44	0.016	1.131
20	9.646	0.104	7.469	0.134	72.052	0.014	1.116

$i = 0.14$

n	F_{PS}	F_{SP}	F_{RP}	F_{PR}	F_{RS}	F_{SR}	F_{PK}
1	1.14	0.877	0.877	1.14	1	1	8.143
2	1.3	0.769	1.647	0.607	2.14	0.467	4.338
3	1.482	0.675	2.322	0.431	3.44	0.291	3.077
4	1.689	0.592	2.914	0.343	4.921	0.203	2.451
5	1.925	0.519	3.433	0.291	6.61	0.151	2.081
6	2.195	0.456	3.889	0.257	8.536	0.117	1.837
7	2.502	0.4	4.288	0.233	10.73	0.093	1.666
8	2.853	0.351	4.639	0.216	13.233	0.076	1.54
9	3.252	0.308	4.946	0.202	16.085	0.062	1.444
10	3.707	0.27	5.216	0.192	19.337	0.052	1.369
11	4.226	0.237	5.453	0.183	23.045	0.043	1.31
12	4.818	0.208	5.66	0.177	27.271	0.037	1.262
13	5.492	0.182	5.842	0.171	32.089	0.031	1.223
14	6.261	0.16	6.002	0.167	37.581	0.027	1.19
15	7.138	0.14	6.142	0.163	43.842	0.023	1.163
16	8.137	0.123	6.265	0.16	50.98	0.02	1.14
17	9.276	0.108	6.373	0.157	59.118	0.017	1.121
18	10.575	0.095	6.467	0.155	68.394	0.015	1.104
19	12.056	0.083	6.55	0.153	78.969	0.013	1.09
20	13.743	0.073	6.623	0.151	91.025	0.011	1.078

$i = 0.16$

n	F_{PS}	F_{SP}	F_{RP}	F_{PR}	F_{RS}	F_{SR}	F_{PK}
1	1.16	0.862	0.862	1.16	1	1	7.25
2	1.346	0.743	1.605	0.623	2.16	0.463	3.894
3	1.561	0.641	2.246	0.445	3.506	0.285	2.783
4	1.811	0.552	2.798	0.357	5.066	0.197	2.234
5	2.1	0.476	3.274	0.305	6.877	0.145	1.909
6	2.436	0.41	3.685	0.271	8.977	0.111	1.696
7	2.826	0.354	4.039	0.248	11.414	0.088	1.548
8	3.278	0.305	4.344	0.23	14.24	0.07	1.439
9	3.803	0.263	4.607	0.217	17.519	0.057	1.357
10	4.411	0.227	4.833	0.207	21.321	0.047	1.293
11	5.117	0.195	5.029	0.199	25.733	0.039	1.243
12	5.936	0.168	5.197	0.192	30.85	0.032	1.203
13	6.886	0.145	5.342	0.187	36.786	0.027	1.17
14	7.988	0.125	5.468	0.183	43.672	0.023	1.143
15	9.266	0.108	5.575	0.179	51.659	0.019	1.121
16	10.748	0.093	5.668	0.176	60.925	0.016	1.103

Table 10.1 (*Continued*).

17	12.468	0.08	5.749	0.174	71.673	0.014	1.087
18	14.463	0.069	5.818	0.172	84.141	0.012	1.074
19	16.777	0.06	5.877	0.17	98.603	0.01	1.063
20	19.461	0.051	5.929	0.169	115.38	0.009	1.054

i = *0.18*

n	F_{PS}	F_{SP}	F_{RP}	F_{PR}	F_{RS}	F_{SR}	F_{PK}
1	1.18	0.847	0.847	1.18	1	1	6.556
2	1.392	0.718	1.566	0.639	2.18	0.459	3.548
3	1.643	0.609	2.174	0.46	3.572	0.28	2.555
4	1.939	0.516	2.69	0.372	5.215	0.192	2.065
5	2.288	0.437	3.127	0.32	7.154	0.14	1.777
6	2.7	0.37	3.498	0.286	9.442	0.106	1.588
7	3.185	0.314	3.812	0.262	12.142	0.082	1.458
8	3.759	0.266	4.078	0.245	15.327	0.065	1.362
9	4.435	0.225	4.303	0.232	19.086	0.052	1.291
10	5.234	0.191	4.494	0.223	23.521	0.043	1.236
11	6.176	0.162	4.656	0.215	28.755	0.035	1.193
12	7.288	0.137	4.793	0.209	34.931	0.029	1.159
13	8.599	0.116	4.91	0.204	42.219	0.024	1.132
14	10.147	0.099	5.008	0.2	50.818	0.02	1.109
15	11.974	0.084	5.092	0.196	60.965	0.016	1.091
16	14.129	0.071	5.162	0.194	72.939	0.014	1.076
17	16.672	0.06	5.222	0.191	87.068	0.011	1.064
18	19.673	0.051	5.273	0.19	103.74	0.01	1.054
19	23.214	0.043	5.316	0.188	123.414	0.008	1.045
20	27.393	0.037	5.353	0.187	146.628	0.007	1.038

i = *0.20*

n	F_{PS}	F_{SP}	F_{RP}	F_{PR}	F_{RS}	F_{SR}	F_{PK}
1	1.2	0.833	0.833	1.2	1	1	6
2	1.44	0.694	1.528	0.655	2.2	0.455	3.273
3	1.728	0.579	2.106	0.475	3.64	0.275	2.374
4	2.074	0.482	2.589	0.386	5.368	0.186	1.931
5	2.488	0.402	2.991	0.334	7.442	0.134	1.672
6	2.986	0.335	3.326	0.301	9.93	0.101	1.504
7	3.583	0.279	3.605	0.277	12.916	0.077	1.387
8	4.3	0.233	3.837	0.261	16.499	0.061	1.303
9	5.16	0.194	4.031	0.248	20.799	0.048	1.24
10	6.192	0.162	4.192	0.239	25.959	0.039	1.193
11	7.43	0.135	4.327	0.231	32.15	0.031	1.156
12	8.916	0.112	4.439	0.225	39.581	0.025	1.126
13	10.699	0.093	4.533	0.221	48.497	0.021	1.103
14	12.839	0.078	4.611	0.217	59.196	0.017	1.084
15	15.407	0.065	4.675	0.214	72.035	0.014	1.069
16	18.488	0.054	4.73	0.211	87.442	0.011	1.057
17	22.186	0.045	4.775	0.209	105.931	0.009	1.047
18	26.623	0.038	4.812	0.208	128.117	0.008	1.039
19	31.948	0.031	4.843	0.206	154.74	0.006	1.032
20	38.338	0.026	4.87	0.205	186.688	0.005	1.027

Similarly, from eqn (10.8b)

$$R_b = \frac{S}{(1+i)} . F_{SR,i,n}$$

Example 10.10

The estimated salvage value of a large-capacity PV/T solar water heater at the end of its useful lifetime of 20 years is 5000. Determine its present worth for a discount rate of 10%.

Solution

From eqn (10.6a)

$$P = S\left[\frac{1}{(1+i)^n}\right]$$

In the present example

$S = USD\ 5000$
$n = 20$ years
$i = 10\%$

Thus the present worth

$$P = 5000[1/(1+0.1)^{20}]$$
$$= USD\ 743.22.$$

10.3 Cash Flow

Cash flow is generally known as the single most pressing concern of any economic analysis. In its simplest form, cash flow is the movement of money into and out of any business and is the life-blood of all growing businesses and the primary indicator of business health. The cash flow is understood graphically on a time scale with the help of a line diagram known as a *cash flow diagram*. The net cash flow is calculated as:

$$\text{Net cash flow} = \text{Receipts (Credits)} - \text{Expenses (Debits)} \tag{10.10}$$

As discussed above this net cash flow can be represented by a cash flow diagram.

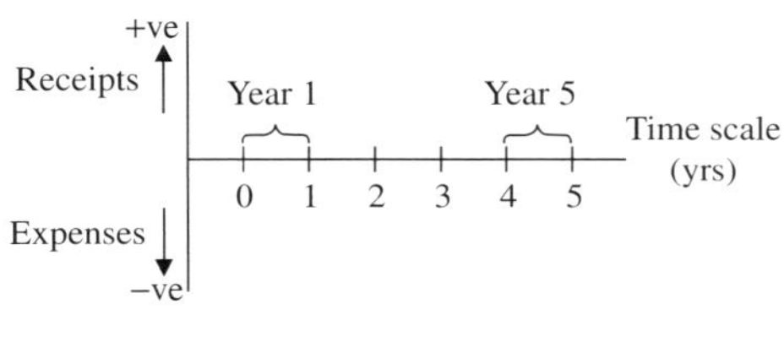

In the above cash flow diagram, a *uniform end-of-year annual amount* (R) can be considered at the end of each year on the time scale. This cash flow diagram will be used in the following examples.

Example 10.11

A person plans to create a forborne annuity by depositing USD 1000 at the end of the year, for 8 years. He wants to withdraw the money at the end of 14 years from now to buy a hybrid solar water heater. Find the accumulated value at the end of the 14th year, if money is worth 10% per year.

Solution

Let X be the amount available at the 14th year which can be considered as a receipt. The cash flow diagram for the payment is

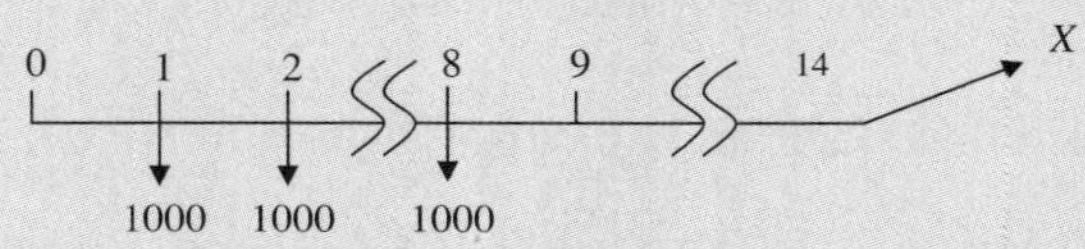

The present value (zero time) can be calculated by using eqn (10.6a), as

$$P = 1000\ F_{RP,10\%,8} = 1000 \times 5.3349 = \text{USD}\,\$\,5334.90$$

If this amount is deposited for 14 years, then the future value at the end of 14 years eqn (10.1a) will be

$$S = 5334.9 \times F_{PS,10\%,14} = 5334.9(3.79) = \text{USD}\,\$\,20259.$$

The above cash flow diagram can also be drawn by considering USD 1000 paid for 14 years less USD 1000 paid as annuity for the last 6 years.

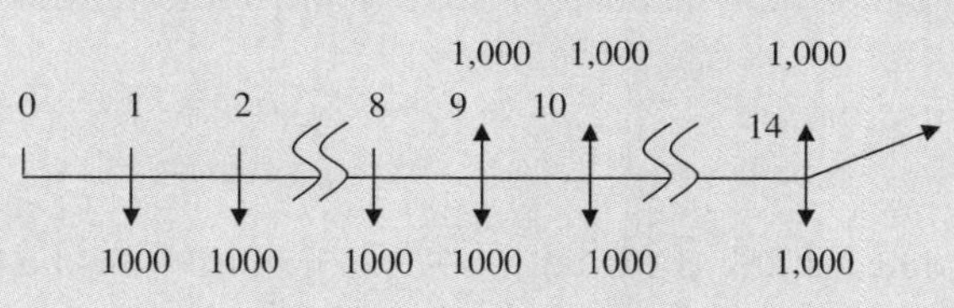

By using eqn (10.9a) we get

$$S = 1000\left[\frac{(1.10)^{14}-1}{0.10} - \frac{(1.10)^{6}-1}{0.10}\right] = 1000\,[27.9 - 7.71]$$

or, $S = \text{USD } 20{,}259.$

Example 10.12

A person wants a down payment of USD 2000 on a hybrid solar system of amount USD 10,000. An annual end-of-year payment (R) of USD 1174.11 is required for 12 years. However, the person elects to pay USD 1000 yearly and a balance payment at the end. Find the balance payment if money is worth 10% interest.

Solution

Let X be the balance payment. The cash flow diagram is

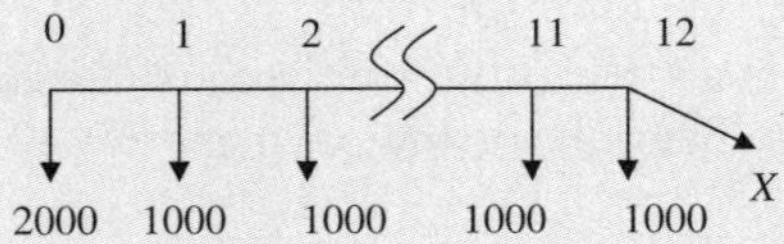

By using the cash flow diagram and eqns (10.4b) and (10.6a), we can write

$$10,000 = 2000 + 1000\,F_{RP,\,10\%,\,12} + XF_{SP,\,10\%,\,12}$$
$$= 2000 + 1000(6.8137) + X(0.31863)$$
$$X = 3723.10$$

The balance payment is USD 3723.10.

Example 10.13

A person decides to spend USD 3000 on the first, second, third and fourth years on energy-efficient equipment and agrees to set aside a certain amount now and each year thereafter until the fourth year. If the contribution forms an arithmetical progression for all years increasing by 20% after the first year; calculate his first contribution if money is worth 10%.

Solution

Let us assume that his first contribution is x. The cash flow diagram can be shown as

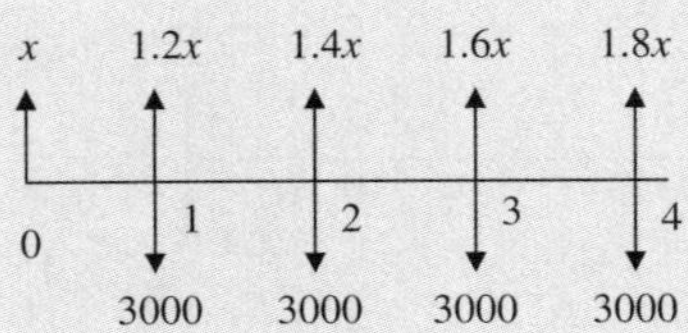

Consider two years from now as the focal point. Now using the time-value conversion relation in the above cash flow diagram, we get

$$x(1.10)^{-2}+1.2x(1.10)^{-1}+1.4x(1.10)^{0}+1.6x(1.10)^{1}+1.8x(1.10)^{2}$$
$$=3000(1.10)^{-1}+3000(1.10)^{0}+3000(1.10)^{1}+3000(1.10)^{2}$$
$$7.2553x=3000\times 4.2191$$
$$x=1744.56.$$

The first contribution would be USD 1744.56.

10.4 Cost Comparisons with Equal Duration

In this section a uniform expense is referred to as a uniform end-of-year cost.

Example 10.14

Two hybrid PV/T solar systems have the following cost comparison. Which system is more economical if the money is worth 10% per year?

Economic components	*System (A)*	*System (B)*
First cost (USD)	30,000	15,000
Uniform end-of-year maintenance per year (USD)	2,000	5,000
Overhaul, end of the third year (USD)	–	3,500
Salvage value (USD)	4,000	1,000
Life of the system (years)	5	5
Benefit from quality control as a uniform end-of-year amount per year (USD)	1,000	–

Solution

The cash flow diagrams for each system have been shown as follows

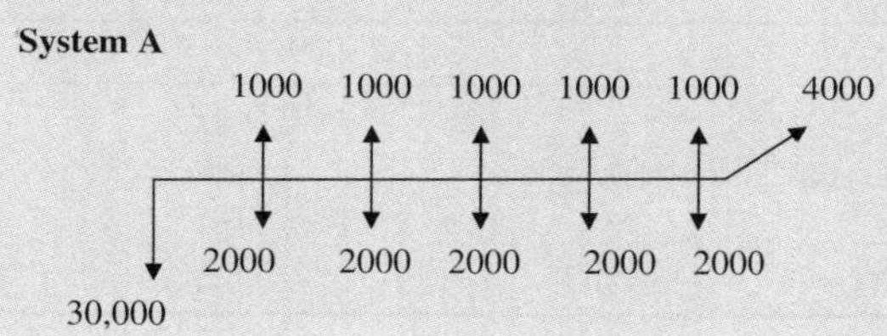

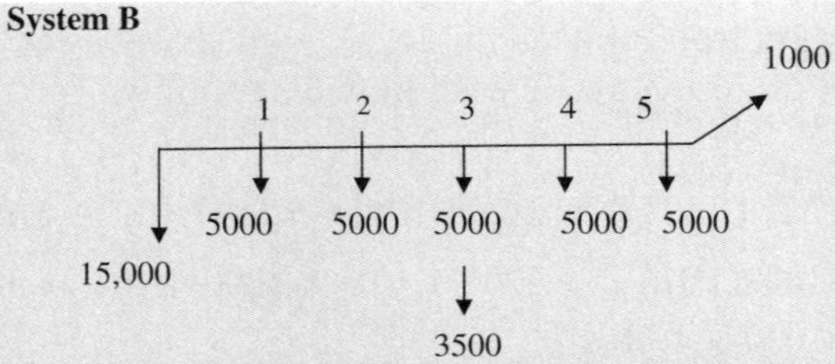

The present value of the costs for system A can be obtained by using eqns (10.4b) and (10.6a) as:

$$P_{AS} = 30,000 + (2000 - 1000)F_{RP,10\%,5} - 4000 \quad F_{SP,10\%,5}$$
$$= 30,000 + 1000(3.7908) - 4000(0.62092) = \text{USD } 31,307.12.$$

The present value of the costs for system B can be obtained by using eqns (10.4b) and (10.6a) as follows:

$$P_{BS} = 15,000 + 5000 \quad F_{RP,10\%,5} + 3500 \quad F_{SP,10\%,3} - 1000F_{SP,10\%,5}$$
$$= 15,000 + 5000 \times 3.7908 + 3500 \times 0.75131 - 1000 \times 0.62092$$
$$= 15,000 + 18,954 + 2629.55 - 620.92 = \text{USD } 35,962.63.$$

From the above calculations, it is clear that system A is more economical than system B.

10.5 Cost Comparisons with Unequal Duration

If two energy-efficient systems have different durations of life, a fair comparison can be made only on the basis of equal duration. One of the methods for comparison is to compare the single present value of costs on the basis of a common denominator of their service lives.

10.5.1 Single Present Value Method

Example 10.15

Two energy-efficient systems have the following cost comparison. Which system is more economical if the money is worth 10% per year?

Cost components (USD)	*System (A)*	*System (B)*
First cost	20,000	30,000
Uniform end-of-year maintenance	4,000	3,000
Salvage value	500	1,500
Service life, years	2	3

Solution

The cash flow diagrams for both systems are first reduced to the single present value of the cost.

System A

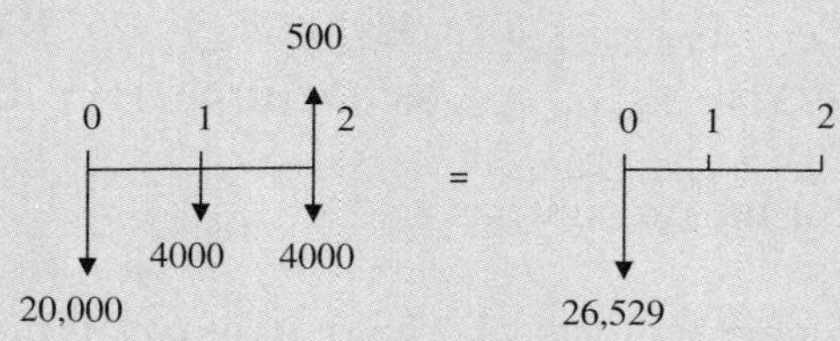

System B

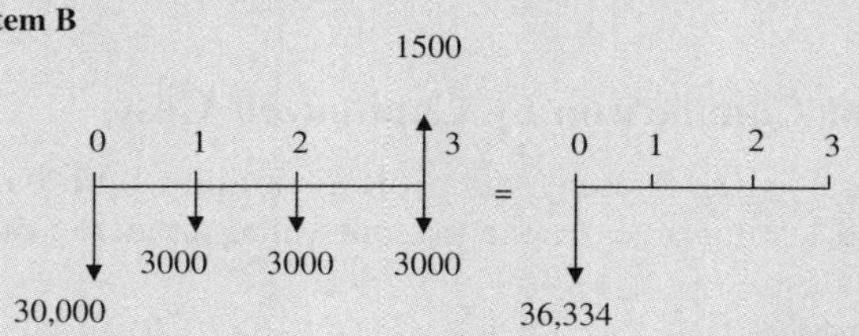

The simplified diagrams are now repeated to obtain a six-year duration. Note that the present value of system A is 26,529 at its time of installation.

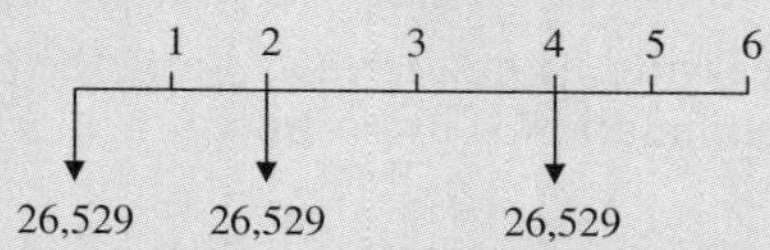

Similarly, the present value of system B is USD 36,334 at the time of installation. The cash flow diagram for a six-year duration is

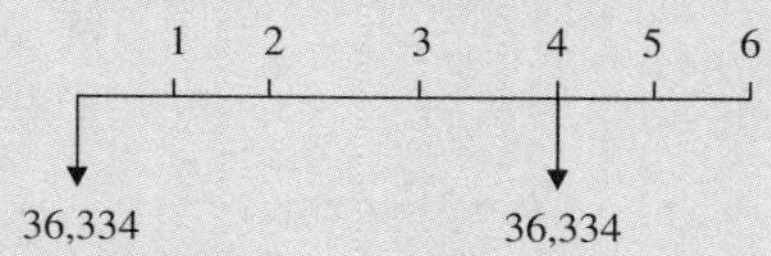

The present value of each of the preceding diagrams at 10% per year is

$$\begin{aligned} P_{A6} &= 25,529 + 26,529 \quad F_{RP,\,10\%,\,2} + 26,529 \quad F_{SP,\,10\%,\,4} \\ &= 26,529 + 26,529(1.10)^{-2} + 26,529(1.10)^{-4} \\ &= 26,529 + 21,924.7 + 18,119.66 = \text{USD}\,\$\,6,573.4 \end{aligned}$$

Similarly, P_{B6} = USD 63,632.2.

The ratio of cost is

$$\frac{P_{A6}}{P_{B6}} = \frac{66,573.45}{63,632.27} = 1.0462$$

Thus, system B is more economical than system A.

10.5.2 Cost Comparison by Annual Cost Method

In this case the uniform end-of-year annual amount will be calculated using eqn (10.7b) for P_{A2} = USD 26,529 and P_{B3} = USD 36,334 of Example 10.11.

$$R_{\mathrm{A}} = \mathrm{P_{A2}} \quad \mathrm{F_{PR,\,10\%,2}} = 26,529(0.57619) = \mathrm{USD}\,\$\,15,285.74$$
$$R_{\mathrm{B}} = 36,334 \quad \mathrm{F_{PR,\,10\%,3}} = 36,334x(0.40211) = \mathrm{USD}\,\$\,14,610.26$$

The unacost for the two systems ais $\frac{R_{\mathrm{A}}}{R_{\mathrm{B}}} = \frac{15,285.74}{14610.26} = 1.0462$

System B is more economical. The ratio of cost is more than one.
System B is more economical than system A as concluded earlier.

10.5.3 Cost Comparison by Capitalized Cost

Capitalized cost is the present value on an infinite time basis. For a system costing P_n and lasting n years, the present value replacing out to infinity is

$$K = P_{\mathrm{n}} \sum_{x=0}^{\mathrm{infinity}} \frac{1}{(1+\mathrm{i})^{\mathrm{xn}}} = P_{\mathrm{n}} \left[1 + \frac{1}{(1+\mathrm{i})^{\mathrm{n}}} + \frac{1}{(1+\mathrm{i})^{2\mathrm{n}}} + \ldots\ldots \right] \tag{10.11}$$

This is a geometric series with the first term as 1 and the ratio of the consecutive terms as $1/(1+i)^n$. Its summation is given by

$$\sum \frac{1}{(1+\mathrm{i})^{\mathrm{xn}}} = 1 \frac{1 - \left(\frac{1}{(1+\mathrm{i})^{\mathrm{n}}}\right)^{\mathrm{infinity}}}{1 - \frac{1}{(1+\mathrm{i})^{\mathrm{n}}}} = \frac{(1+\mathrm{i})^{\mathrm{n}}}{(1+\mathrm{i})^{\mathrm{n}} - 1} \tag{10.12a}$$

Equation (10.12) becomes

$$K = \mathrm{P_n} F_{\mathrm{PK,i,n}} \tag{10.12b}$$

or, *Capitalized cost = Present value basis n years duration* × (*capitalized cost factor*)

where K is the capitalized cost and $F_{\mathrm{PK,i,n}}$ the factor that converts a present value to capitalized cost (Table 10.1), also known as the capitalized cost factor and given as

$$F_{\mathrm{PK,i,n}} = \frac{\mathrm{i}(1+\mathrm{i})^{\mathrm{n}}}{(1+\mathrm{i})^{\mathrm{n}} - 1} \tag{10.13a}$$

From eqns (10.7b), $F_{\mathrm{PR,i,n}}$ is given as

$$F_{\mathrm{PR,i,n}} = \frac{\mathrm{i}(1+\mathrm{i})^{\mathrm{n}}}{(1+\mathrm{i})^{\mathrm{n}} - 1} \tag{10.13b}$$

Comparing eqns (10.13a) and (10.13b), $F_{PR,i,n}$ and $F_{PK,i,n}$ give the relationship

$$F_{PR,i,n} = iF_{PK,i,n} \qquad (10.13c)$$

or, *Capital recovery factor = rate of return × (capitalized cost factor)*

Similarly, from eqns (10.7a) and (10.12b), R and K are related as

$$R = iK \qquad (10.14)$$

or, *Unacost = rate of return × (capitalized cost factor).*

In this case, we solve Example 10.11 by using the capitalized cost method.

From Example 10.11, we have, P_{A2} = USD 26,529 and P_{B3} = USD 36,334.

By using eqn (10.12b), we get

$$\begin{aligned} K_A &= P_{A2}, F_{PK,\,10\%,2} = 26,529 \text{ x } (5.7619) = \text{USD}\,\$\,152,857.45 \\ K_B &= 36,334\ F_{PK,\,10\%,3} = 36,334 \text{ x } (4.0211) = \text{USD}\,\$\,146,102.65 \end{aligned}$$

The ratio of cost is

$$\frac{K_A}{K_B} = \frac{152,857.45}{146,102.65} = 1.0462$$

It is clear from the above calculation that the results obtained are the same as in the earlier solution. From this calculation, we can conclude that system B is more economical than system A.

As a matter of fact, it is possible to convert a present value P_{n1}, of n_1 years duration to an equivalent present value P_{n2} of n_2 years duration.

Hence, applying eqn (10.12b) gives

$$\begin{aligned} P_{n1}F_{PR,i,n} &= P_{n2}F_{PR,i,n2} \\ P_{n2} &= P_{n1}\frac{F_{PR,i,n}}{F_{PR,i,n2}} \end{aligned} \qquad (10.15)$$

As discussed earlier

$$\begin{aligned} P_{A2} &= \text{2-year duration} = \text{USD}\,26,529 \\ P_{B3} &= \text{3-year duration} = \text{USD}\,36,334. \end{aligned}$$

Convert the present value of system B to an equivalent value for 2 years duration using eqn (10.15).

$$P_{B2} = P_{B3}\frac{F_{PR,10\%,3}}{F_{PR,10\%,2}} = 36,334 \quad \frac{0.40211}{0.57619} = \text{USD}\,\$\,25,356.68$$

$$\frac{P_{A2}}{P_{B2}} = \frac{26529}{25356} = 1.0462$$

Further, the result of cost ratio is the same as that obtained by various methods discussed earlier. Hence, system B is more economical.

10.6 Analytical Expression for Payout Time

The pay back period (n), the number of years necessary to exactly recover the initial investment P, is computed by summing the annual cash flow values and estimating n through the relation:

$$0 = -\text{initial investment} + \text{sum of annual cash flows}$$

For equal annual savings

$$\text{Payback period(yrs)} = \frac{\text{Initial capital cost}}{\text{Annual operating cash flow}} \tag{10.16a}$$

For unequal annual savings

$$0 = -P + \sum_{t=1}^{n} CF_t\left(F_{SP,i\%,t}\right) \tag{10.16b}$$

where CF_t is the net cash flow at the end of year t. If the cash flow is same each year, the F_{RP} factor may be used in the above relation:

$$0 = -P + CF_t\left(F_{RP,i\%,n}\right) \tag{10.17}$$

i.e. after n years, the cash flow will recover the investment and a return of $i\%$. If the expected retention period (life) of the asset/project is less than n years, then investment is not advisable.

Considering i to be zero, eqn (10.16b) becomes

$$0 = -P + \sum_{t=1}^{n} CF_t \tag{10.18}$$

and if CF_t values are assumed equal, then

$$n = \frac{P}{CF}(i.e.\ P = n \times CF) \tag{10.19a}$$

There is little value in techno-economic study for n computed from eqns (10.18) and (10.19). When $i\% > 0$ is used to estimate n, the results incorporate the risk considered in the project undertaken.

Using eqns (10.7c) and (10.17), the expression for the pay back period for unequal annual savings can be written as

$$n = \frac{\ln\left[\frac{CF}{CF - P \times i}\right]}{\ln[1 + i]} \tag{10.19b}$$

Example 10.16

Energy-efficient systems purchased for USD 18,000 are expected to generate annual revenues of USD 3000, and have salvage a value of USD 3000 at any time during 10 years of anticipated ownership. If a 15% per year required return is imposed on the purchase, compute the pay back period.

Solution

The cash flow for each year is USD 3000 (P) with an additional revenue of USD 3000 in year n. The cash flow diagram has been shown below:

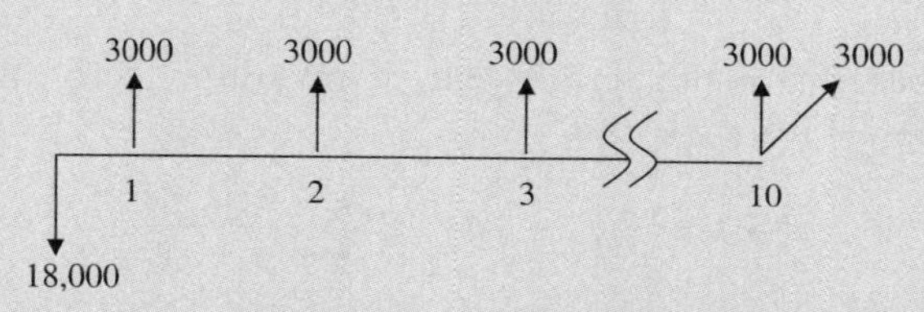

After using eqns (10.16b) and (10.17) for the above cash flow, one gets

$$0 = -P + C\ F_t(F_{RP,15\%,n}) + S.V(F_{SP,15\%,n})$$

or $$0 = -18,000 + 3000\ F_{RP,15\%,n} + 3000\ F_{SP,15\%,n}$$

The resulting payout time can be evaluated after further using eqns (10.4b) and (10.6c) for F_{SP} and F_{RP}, respectively, and we get $n = 15.3$ years, which is not economical with such high interest.

For $i = 0$, eqn (10.18) can be used and we get

$$0 = -18,000 + n(3000) + 3000$$

The resulting payout time is 5 years, which is most economical, without interest rate.

10.7 Net Present Value

The difference between the present value of the benefits and the costs resulting from an investment is the *net present value (NPV)* of the investment. A positive NPV means a positive surplus indicating that the financial position of the investor will be improved by undertaking the project. Obviously, a negative NPV would indicate a financial loss. An NPV of zero would mean that the present value of all benefits over the useful lifetime is equal to the present value of all the costs. In mathematical terms

$$NPV = \sum_{j=0}^{n} \frac{B_j - C_j}{(1+i)^j} \tag{10.20}$$

where B_j stands for benefits at the end of the period j, C_j for costs at the end of period j, n the useful life of the project and i the interest rate. Equation (10.20) involves subtracting the cost from the benefits at any period and then multiplying the result by the single payment present worth factor for that period. Finally, the NPV is determined by algebraically adding the results for all the periods under consideration.

It often happens that $(B_j - C_j)$ is constant for all j except for j = 0. In such a case, eqn (10.20) can be modified as

$$\text{NPV} = (B_0 - C_0) + \sum_{j=1}^{n} \frac{B_j - C_j}{(1+i)^j}$$

Since B_0, the benefits in the zeroth year, is invariably zero and $(B_j - C_j)$ is constant (= B – C) for j = 1 to n,

$$\text{NPV} = -C_0 + (B - C) \sum_{j=1}^{n} \frac{1}{(1+i)^n}$$

or $$\text{NPV} = -C_0 + (B - C)\left[\frac{(1+i)^n - 1}{i(1+i)^n}\right] \quad (10.21)$$

with C_0 representing the initial capital investment in the project.

Example 10.17

A PV system for water pumping costs USD 10,000 to purchase and install on the field of a farmer. It is expected to save USD 1200 worth of diesel annually to the farmer and its annual maintenance cost is estimated at USD 100. Calculate the NPV of the investment on the PV system if the useful life of the system is 30 years and the interest rate is 8%.

Solution

Net annual benefits of using a PV system = 1200 – 100 = USD 1100.

Since the amount of net annual benefits is constant over the useful life of the system, eqn (10.21) can be used for determining the NPV.

$$i.e.\ \text{NPV} = -C_0 + (B - C)\left[\frac{(1+i)^n - 1}{i(1+i)^n}\right]$$

$$= -10,000 + (1100)\left[\frac{(1+0.08)^{30} - 1}{0.08(1+0.08)^{30}}\right] = -10,000 + 12,384 = 2,384$$

Therefore, the investment in the PV system is a financially viable investment for the farmer.

Equations (10.20) and (10.21) are based on the assumption that the interest rate i remains constant over time. The NPV could also be calculated with different rates of interest rate over the jth period, eqn (10.20) for NPV can be modified as

$$\begin{aligned} \text{NPV} =& (B_0 - C_0) + \frac{B_1 - C_1}{(1+i_1)} + \frac{B_2 - C_2}{(1+i_1)(1+i_2)} + \ldots\ldots \\ &+ \frac{B_j - C_j}{(1+i_1)(1+i_2)\ldots\ldots(1+i_j)} + \ldots\ldots \\ &+ \frac{B_n - C_n}{(1+i_1)(1+i_2)\ldots\ldots(1+i_n)} \end{aligned} \tag{10.22}$$

The acceptance criteria of an investment project, as evaluated from the NPV method are:

a) NPV > 0, accept the project;
b) NPV = 0, remain indifferent;
c) NPV < 0, reject the project.

As mentioned earlier, a positive NPV represents a positive surplus and therefore the project may be accepted subject to availability of funds. A project with negative NPV should be rejected as the funds may be advantageously invested in the other projects. Thus, unless the project is mandatory only those investments having positive NPV may be accepted. In case of mutually exclusive alternative investments the project with highest positive NPV should be chosen.

Limitations of NPV method

As regards the limitations of the NPV method, the following points are worth mentioning:

(a) The NPV method focuses only on benefits and does not distinguish between an investment involving relatively large costs and benefits and one involving much smaller costs and benefits as long as the two projects result in equal NPVs. Thus, it does not give any indication of the scale of efforts required to achieve the results.
(b) The results of the NPV method are quite sensitive to the interest/discount rate chosen. Thus, failure to select an appropriate value of the interest rate used in the computation of NPV may alter or even reverse the relative ranking of different alternatives being compared using this method. For example, with a very low value of interest rate, an alternative with benefits spread far into the future may unjustifiably appear more profitable than an alternative whose benefits are more quickly realized but is of a lower amount in undiscounted terms.

From eqn (10.20) it may be noted that as the interest rate i is increased, every cash flow in the future is discounted to the present by a factor of the general form $1/(1+i)^j$. As i approaches infinity, $1/(1+i)^j$ approaches zero. Mathematically, for two extreme values of the interest rate, eqn (10.20) gives

$$NPV = \sum_{j=0}^{n} B_j - C_j \quad \text{for } i = 0$$

$$\text{and} \quad \text{NPV} = -C_0 \quad \text{for } I = \infty.$$

Example 10.18

The cost of a BIPV air-circulating collector is USD 35,000. During its useful life of 20 years, besides other routine maintenance costs of USD 300 each year, replacement of the wooden duct in the 10th year is expected to cost USD 4000. Determine the equivalent annual cost of the system for an interest rate of 10%.

Solution

Present values of all the costs associated with the BIPV air collector:

$$= 35,000 + 300[1/(1+0.1) + 1/(1+0.1)^2 + \ldots + 1/(1+0.1)^{10}] + 4000/(1+0.1)^{10}$$

$$= 35,000 + 300\left[\frac{(1+0.1)^{20}-1}{0.1(1+0.1)^{20}}\right] + \left[\frac{4000}{(1+0.1)^{10}}\right]$$

$$= 35,000 + 300(8.51) + 1542.2$$

$$= \text{USD } 39,095.4.$$

Hence, the equivalent annual cost of the BIPV system is

$$= 39,095.4\left[\frac{0.1(1+0.1)^{20}}{(1+0.1)^{20}-1}\right]$$

$$= 39,095.4(0.117)$$

$$= \text{USD } 4574.1.$$

10.8 Benefit-Cost Analysis

The benefit-cost ratio is another method of analysing and making a decision on investments. As its name suggests, the benefit-cost (B-C) ratio method of analysis is based on the ratio of the benefits to costs associated with a particular project. The ratio of benefits to costs as a measure of financial or economic efficiency is conceptually simple and quite versatile and it measures cost efficiency. Obviously, the first step in a B-C ratio analysis is to identify the costs and benefits separately. In general, the benefits are advantages (fuel saving in the case of energy projects) expressed in monetary terms and the disadvantages are the associated disbenefits. The costs are the anticipated expenditures for construction, installation, operation, maintenance, *etc*. The B-C ratio method has frequently been used by government agencies for projects whose benefits are reaped by the common public and the costs are incurred by the government. Therefore, the determination of whether an item is to be considered as a benefit, disbenefit or cost depends on who is affected by the consequences of the project implementation.

A project is considered to be attractive when the benefits derived from its execution exceed its associated costs.

The conventional B/C ratio is calculated as

$$B/C = (\text{Benefits} - \text{Disbenefits})/\text{cost} = (B - D)/C$$

The modified B/C ratio, which is gaining support, includes operation and maintenance (O&M) costs in the numerator and treats them in a manner similar to disbenefits, and is given by

$$B/C = (\text{Benefits} - \text{Disbenefits} - \text{O\&M cost})/(\text{Initial investment})$$

The salvage value can also be considered in the denominator.

The B/C ratio influences the decision on the project approval.

$$\text{If} \quad B/C > 1 \quad \text{accept the project}$$
$$B/C < 1 \quad \text{reject the project.}$$

Thus, in the case of mutually exclusive projects, the B/C ratio gives a method to compare them against each other.

Benefits (B): Benefits are the advantages to the owner.
Disbenefits (D): When the project under consideration involves disadvantages to the owner.
Costs: The anticipated expenditures for construction, operation, maintenance, *etc*.
Owner: Public: One who incurs the costs as the government.

Let B and C be the present values of the cash inflows (benefits) and outflows (costs) defined as

$$B = \sum_{j=0}^{n} \frac{B_j}{(1+i)^j} \tag{10.23}$$

$$C = \sum_{j=0}^{n} \frac{C_j}{(1+i)^j} \tag{10.24}$$

where B_j and C_j respectively represent the benefits and costs at the end of the jth period and n is the useful life of the project.

The equivalent present value cost C (eqn (10.24)) may be split into two components – (i) the initial capital expenditure and (ii) the annual costs accrued in each successive period. If it is assumed that the initial investment is required in the first m periods and that the annual costs accrue in each of the following periods till the end of the useful life of n periods, the above two components of the equivalent present value cost C may be expressed as

$$C_0 = \sum_{j=0}^{m} \frac{C_j}{(1+i)^j} \tag{10.25}$$

and $$C'' = \sum_{j=m+1}^{n} \frac{C_j}{(1+i)^j} \tag{10.26}$$

with $$C = C_0 + C'' \tag{10.27}$$

Using the above three expressions (eqns (10.25) to (10.27)), the following three types of benefits-cost rates are usually defined:

(i) *Aggregate B-C ratio*
This is the ratio of the present value of total benefits to total costs.

$$\left(\frac{B}{C}\right)_{\text{aggregate}} = \frac{B}{C} = \frac{B}{C_0 + C''}, \quad C > 0 \text{ (or } C_0 + C'' > 0) \tag{10.28}$$

or $$\left(\frac{B}{C}\right)_{\text{aggregate}} = \frac{\sum\limits_{j=0}^{n} \frac{B_j}{(1+j)^n}}{\sum\limits_{j=0}^{n} \frac{C_j}{(1+j)^n} + \sum\limits_{j=m+1}^{n} \frac{C_j}{(1+j)^n}} \tag{10.29}$$

Obviously, to accept a project the ratio $\left(\frac{B}{C}\right)_{\text{aggregate}}$ must be greater than 1.

(ii) *Net B-C ratio*
In another definition of the B-C ratio, only the initial capital expenditure is considered as a cash outlay, and equivalent benefits become net benefits (*i.e.* annual revenues minus annual outlays).

The net benefit-cost ratio is expressed as

$$\left(\frac{B}{C}\right)_{\text{net}} = \frac{B - C'}{C_0}, \quad C_o > 0 \tag{10.30}$$

Once again, for a project to be viable, the ratio $\left(\frac{B}{C}\right)_{\text{net}}$ must be greater than 1. The benefit-cost ratio defined in this manner essentially provides an index which indicates the benefits expected per unit of capital investment and can hence be used as a profitability index. It may be noted that it is simply a comparison of the present value of net revenues with the present value of capital investment. Thus $\left(\frac{B}{C}\right)_{net}$ ensures that there is a surplus at time zero and the project is favourable.

Advantages and limitations of B-C ratio:

The benefits-cost ratio method offers the following advantages over other measures of evaluating different alternatives.

(a) It compares alternatives on a common scale and permits evaluation of different-sized alternatives.
(b) It can be used to rank alternative projects to determine the most profitable alternative for an investor with a limited budget.
(c) It directly provides an indication of whether a project is worthwhile.
(d) It can also be used to determine the optimal size of a project if it is computed for increment in the investment size.

The shortcomings of the benefits-cost ratio include:

(a) The benefit-cost ratio is influenced by the decision as to whether an item is classified as a cost or a disbenefit, *i.e.* whether it appears in the denominator or the numerator of the ratio. Often it may be an arbitrary decision but can lead to inefficient ranking of investment alternatives.
(b) The simple benefit-cost ratio cannot be used to determine the efficient scale of a given project. Incremental analysis is required to be undertaken for this purpose.

Example 10.19

The latest building regulation in a city stipulates that all new student hostels must use solar energy for water heating. The manager of a hostel under construction is considering two hybrid solar water-heating systems to supplement a natural gas-fired water heating system. One of the day systems (alternative X) is based on double-glazed flat-plate collectors and the other (alternative Y) uses evacuated tubular collectors. Both the options have a useful life of 20 years and the associated costs and benefits are tabulated below:

	Amount (USD)	
	Alternative X	Alternative Y
Capital cost	3,200,000	2,700,000
Annual maintenance cost	50,000	80,000
Annual benefits due to fuel savings	600,000	560,000

Which option should be preferred on the basis of incremental net benefit-cost ratio? Use an interest rate of 10% and also assume that salvage value is negligible for both alternatives. What if the benefit-cost ratio for each alternative is computed and the alternatives with higher benefit-cost ratio are selected?

Solution

As is stipulated by the latest building regulations in the city, one of the two alternatives is sure to be chosen. Thus the lower (Alternative Y) need not be compared with the 'do nothing' alternative. Instead, alternatives X and Y are compared with each other in terms of their incremental costs and benefits. The incremental capital cost of alternative X over alternative Y is USD 5,00,000 (USD 32,00,000 – USD 27,00,000). Similarly the incremental net annual benefits of alternative X over the net annual benefits of alternative Y are

(600,000 – 50,000) – (560,000 – 80,000) = 550,000 – 480,000 = USD 70,000.

The cumulative present worth of the incremental benefits over 20 years of useful life of alternative X over alternative Y is

$$= \sum_{j=1}^{n} \frac{70000}{(1+i)^j}$$

$$= 70,000 \left[\frac{(1+0.1)^{20} - 1}{0.1(1+0.1)^{20}} \right]$$

$$= 70,000\,(8.51)$$

$$= \text{USD } 595,949.4$$

Thus the net incremental benefits to cost ratio = 5,95,949.4/5,00,000 ≈ 1.19.

A value greater than one for the ratio of net incremental benefits to incremental capital cost implies that the additional discounted benefits more than justify the extra capital cost of alternative X compared to alternative Y. Therefore, alternative X should be selected for installation on the hostel.

The computation for the net benefit to cost ratio for each alternative independent of each other are given below.

The net benefit to cost ratio for alternative X is

$$= \frac{(60,000 - 50,000)\left[\frac{(1+0.1)^{20} - 1}{0.1(1+0.1)^{20}}\right]}{3,200,000}$$

$$= \frac{50,000 \times 8.51}{3,200,000}$$

$$= 1.463$$

Similarly, the net benefits to cost ratio for alternative Y is

$$= \frac{(5,600,000 - 80,000)\left[\frac{(1+0.1)^{20}-1}{0.1(1+0.1)^{20}}\right]}{2,700,000}$$
$$= \frac{480,000 \times 8.51}{2,700,000}$$
$$= 1.513$$

It may be noted that an appraisal of the two alternatives using their net benefit to cost ratios would suggest that the alternative Y is selected.

As the results obtained with the two methods do not match, the net present values of both the alternatives are determined to identify the correct method.

NPV of alternative X is

$$\text{NPVX} = -3,200,000 + (600,000 - 50,000)\left[\frac{(1+0.1)^{20}-1}{0.1(1+0.1)^{20}}\right]$$
$$= -3,200,000 + 550,000(8.51)$$
$$= \text{USD}\ 1,482,460.$$

NPV of alternative Y is

$$\text{NPVY} = -2,700,000 + (560,000 - 80,000)\left[\frac{(1+0.1)^{20}-1}{0.1(1+0.1)^{20}}\right]$$
$$= -2,700,000 + 480,000(8.51)$$
$$= \text{USD}\ 1,386,510$$

i.e. $\text{NPV}_X\text{s} > \text{NPV}_Y$.

Thus, the appraisal based on incremental costs and benefits is correct.

10.9 Internal Rate of Return

The internal rate of return (IRR) is a widely accepted discounted measure of investment worth and is used as an index of profitability for the appraisal of projects. The IRR is defined as the rate of interest that equates the present value of a series of cash flows to zero. Mathematically, the internal rate of return is the interest rate i_{IRR} that satisfies the equation

$$\text{NPV}(i_{\text{IRR}}) = \sum_{j=0}^{n} \frac{B_j - C_j}{(1 + i_{\text{IRR}})^j} = 0 \quad (10.31)$$

Alternatively, the internal rate of return is the interest rate that causes the discounted present value of the benefits in a cash flow to be equal to the present value of the costs, *i.e.*

$$\sum_{j=0}^{n} \frac{B_j}{(1+i_{\text{IRR}})^j} = \sum_{j=0}^{n} \frac{C_j}{(1+i_{\text{IRR}})^j} \tag{10.32}$$

Multiply both sides of eqn (10.31) by $(1 + i_{\text{IRR}})^n$

$$(1+i_{\text{IRR}})^n NPV(i_{\text{IRR}}) = \sum_{j=0}^{n} \left\{ \frac{B_j - C_j}{(1+i_{\text{IRR}})^j} \right\} (1+i_{\text{IRR}})^n$$

$$\text{or,} \quad \text{NPV}(\text{i}_{\text{IRR}})(1+\text{i}_{\text{IRR}})^{\text{n}} = \sum_{\text{j}=0}^{\text{n}} (\text{B}_\text{j} - \text{C}_\text{j})(1+\text{i}_{\text{IRR}})^{\text{n}-\text{j}} = 0 \tag{10.33}$$

IRR is widely used in the appraisal of projects because (i) the IRR on a project is its expected rate of return, (ii) it employs a percentage rate of return as the decision variable which suits the banking community and (iii) for situations in which IRR exceeds the cost of the funds used to finance the project, a surplus would remain after paying for the capital.

Iterative procedure for computation of IRR

The following step-by-step procedure is suggested for computation of IRR by the iterative approach.

Step 1: Make a guess at a trial rate of interest.

Step 2: Using the guessed rate of interest, calculate the NPV of all disbursements and receipts.

Step 3: If the calculated value of NPV is positive then the receipts from the investments are worth more than the disbursements of the investments and the actual value of IRR would be more than the trial rate. On the other hand, if NPV is negative the actual value of IRR would be less than the trial rate of interest. Adjust the estimate of the trial rate of return accordingly.

Step 4: Proceed with Steps 2 and 3 again until one value of i ($= i_1$) is found that results in a positive (+) NPV and the next higher value of i ($= i_2$) is found with a negative NPV.

Step 5: Solve for the value of IRR by interpolation using the values of i_1 and i_2 as obtained in Step 4 (Figure 10.1).

$$\text{IRR} = i_1 \left(\frac{i_2 - i_1}{NPV_1 - NPV_2} \right) \text{NPV}_1 \tag{10.34}$$

An important aspect of the iterative method of computing IRR is making the initial estimate. If the initial estimate is too far from the actual value of IRR, a large number of trials will have to be made to obtain the two consecutive values of interest rate(i_1 and i_2) to permit accurate interpolation. It should be noted that the initial estimate of the IRR will always be somewhat in error and several iterations will normally be required to determine i_1 and i_2. A simple approach for making a guess of the first trial rate of return is given below.

The NPV of a capital investment C_0 resulting in uniform net annual cash flows of amount A for an infinite time horizon can be expressed as

$$\mathrm{NPV} = -\mathrm{C}_0 + \left[\frac{\mathrm{A}}{(1+\mathrm{i})} + \frac{\mathrm{A}}{(1+\mathrm{i})^2} + \ldots\ldots\right]$$

where i is the interest rate.

$$\text{Thus,}\quad \mathrm{NPV} = -\mathrm{C}_0 + \frac{\mathrm{A}}{(1+\mathrm{i})}\left[\frac{1}{1-\frac{1}{(1+\mathrm{i})}}\right] = -\mathrm{C}_0 + \frac{\mathrm{A}}{\mathrm{i}}$$

since NPV = 0 at i = IRR, we have

$$-C_0 + \frac{A}{\mathrm{IRR}} = 0$$

$$\text{or,}\qquad \mathrm{IRR} = \frac{A}{C_0}$$

In actual practice, for investment projects with finite life the IRR shall be less than A/C_0. However, to begin with, for cases with uniform periodic cash flows, the figure A/C_0 or a value close to it may be used as the trial rate of return in the iterative procedure used for determining IRR.

The above interpolation between two consecutive values of interest rates that bracket the IRR always overestimates its true value. This is because of the fact that the linear interpolation technique makes an implicit assumption that between two interest rates i_1 and i_2 the IRR changes, following a straight line, whereas the true value of IRR follows a concave curvilinear function between the two values. However, the error introduced by interpolation is usually very small. Referring to Figure 10.1, the true value of IRR is that value of i for which the NPV (i) function intersects the horizontal axis, whereas the interpolated value of IRR is somewhat higher than the true value. Obviously, the interpolation error would become less and less as the incremental change in the trial values of i used in iteration is made smaller and smaller.

Example 10.20

Calculate the internal rate of return for the investment in a heat exchanger which will costs USD 500,000 to purchase and install, will last 10 years and will result in fuel savings of USD 145,000 per year. Also assume that the salvage value of the heat exchanger at the end of 10 years is negligible.

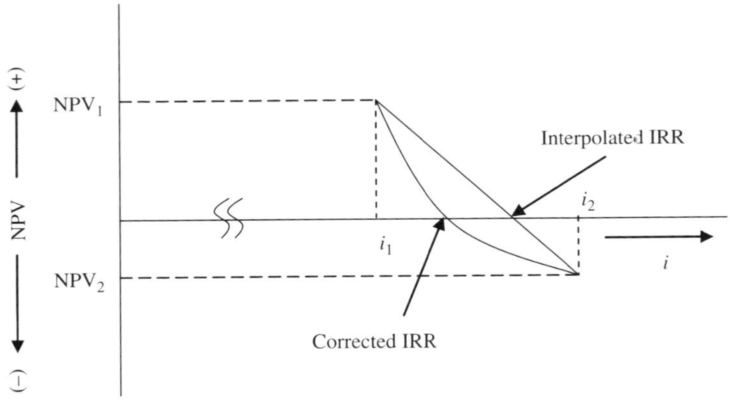

Figure 10.1 Interpolation of IRR.

Solution

Let the first guess at the value of IRR be 25%.

$$\begin{aligned}\text{NPV at } 25\% &= 145,000\left[\frac{(1+0.25)^{10}-1}{0.25(1+0.25)^{10}}\right] - 500,000\\ &= 145,000(3.57) - 500,000\\ &= \text{USD}17,722.\end{aligned}$$

Since the NPV at 25% is positive, the IRR shall be greater than 25%. If the next trial value is chosen at 30%, then

$$\begin{aligned}\text{NPV at } 30\% &= 145,000\left[\frac{(1+0.3)^{10}-1}{0.3(1+0.3)^{10}}\right] - 500,000\\ &= 145,000(3.09) - 500,000\\ &= \text{USD} - 51,724.\end{aligned}$$

Obviously, the true IRR lies between 25% and 30%. By interpolating between the two, the IRR can be estimated as

$$\begin{aligned}\text{IRR} &= 0.25 + \left(\frac{0.3-0.25}{17,722+51,724}\right)17,722\\ &= 0.26275\\ \text{or,}\qquad \text{IRR} &= 26.275\%.\end{aligned}$$

A better estimate of the true IRR may be obtained by using smaller incremental changes in the interest rate.

Example 10.21

Installation of a USD 50,00,000 energy management system in an industry is expected to result in a 25% reduction in electricity use and a 40% saving in process heating costs. This translates to net yearly savings of USD 600,000 and USD 750,000 respectively. If the energy management system has an expected useful life of 20 years, determine the internal rate of return on the investment. Salvage value need not be considered in the analysis.

Solution

$$\text{Total annual benefits} = \text{USD } 600,000 + \text{USD } 750,000 = \text{USD } 1,350,000.$$

$$\text{NPV of the investment} = -5,000,000 + 1,350,000\left[\frac{(1+i)^{20}-1}{i(1+i)^{20}}\right]$$

NPV at $i = 0.27$

$$= -5,000,000 + 1,350,000\left[\frac{(1+0.27)^{20}-1}{0.27(1+0.27)^{20}}\right]$$
$$= -5,000,000 + 4,958,034$$
$$= -41,965.$$

NPV at $i = 0.26$

$$= 50,00,000 + 13,50,000\left[\frac{(1+0.26)^{20}-1}{0.27(1+0.26)^{20}}\right]$$
$$= 50,00,000 + 51,41,263$$
$$= 1,41,263.$$

Thus, the IRR can be obtained by interpolating between $i = 0.26$ and $i = 0.27$ in the following manner:

$$\text{IRR} = 0.26 + \left(\frac{0.27 - 0.26}{1,41,263 + 41,965}\right)141,263 = 0.2677$$

i.e. the internal rate of return is 26.77%.

Multiple values of IRR

The NPV of a set of cash receipts and disbursements can be expressed as an nth degree polynomial of the form

$$\text{NPV}(i_{\text{IRR}}) = 0 = F_0 + F_1x + F_2x^2 + \ldots + F_nx^n \qquad (10.35)$$

where $x = 1/(1+i)$ and F_i's are coefficients of the n terms in the polynomial.

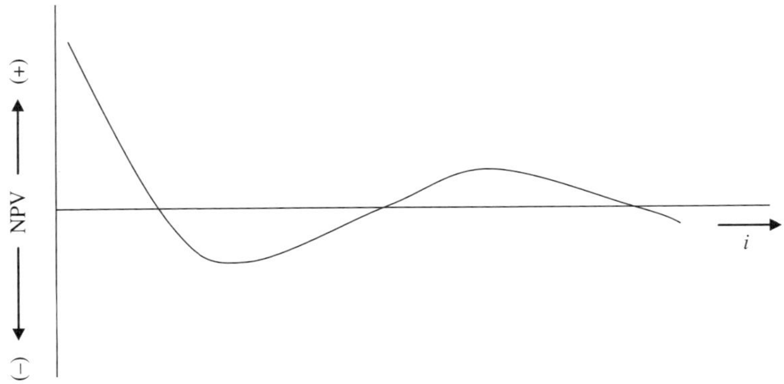

Figure 10.2 Multiple values of IRR.

For the above polynomial, in principle, there may be *n* different roots or values of x which satisfy eqn (10.35). Thus, it is possible that the NPV (*i*) function crosses the *i* axis several times as shown in Figure 10.2.

It may be noted that a unique value of IRR of special interest in applying the IRR method and consequently multiple values of IRR essentially hinder the application of the IRR criterion. In fact, in the case with multiple IRR values, use of the IRR criterion is normally not recommended.

10.10 Effect of Depreciation

Initial cost (C_i)*:* Also referred to as *first cost* or *initial value* or *single amount*, it is the installed cost of the system. The cost includes the purchase price, delivery and installation fee and other depreciable direct costs (defined later) incurred to ready the asset for use.

Salvage value (C_{sal})*:* This is the expected market value at the end of the useful life of the asset. It is negative if dismantling cost or carrying away cost is anticipated. It can be zero also. For example, the window glass has zero salvage value.

Depreciation (C_d)*:* An expenditure that decreases in value with time. This must be apportioned over its lifetime. The term used to describe this loss in value is known as depreciation.

$$C_{\mathrm{d}} = C_{\mathrm{i}} - C_{\mathrm{sal}}$$

Book value (B): This represents the remaining undepreciated investment on corporate books. It can be obtained after the total amount of annual depreciation charges to date has been subtracted from the first cost (present value/initial cost). The book value is usually determined at the end of each year.

Book value = *initial cost (first cost)* − *accumulated cost*

or, *Book value* = *salvage value* + *future depreciation*

Depreciation rate (D_t): This is the fraction of first cost removed through depreciation from corporate books. This rate may be the same *i.e.* straight-line (SL) rate or different for each year of the recovery period.

Mathematically, it can be written for straight-line (SL) depreciation as follows:

$$D_t = \frac{C_i - C_{\mathrm{sal}}}{n} \tag{10.36}$$

The book value at the *n*th year can be expressed as

$$B_n = C_{\mathrm{i}} - nD_{\mathrm{t}} \tag{10.37}$$

Recovery period (n): This is the life of the asset (in years) for depreciation and tax purposes. It is also referred to as the *expected life* of the asset in years.

Market value: This is the actual amount that could be obtained after selling the asset in the open market. For example, (i) the market value of a commercial building tends to increase with period in the open market but the book value will decrease as depreciation charges are taken in to account and (ii) an electronic equipment (computer system) may have a market value much lower than the book value due to the rapid change of technology.

Present value of Re. 1 of depreciation ($C_{\mathrm{d}} =$ Re. 1) is

$$F_{SLP,i,n} = \frac{1}{n}\left[\frac{(1+i)^n - 1}{i(1+i)^n}\right] = \frac{1}{n} F_{RP,i,n} \tag{10.38}$$

An expression for the conversion factor from straight-line depreciation to the present value with tax is given by

$$F_{SLP,r,n} = \frac{1}{n}\left[\frac{(1+r)^n - 1}{r(1+r)^n}\right] \tag{10.39}$$

10.11 Cost Comparisons of Solar Dryers with Duration

Hossain *et al.*[1] reported that the pay back period of a solar tunnel dryer is 4 years for a basic-mode dryer and those for optimum-mode dryers are 4 years and about 3 years (Section 7.5.1). On the basis of sensitivity analysis, they showed that the design geometry was not very sensitive to minor material costs, fixed costs and operating costs. It is sensitive to costs of major construction materials of the collector, solar radiation and air velocity in the dryer.

For a solar grain dryer incorporating photovoltaic powered air circulation, the variation of pay back period with respect to PV area to air-heater area ratio

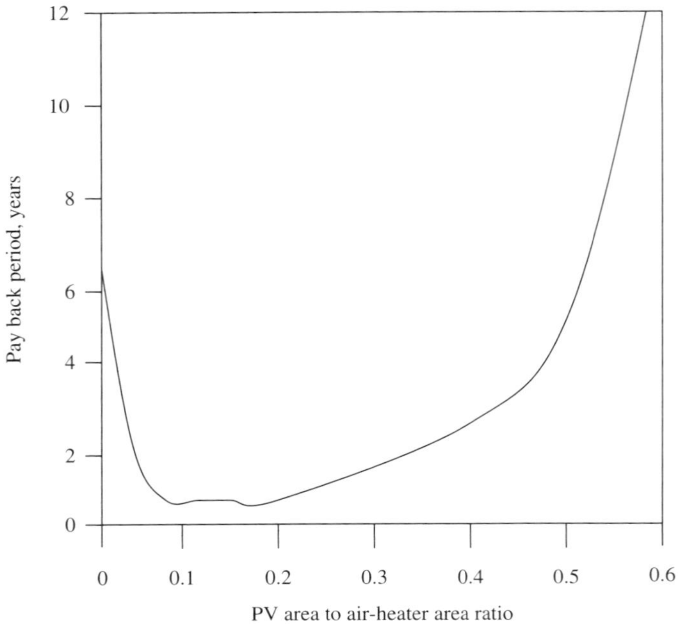

Figure 10.3 Comparison of annual cost (Rs.) with respect to life of the system (years).

is shown in Figure 10.3.[2] It is clear from the figure that the optimum PV area to air-heater area ratio is 0.22 for a pay back period of 0.5 year, which is less than a year, if used to dry surplus grain for selling at the markets.

Kumar and Kandpal[3] have estimated the potential for solar drying of selected cash crops, namely tobacco, tea, coffee, grapes, raisins, small cardamom, chilli, coriander seeds, ginger, turmeric, black pepper and onion flakes *etc.* for Indian conditions. They also estimated the potential of net fossil CO_2 emissions mitigation due to the amounts of different fuels that would be saved, along with the unit cost of CO_2 emissions mitigation.

Table 10.2 gives the cost of different types of solar dryers constructed at the Indian Institute of Technology, New Delhi, India.

Table 10.3 gives the cost of plastic and conventional solar collectors, having 560 m^2 gross collector area, to deliver useful energy of 203 MWh yr^{-1}.[4] Figures 10.4 and 10.5 show the comparison of annual cost (USD) and useful energy cost (USD kWh^{-1}), respectively, for plastic and conventional solar collectors with respect to life of the system.[4]

Problems

10.1 Calculate future (F_{PS}) and present (F_{SP}) value factors for a given life of a solar system for 10% rate of interest and show that $F_{PS} \cdot F_{SP} = 1$ for

Table 10.2 Cost of different types of solar dryers (In India).[6]

Type of solar dryer	*Initial investment (Rs)*	*Salvage value (Rs.)*	*O&M[a] cost year^{-1}(Rs)*	*Life (yrs)*
Cabinet dryer	5,000	–	200	5
Greenhouse crop dryer (natural)	2,000	–	–	5
Reverse absorber cabinet dryer	8,000	–	200	5
Conventional active solar dryer	15,000	2,000	200	5
Hybrid PV/T Integrated greenhouse dryer	43,000	10,000	1,000[b]	35
Hybrid PV/T solar dryer	39,000	5,000	200[c]	30

[a]*O&M represents operation and maintenance.*
[b]*per five years for UV polyethylene sheet replacement.*
[c]*per three years for glass replacement.*

Table 10.3 Cost of plastic and conventional solar collectors (area, 560 m^2) (In India).

Dryer	*Initial investment (Rs)*	*Salvage value (Rs)*	*O&M cost year^{-1} (Rs)*	*Life (yrs)*
Plastic solar collectors	560,000	56,000	14,772.8	5
	560,000	56,000	9,116.8	10
	560,000	56,000	7,362.3	15
Conventional solar collectors	1,120,000	112,000	18,233.6	10
	1,120,000	112,000	13,160	20
	1,120,000	112,000	11,883.2	30

each case. Hint: use eqns (10.1b) and (10.4b) for $n = 0$, 2, 4, 6, 8, 10 and 12.

10.2 Calculate the effective rate of return for different values of p for 10% rate of interest. Hint: use eqn (10.2) for $p = 1$, 2, 3, 4 and 6.

10.3 Calculate the capital recovery (F_{PR}) and sinking fund (F_{SR}) factors for different numbers of years ($n = 1$, 5, 10, 15 and 20) for a given rate of interest ($i = 0.05$, 0.10, 0.15 and 0.20 percentages). Hint: use eqns (10.6c) and (10.8b).

10.4 Draw the curve between F_{PR} and n for different values of 'i' of Problem 10.3.

10.5 Prove that $F_{SR} \cdot F_{RS} = 1$. Hint: use eqn (10.9c).

10.6 A hybrid solar dryer purchased for USD 1200 is expected to generate annual revenues of USD 150 and have a salvage value of USD 400 at the end of 15 years. If 18% per year required return is imposed on the

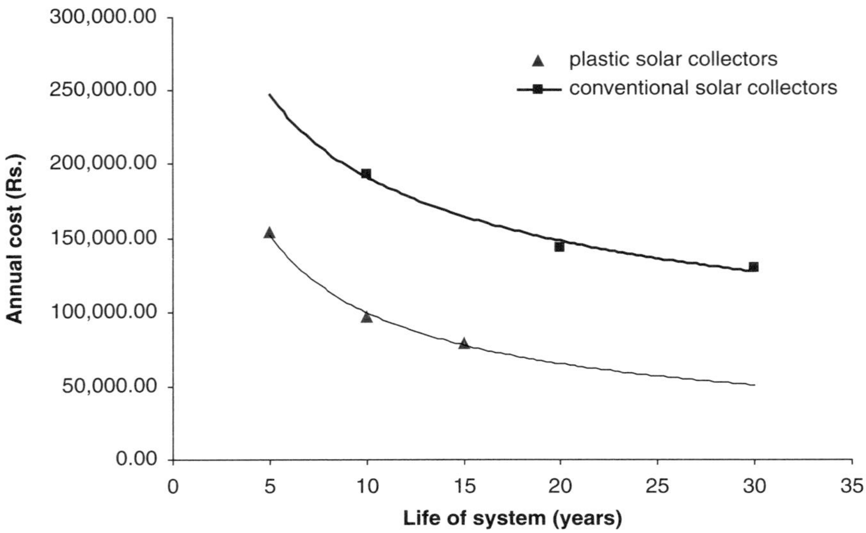

Figure 10.4 Comparison of useful energy cost (Rs. kWh^{-1}) with respect to life of the system (years).

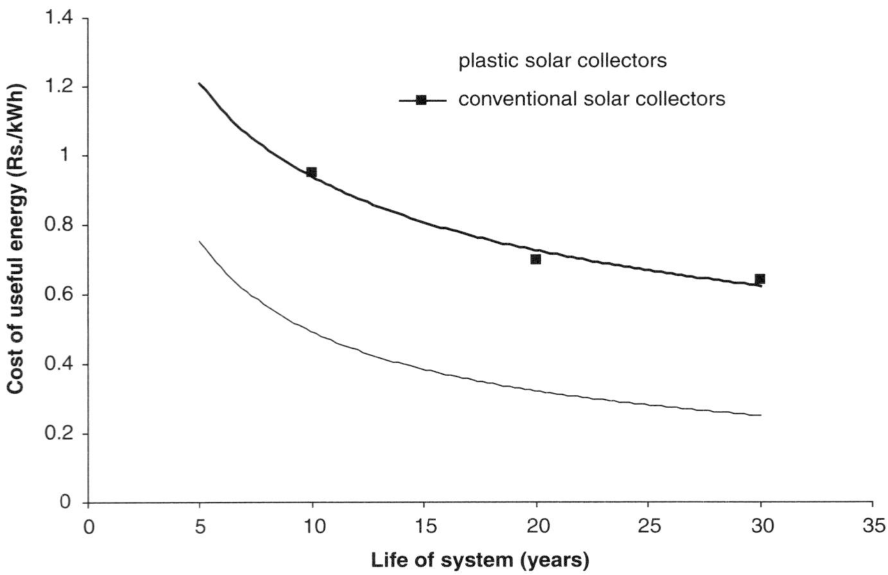

Figure 10.5 Variation of the dryer pay back time with the ratio of PV area to the solar air-heater area.

purchase, compute the pay back period. Hint: Solve the problem with a cash flow diagram.

10.7 Two solar dryers have been heated by solar energy and the cost comparison is given in Table 10.2. Find out which system is more economical if the money is worth 12% per year. Draw a cash flow diagram of both the systems. Hint: see Example 10.8.

10.8 Repeat Problem 10.7 and compare the cost for unequal life of the plastic and conventional solar collectors given in Table 10.3.

10.9 Derive an expression for the present value P for a uniform end-of-year cost R occurring simultaneously with the tax instant t.

10.10 Derive an expression for the present value P for a given salvage C_{sal} value at the end of the nth year by treating as a non-depreciable first cost, an expense.

10.11 A non-profit organization is contemplating an investment of USD 1,00,000 to install a hybrid solar water-heating system. The grant would extend over a 10-year period and would create an estimated saving of USD 20,000 per year. The organization uses a rate of return of 6% per year on all grant investments. An estimated USD 4000 a year would have to be released, from other sources, for expenses. In order to make this program successful, a USD 2000 per year operating expense will be incurred by the organization from its regular O&M budget. Use the following analysis methods to determine whether the program is justified over a 10-year period: (a) Conventional B/C (b) modified B/C and B-C analysis. Hint: use eqn (10.20) and (10.21).

10.12 Two swimming pools have been heated by PV/T solar water-heating systems which have the following cost comparison. Find out which system is more economical if the money is worth 12% per year. Hint: see Example 10.11.

Economic components	*System I*	*System II*
First cost (USD)	60,000	30,000
Uniform end-of-year maintenance per year (USD)	3,500	7,000
Overall, end of the fifth year (USD)	3,000	2,500
Salvage value (USD)	10,000	10,000
Life of the system (years)	25	25

References

1. M. A. Hossain, J. L. Woods and B. K. Bala, *Optimisation Renewable Energy*, 2005, **30**, 729–742.
2. J. Mumba, *Energ. Convers. Manag.*, 1996, **37**(5), 615–621.

3. A. Kumar and T. C. Kandpal, *Sol. Energ.*, 2005, **78**(2), 321–329.
4. M. S. Sodha, R. Chandra, K. Pathak, N. P. Singh and N. K. Bansal, *Energ. Convers. Manag.*, 1991, **31**(6), 509–513.
5. G. N. Tiwari, *Solar Energy: Fundamentals, Design, Modeling and Applications*, Narosa Publishing House, New Delhi, India, 2004.
6. P. Barnwal and A. Tiwari, *Int. J. Agr. Res.*, 2008, **3**(2), 110–120.

APPENDIX I

Conversion of Units

i) Length, m

1 yd (yard) = 3 ft = 36 in (inches) = 9144 m
1 m = 39.3701 in = 3.280839 ft = 1.093613 yd = 1650763.73 wavelength
1 ft = 12 in = 0.3048 m
1 in = 2.54 cm = 25.4 mm
1 mil = 2.54×10^{-3} cm
1 μm = 10^{-6} m
1 nm = 10^{-9} m = 10^{-3} μm

ii) Area, m^2

1 ft^2 = 0.0929 m^2
1 in^2 = 6.452 cm^2 = 0.00064516 m^2
1 cm^2 = 10^{-4} m^2 = 10.764×10^{-4} ft^2 = 0.1550 in^2
1 ha = 10,000 m^2

iii) Volume, m^3

1 ft^3 = 0.02832 m^3 = 28.3168 l (litre)
1 in^3 = 16.39 cm^3 = 1.639×10^2 l
1 yd^3 = 0.764555 m^3 = 7.646×10^2 l
1 UK gallon = 4.54609 l
1 US gallon = 3.785 l = 0.1337 ft^3
1 m^3 = 1.000×10^6 cm^3 = 2.642×10^{12} US gallons = 109 l
1 l = 10^{-3} m^3
1 fluid ounce = 28.41 cm^3

iv) Mass, kg

1 kg = 2.20462 lb = 0.068522 slug
1 ton (short) = 2000 lb (pounds) = 907.184 kg
1 ton (long) = 1016.05 kg
1 lb = 16 oz (ounces) = 0.4536 kg
1 oz = 28.3495 g
1 quintal = 100 kg

RSC Energy Series No. 2
Fundamentals of Photovoltaic Modules and Their Applications
By G. N. Tiwari and Swapnil Dubey

Published by the Royal Society of Chemistry, www.rsc.org

1 kg = 1000 g = 10,000 mg
1 μg = 10^{-6} g
1 ng = 10^{-9} g

v) Density and specific volumes, kg m^{-3}, m^3 kg^{-1}
1 lb ft^{-3} = 16.0185 kg m^{-3} = 5.787 × 10^{-4} lb in^{-3}
1 g cm^{-3} = 10^3 kg m^{-3} = 62.43 lb ft^{-3}
1 lb ft^{-3} = 0.016 g cm^{-3} = 16 kg m^{-3}
1 ft^3 (air) = 0.08009 lb = 36.5 g at N.T.P.
1 gallon lb^{-1} = 0.010 cm^3 kg^{-1}
1 μg m^{-3} = 10^{-6} g m^{-3}

vi) Pressure, Pa (Pascal)
1 lb ft^{-2} = 4.88 kg m^{-2} = 47.88 Pa
1 lb in^{-2} = 702.7 kg m^{-2} = 51.71 mm Hg = 6.894757 × 10^3 Pa = 6.894757× 10^3 N m^{-2}
1 atm = 1.013 × 10^5 N m^{-2} = 760 mm Hg = 101.325 kPa
1 in H_2O = 2.491 × 10^2 N m^{-2} = 248.8 Pa = 0.036 lb in^{-2}
1 bar = 0.987 atm = 1.000 × 10^6 dynes cm^{-2} = 1.020 020 kgf cm^{-2} = 14.50 50 lbf in^{-2} = 10^5 N m^{-2} = 100 kPa
1 torr (mm Hg 0 °C) = 133 Pa
1 Pascal (Pa) = 1 N m^{-2} = 1.89476 kg
1 inch of Hg = 3.377 kPa = 0.489 lb in^{-2}

vii) Velocity, m s^{-1}
1 ft s^{-1} = 0.3041 m s^{-1}
1 mile h^{-1} = 0.447 m s^{-1} = 1.4667 ft s^{-1} = 0.8690 knots
1 km h^{-1} = 0.2778 m s^{-1}
1 ft min^{-1} = 0.00508 m s^{-1}

viii) Force, N
1 N (Newton) = 10^5 dynes = 0.22481 lb wt = 0.224 lb f
1 pdl (poundal) = 0.138255 N (Newton) = 13.83 dynes = 14.10 gf
1 lbf (*i.e.* wt of 1 lb mass) = 4.448222 N = 444.8222 dynes
1 ton = 9.964 × 10^3 N
1 bar = 10^5 Pa (Pascal)
1 ft of H_2O = 2.950 × 10^{-2} atm = 9.807 × 10^3 N m^{-2}
1 in H_2O = 249.089 Pa
1 mm H_2O = 9.80665 Pa
1 dyne = 1.020 × 10^{-6} kg f = 2.2481 × 10^{-6} lb f = 7.2330 × 10^{-5} pdl = 10^{-5} N
1 mm of Hg = 133.3 Pa
1 atm = 1 kg f cm^{-2} = 98.0665 k Pa
1 Pa (Pascal) = 1 N m^{-2}

ix) Mass flow rate and discharge, kg s^{-1}, m^3 s^{-1}
1 lb s^{-1} = 0.4536 kg s^{-1}
1 ft^3 min^{-1} = 0.4720 1 s^{-1} = 4.179 × 10^{-4} m^3 s^{-1}
1 m^3 s^{-1} = 3.6 × 10^6 1 h^{-1}
1 g cm^{-3} = 10^3 kg m^{-3}

$1\ lb\ h^{-1}\ ft^{2} = 0.001356\ kg\ s^{-1}\ m^{2}$
$1\ lb\ ft^{-3} = 16.2\ kg\ m^{-2}$
$1\ litre\ s^{-1}\ (l\ s^{-1}) = 10^{-3}\ m^{3}\ s^{-1}$

x) Energy, J
1 cal = 4.187 J (Joules)
$1\ kcal = 3.97\ Btu = 12 \times 10^{-4}\ kWh = 4.187 \times 10^{3}\ J$
$1\ watt = 1.0\ J\ s^{-1}$
$1\ Btu = 0.252\ kcal = 2.93 \times 10^{-4}\ kWh = 1.022 \times 10^{3}\ J$
1 hp = 632.34 kcal = 0.736 kWh
$1\ kWh = 3.6 \times 10^{6}\ J = 1\ unit$
$1\ J = 2.390 \times 10^{-4}\ kcal = 2.778 \times 10^{-4}\ Wh$
1 kWh = 860 kcal = 3413 Btu
$1\ erg = 1.0 \times 10^{-7}\ J = 1.0 \times 10^{-7}\ Nm = 1.0\ dyne\ cm$
1 J = 1 Ws = 1 Nm
$1\ eV = 1.602 \times 10^{-19}\ J$
$1\ GJ = 10^{9}\ J$
$1\ MJ = 10^{6}\ J$
$1\ TJ\ (Terajoules) = 10^{12}\ J$
$1\ EJ\ (Exajoules) = 10^{18}\ J$

xi) Power, watt ($J\ s^{-1}$)
$1\ Btu\ h^{-1} = 0.293071\ W = 0.252\ kcal\ h^{-1}$
$1\ Btu\ h^{-1} = 1.163\ W = 3.97\ Btu\ h^{-1}$
$1\ W = 1.0\ J\ s^{-1} = 1.341 \times 10^{-3}\ hp = 0.0569\ Btu\ min^{-1} = 0.01433\ kcal\ min^{-1}$
$1\ hp\ (F.P.S.) = 550\ ft\ lb\ f\ s^{-1} = 746\ W = 596\ kcal\ h^{-1} = 1.015\ hp\ (M.K.S.)$
$1\ hp\ (M.K.S.) = 75\ mm\ kg\ f\ s^{-1} = 0.17569\ kcal\ s^{-1} = 735.3\ W$
$1\ W\ ft^{-2} = 10.76\ W\ m^{-2}$
1 ton (refrigeration) = 3.5 kW
1 kW = 1000 W
$1\ GW = 10^{9}\ W$
$1\ W\ m^{-2} = 100\ lux$

xii) Specific heat, $J\ kg^{-1}\ {}^{\circ}C^{-1}$
$1\ Btu\ lb^{-1}\ {}^{\circ}F = 1.0\ kcal\ kg^{-1}\ {}^{\circ}C^{-1} = 4.187 \times 10^{3}\ J\ kg^{-1}\ {}^{\circ}C^{-1}$
$1\ Btu\ lb^{-1} = 2.326\ kJ\ kg^{-1}$

xiii) Temperature, °C and K used in SI
$T_{(Celsius,\ {}^{\circ}C)} = (5/9)\ [T_{(Fahrenheit,\ {}^{\circ}F)} + 40] - 40$
$T_{({}^{\circ}F)} = (9/5)\ [T_{({}^{\circ}C)} + 40] - 40$
$T_{(Rankine,\ {}^{\circ}R)} = 460 + T_{({}^{\circ}F)}$
$T_{(Kelvin,\ K)} = (5/9)\ T_{({}^{\circ}R)}$
$T_{(Kelvin,\ K)} = 273.15 + T_{({}^{\circ}C)}$
$T_{({}^{\circ}C)} = T_{({}^{\circ}F)}/1.8 = (5/9)\ T_{({}^{\circ}F)}$

xiv) Rate of heat flow per unit area or heat flux, $W\ m^{-2}$
$1\ Btu\ ft^{-2}\ h = 2.713\ kcal\ m^{-2}\ h = 3.1552\ W\ m^{-2}$
$1\ kcal\ m^{-2}\ h = 0.3690\ Btu\ ft^{-2}\ h = 1.163\ W\ m^{-2} = 27.78 \times 10^{-6}\ cal\ s^{-1}\ cm^{2}$

$1\ cal\ cm^{-2}\ min = 221.4\ Btu\ ft^{-2}\ h$
$1\ W\ ft^{-2} = 10.76\ W\ m^{-2}$
$1\ W\ m^{-2} = 0.86\ kcal\ hm^{-2} = 0.23901 \times 10^{-4}\ cal\ s^{-1}\ cm^{2} = 0.137\ Btu\ h^{-1}\ ft^{2}$
$1\ Btu\ h^{-1}\ ft = 0.96128\ W\ m^{-1}$

xv) Heat transfer coefficient, $W\ m^{-2}\ {}^{\circ}C^{-1}$
$1\ Btu\ ft^{-2} h\ {}^{\circ}F = 4.882\ kcal\ m^{-2} h\ {}^{\circ}C^{-1} = 1.3571 \times 10^{-4}\ cal\ cm^{-2}\ s\ {}^{\circ}C^{-1}$
$1\ Btu\ ft^{-2} h\ {}^{\circ}F = 5.678\ W\ m^{-2}\ {}^{\circ}C^{-1}$
$1\ kcal\ m^{-2} h\ {}^{\circ}C^{-1} = 0.2048\ Btu\ ft^{-2}\ h\ {}^{\circ}F = 1.163\ W\ m^{-2}\ {}^{\circ}C^{-1}$
$1\ W\ m^{-2}\ K^{-1} = 2.3901 \times 10^{-5}\ cal\ cm^{-2}\ s\ K = 1.7611 \times 10^{-1}\ Btu\ ft^{-2}\ {}^{\circ}F$
$= 0.86\ kcal\ m^{-2}\ h\ {}^{\circ}C^{-1}$

xvi) Thermal conductivity, $W\ m^{-1}\ {}^{\circ}C^{-1}$
$1\ Btu\ ft^{-1}\ h\ {}^{\circ}F^{-1} = 1.488\ kcal\ m^{-1}\ h\ {}^{\circ}C^{-1} = 1.73073\ W\ m^{-1}\ {}^{\circ}C^{-1}$
$1\ kcal\ m^{-1}\ h\ {}^{\circ}C^{-1} = 0.6720\ Btu\ ft^{-1}\ h\ {}^{\circ}F^{-1} = 1.1631\ W\ m^{-1}\ {}^{\circ}C^{-1}$
$1\ Btu\ in^{-1}\ ft^{-2}\ h\ {}^{\circ}F^{-1} = 0.124\ kcal\ mh^{-1}\ {}^{\circ}C^{-1} = 0.144228\ W\ m^{-1}\ {}^{\circ}C^{-1}$
$1\ Btu\ in^{-1}\ h\ {}^{\circ}F^{-1} = 17.88\ kcal\ mh^{-1}\ {}^{\circ}C^{-1}$
$1\ cal\ cm^{-1}\ s\ {}^{\circ}F^{-1} = 4.187 \times 10^{2}\ W\ m^{-1}\ {}^{\circ}C^{-1} = 242\ Btu\ h^{-1}\ ft\ {}^{\circ}F^{-1}$
$1\ W\ cm^{-1}\ {}^{\circ}C^{-1} = 57.79\ Btu\ h^{-1}\ ft\ {}^{\circ}F^{-1}$

xvii) Angle, rad
2π rad (radian) = 360° (degree)
1° (degree) = 0.0174533 rad = 60′ (minutes)
$1' = 0.290888 \times 10^{-3}$ rad = 60″ (seconds)
$1'' = 4.84814 \times 10^{-6}$ rad
1° (hour angle) = 4 minutes (time)

xviii) Illumination
1 lx (lux) = 1.0 lm (lumen) m^{-2}
1 lm ft^{-2} = 1.0 foot candle
1 foot candle = 10.7639 lx
100 lux = 1 W m^{-2}

xix) Time, h
1 week = 7 days = 168 h = 10,080 minutes = 604,800 s
1 mean solar day = 1440 minutes = 86,400 s
1 calendar year = 365 days = 8760 h = 5.256×10^{5} minutes
1 tropical mean solar year = 365.2422 days
1 sidereal year = 365.2564 days (mean solar)
1 s (second) = $9.192631770 \times 10^{9}$ Hertz (Hz)
1 day = 24 hours = 360° (hour angle)

xx) Concentration, $kg\ m^{-3}$ and $g\ m^{-3}$
$1\ g\ l^{-1} = 1\ kg\ m^{-3}$
$1\ lb\ ft^{-3} = 6.236\ kg\ m^{-3}$

xxi) Diffusivity, $m^{2}\ s^{-1}$
$1\ ft^{2}\ h^{-1} = 25.81 \times 10^{-6}\ m^{2}\ s^{-1}$

APPENDIX II

Parameters on Horizontal Surface for Sunshine Hours = 10 for All Four Weather Types of Days for Different Indian Climates

RSC Energy Series No. 2
Fundamentals of Photovoltaic Modules and Their Applications
By G. N. Tiwari and Swapnil Dubey

Published by the Royal Society of Chemistry, www.rsc.org

(a) New Delhi

Type of day	*Month▶* / *Parameters▼*	*January*	*February*	*March*	*April*	*May*	*June*	*July*	*August*	*September*	*October*	*November*	*December*
a	T_R	2.25	2.79	2.85	2.72	3.54	2.47	2.73	2.58	2.53	1.38	0.62	0.72
	α	0.07	0.10	0.17	0.23	0.16	0.28	0.37	0.41	0.29	0.47	0.59	0.54
	K_1	0.47	0.39	0.33	0.28	0.20	0.27	0.41	0.40	0.23	0.21	0.21	0.28
	K_2	−13.17	−6.25	5.61	38.32	65.04	31.86	−40.57	−55.08	39.92	32.77	30.62	9.73
b	T_R	2.28	2.78	2.89	3.15	5.44	4.72	5.58	5.43	3.23	4.56	0.19	1.83
	α	0.15	0.13	0.14	0.17	0.16	0.20	0.24	0.18	0.31	0.22	1.14	0.42
	K_1	0.51	0.54	0.49	0.46	0.45	0.45	0.53	0.39	0.37	0.42	0.35	0.40
	K_2	−21.77	−28.26	−9.22	−11.55	1.54	23.99	−51.61	9.46	14.07	−9.50	17.47	−0.07
c	T_R	5.88	6.36	6.11	7.77	9.20	10.54	7.13	7.97	5.51	5.01	4.93	3.23
	α	0.27	0.37	0.37	0.31	0.07	0.06	0.41	0.51	0.49	1.26	1.06	0.64
	K_1	0.39	0.36	0.33	0.35	0.56	0.48	0.47	0.35	0.39	0.36	0.31	0.43
	K_2	−14.73	−7.97	10.87	20.45	−56.00	−0.37	−52.27	47.70	35.64	−0.68	13.06	−7.04
d	T_R	7.47	8.97	10.77	11.18	13.69	12.47	8.21	8.58	9.40	7.24	4.30	4.02
	α	0.96	1.04	0.24	0.07	0.07	0.61	1.26	1.10	0.84	1.29	1.43	1.70
	K_1	0.35	0.30	0.43	0.49	0.48	0.46	0.43	0.43	0.41	0.36	0.31	0.38
	K_2	−25.89	−6.48	−36.46	−44.07	−42.58	−62.66	−56.75	−61.08	−27.09	3.90	20.10	−11.78

(b) Bangalore

Type of day	Month▶ / Parameters▼	January	February	March	April	May	June	July	August	September	October	November	December
a	T_R	3.36	3.27	3.63	5.05	4.24	4.32	5.18	4.75	4.10	2.28	1.66	1.65
	α	0.07	0.13	0.06	–0.06	0.10	0.19	0.10	0.18	0.13	0.33	0.35	0.36
	K_1	0.33	0.35	0.33	0.29	0.21	0.25	0.32	0.23	0.20	0.05	0.03	0.12
	K_2	–18.05	–22.11	–5.44	14.54	47.81	22.40	–26.04	10.14	38.54	107.04	103.64	47.70
b	T_R	3.24	5.25	6.21	5.72	5.90	7.35	4.12	5.27	4.83	2.43	1.89	3.68
	α	0.31	0.24	0.21	0.19	0.25	0.17	0.51	0.44	0.62	0.56	0.78	0.39
	K_1	0.50	0.45	0.48	0.50	0.41	0.50	0.46	0.50	0.33	0.26	0.37	0.41
	K_2	–60.12	–60.50	–80.04	–75.59	–28.55	–103.35	–90.54	–115.27	13.80	69.14	9.08	–33.76
c	T_R	3.70	4.51	7.74	5.83	4.95	4.39	5.68	2.67	6.64	4.71	5.68	2.02
	α	0.96	0.94	0.63	0.98	0.96	1.12	1.07	1.35	0.78	1.03	0.93	1.44
	K_1	0.46	0.57	0.36	0.50	0.53	0.58	0.50	0.55	0.48	0.43	0.36	0.43
	K_2	–63.02	–129.68	–20.76	–61.13	–103.14	–156.14	–108.34	–161.61	–52.93	–26.53	–15.95	–47.21
d	T_R	6.13	7.49	7.35	6.86	6.33	4.84	4.45	6.68	3.94	3.91	3.84	2.80
	α	1.61	1.31	1.41	1.48	1.59	2.00	2.32	1.69	2.16	2.00	2.04	2.58
	K_1	0.29	0.30	0.40	0.45	0.53	0.61	0.41	0.50	0.38	0.42	0.55	0.27
	K_2	36.80	83.73	–39.85	–72.22	–99.52	–213.29	–79.79	–146.94	–88.62	–125.35	–177.28	–12.29

(c) Jodhpur

Type of day	Month▶ / Parameters▼	January	February	March	April	May	June	July	August	September	October	November	December
a	T_R	1.26	1.33	1.59	2.82	3.72	3.87	3.25	3.39	3.20	2.26	1.56	1.54
	α	0.37	0.38	0.37	0.27	0.21	0.21	0.27	0.28	0.27	0.33	0.39	0.31
	K_1	0.22	0.14	0.18	0.21	0.20	0.13	0.10	0.17	0.26	0.24	0.23	0.26
	K_2	30.67	63.90	56.40	47.66	50.84	87.88	105.23	59.41	14.42	27.40	22.71	9.48
b	T_R	2.34	2.03	3.00	4.07	5.21	5.50	5.07	4.73	3.81	2.90	2.28	3.43
	α	0.46	0.55	0.42	0.31	0.23	0.28	0.37	0.40	0.35	0.38	0.46	0.24
	K_1	0.33	0.29	0.31	0.34	0.33	0.33	0.34	0.33	0.34	0.30	0.33	0.40
	K_2	12.89	43.13	42.22	23.50	31.22	33.40	35.81	29.57	8.71	24.12	12.35	–11.64
c	T_R	3.81	4.78	4.04	4.97	6.87	5.58	4.90	5.10	3.40	3.71	3.28	4.23
	α	0.93	1.32	0.98	0.64	0.61	0.67	1.02	0.88	0.97	2.05	1.31	1.06
	K_1	0.43	0.40	0.42	0.47	0.47	0.46	0.41	0.50	0.48	0.53	0.44	0.44
	K_2	–33.72	12.44	–19.11	–26.93	–44.76	–35.15	2.06	–60.42	–26.96	–62.06	–35.85	–32.84
d	T_R	2.25	5.20	7.09	9.33	8.01	3.52	9.62	3.17	1.63	7.67	1.71	1.94
	α	1.89	1.64	2.03	1.59	1.66	2.37	2.37	2.77	3.24	0.86	2.89	2.03
	K_1	0.44	0.46	0.42	0.44	0.43	0.28	0.52	0.44	0.44	0.52	0.36	0.39
	K_2	–19.31	–45.44	–89.92	–149.27	–117.01	60.69	–221.29	–87.34	–77.55	–26.47	–15.46	–14.88

(d) Mumbai

Type of day	Month▶ / Parameters▼	January	February	March	April	May	June	July	August	September	October	November	December
a	T_R	1.95	1.80	2.88	3.95	5.40	3.20	3.31	4.25	4.22	3.16	2.97	3.27
	α	0.34	0.37	0.23	0.14	–0.02	0.16	0.61	0.33	0.15	0.30	0.23	0.18
	K_1	0.26	0.19	0.28	0.34	0.28	0.25	0.09	0.12	0.24	0.24	0.26	0.30
	K_2	19.77	53.96	27.13	–0.75	30.06	4.55	27.28	47.27	30.02	15.87	9.11	–4.81
b	T_R	2.96	2.68	3.57	4.98	6.25	6.08	7.74	6.70	4.78	3.93	3.40	4.21
	α	0.43	0.49	0.37	0.25	0.15	0.19	0.20	0.37	0.47	0.47	0.45	0.24
	K_1	0.35	0.31	0.35	0.40	0.42	0.44	0.31	0.39	0.41	0.36	0.34	0.37
	K_2	–0.14	24.17	11.73	–13.57	–13.69	–19.52	61.35	22.16	–14.71	5.99	0.60	–14.17
c	T_R	3.06	2.26	3.24	4.39	5.91	5.97	8.17	4.24	5.36	3.16	2.97	3.75
	α	1.14	1.18	1.10	1.00	0.79	0.86	0.62	1.26	0.98	1.13	1.10	0.91
	K_1	0.59	0.58	0.52	0.54	0.60	0.52	0.54	0.43	0.44	0.47	0.57	0.54
	K_2	–59.86	–47.12	–58.09	–78.37	–111.97	–81.79	–95.21	–34.40	–39.31	–28.02	–48.41	–52.45
d	T_R	3.38	7.42	4.45	2.30	4.71	4.71	6.41	7.40	7.46	3.22	5.13	3.05
	α	1.71	1.73	2.29	2.08	2.95	2.66	2.68	1.81	2.14	2.15	1.53	1.51
	K_1	0.52	0.56	0.50	0.35	0.41	0.38	0.32	0.47	0.34	0.42	0.57	0.53
	K_2	–59.78	–26.16	–82.34	63.52	–101.81	–87.19	–61.50	–108.37	–38.68	–25.89	–78.03	–40.51

(e) Srinagar

Type of day	Month▶ / Parameters▼	January	February	March	April	May	June	July	August	September	October	November	December
a	T_R	1.45	5.37	3.31	4.25	5.41	3.63	5.77	6.45	4.06	2.61	4.03	0.72
	α	0.33	–0.36	–0.03	–0.03	–0.12	0.08	–0.09	–0.23	0.03	0.20	–0.37	0.53
	K_1	0.37	0.63	0.69	0.37	0.51	0.33	0.17	0.37	0.46	0.43	0.66	0.33
	K_2	–6.14	–82.86	–94.01	–10.95	–79.57	–13.73	68.06	–42.79	–60.27	–47.83	–37.00	–6.60
b	T_R	3.09	6.98	4.65	6.92	5.86	6.82	7.40	7.58	6.41	4.04	0.04	0.35
	α	0.38	–0.48	0.23	0.06	0.29	0.11	0.00	–0.13	–0.04	0.19	1.16	1.00
	K_1	0.39	0.83	0.59	0.42	0.32	0.63	0.48	0.38	0.48	0.52	0.37	0.41
	K_2	–23.08	–110.23	–107.74	–49.61	0.26	–167.86	–80.06	–13.91	–66.64	–62.52	–14.63	–12.20
c	T_R	2.35	6.59	6.31	7.57	8.69	8.00	9.72	8.23	7.36	5.02	1.86	0.76
	α	1.64	0.86	1.35	0.57	0.61	0.81	0.69	0.90	0.99	1.49	1.47	1.98
	K_1	0.41	0.42	0.48	0.54	0.50	0.39	0.56	0.49	0.44	0.52	0.41	0.31
	K_2	–37.87	–85.68	–180.45	–120.38	–146.97	–87.44	–228.91	–147.96	–62.10	–93.64	–40.07	–12.15
d	T_R	1.69	1.36	7.52	9.09	9.48	10.79	10.93	8.54	8.16	7.75	3.78	2.44
	α	2.63	2.97	1.87	1.35	1.13	1.56	3.08	1.71	3.15	1.70	1.74	2.04
	K_1	0.43	0.36	0.35	0.62	0.92	0.80	0.45	0.75	0.67	0.55	0.48	0.63
	K_2	–41.27	–44.68	–65.17	–254.24	–467.30	–421.63	–129.49	–356.92	–261.85	–119.53	–49.16	–64.02

APPENDIX III

Specifications of Solar Cell Material (at Solar Intensity 1000 W m^{-2} and Cell Temperature 25 °C) and Cost

Cell technology	*Efficiency (%)*	*Fill factor (FF)*	*Aperture area (10^{-4} × m^{-2})*	*Lifetime*[a] *(years)*	*Manufacturing cost (\$ kWp^{-1} in 2007)*	*Selling price (\$ kWp^{-1} in 2007)*
Monocrystalline silicon	24.7 ± 0.5	0.828	4.0	30	2.5	3.7
Multicrystalline silicon	19.8 ± 0.5	0.795	1.09	30	2.4	3.5
Copper indium diselenide (CIS/CIGS)	18.4 ± 0.5	0.77	1.04	5	1.5	2.5
Thin silicon cell	16.6 ± 0.4	0.782	4.02	25	2.0	3.3
Cadmium telluride (CdTe)	16.5 ± 0.5	0.755	1.03	15	1.5	2.5
Amorphous silicon (a-Si)	10.1 ± 0.2	0.766	1.2	20	1.5	2.5

[a]*Based on experience.*
Source: B. Agarwal, G.N. Tiwari, Development in environmental durability for photovoltaics, Pira International Ltd, UK, 2008.
Courtesy : L.L.Kazmerski, NREL, http : //en.wikipedia.org/wiki/File : PVeff(rev110707)d.png.

RSC Energy Series No. 2
Fundamentals of Photovoltaic Modules and Their Applications
By G. N. Tiwari and Swapnil Dubey

Published by the Royal Society of Chemistry, www.rsc.org

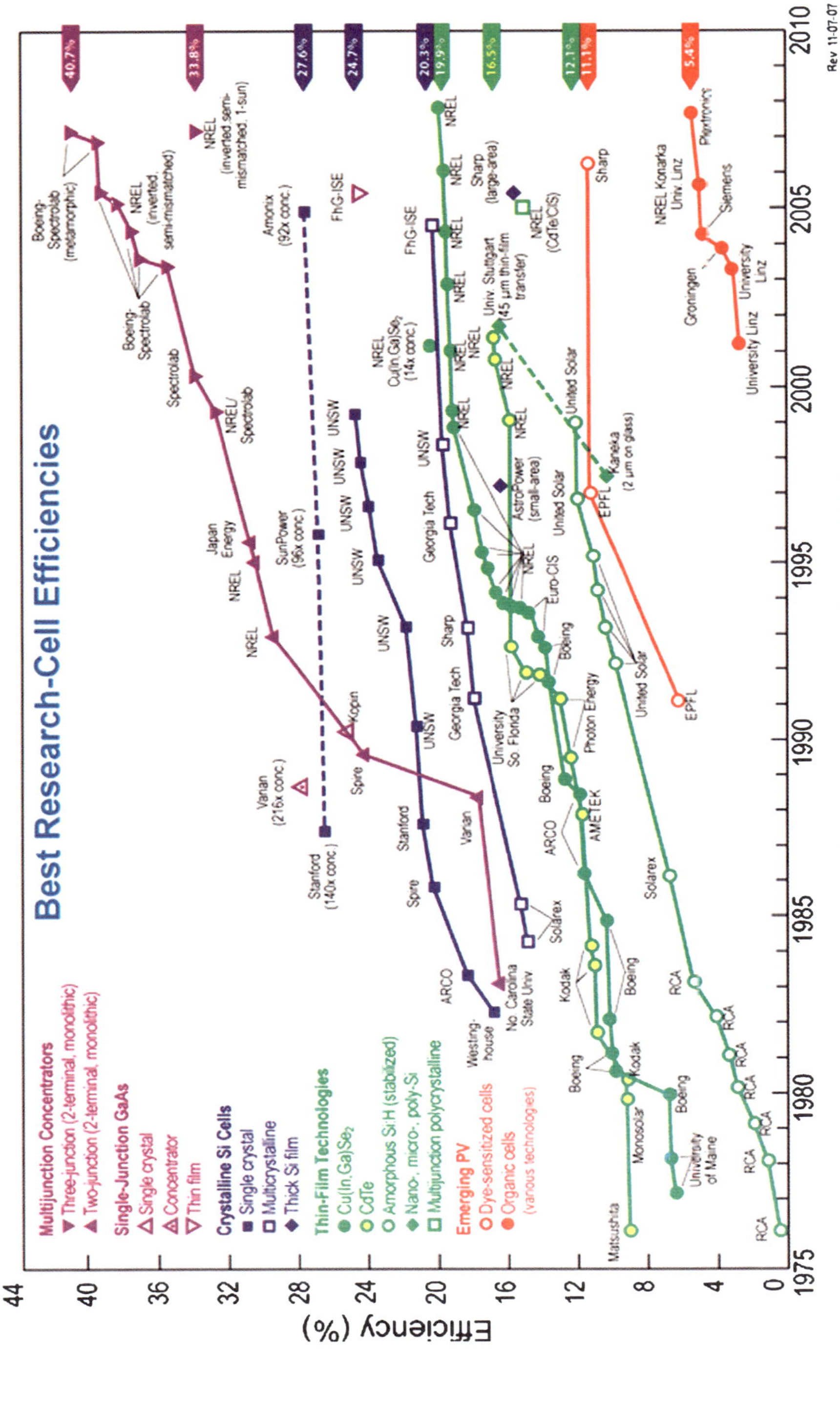

Courtesy: L.L. Kazmerski, NREL, http://en.wikipedia.org/wiki/File:PVeff(rev110707)d.png.

APPENDIX IV

List of Embodied Energy Coefficients

Material	$MJ\,kg^{-1}$	$MJ\,m^{-3}$
Aggregate, general	0.10	150
Virgin rock	0.04	63
River	0.02	36
Aluminium, virgin	191	515 700
Extruded	201	542 700
Extruded, anodized	227	612 900
Extruded, factory painted	218	588 600
Foil	204	550 800
Sheet	199	537 300
Aluminium, recycled	8.1	21 870
Extruded	17.3	46 710
Extruded, anodized	42.9	115 830
Extruded, factory painted	34.3	92 610
Foil	20.1	54 270
Sheet	14.8	39 960
Asphalt (paving)	3.4	7 140
Bitumen	44.1	45 420
Brass	62.0	519 560
Carpet	72.4	–
Felt underlay	18.6	–
Nylon	148	–
Polyester	53.7	–
Polyethylterepthalate (PET)	107	–
Polypropylene	95.4	–
Wool	106	–

RSC Energy Series No. 2
Fundamentals of Photovoltaic Modules and Their Applications
By G. N. Tiwari and Swapnil Dubey

Published by the Royal Society of Chemistry, www.rsc.org

(*Continued*).

Material	$MJ\,kg^{-1}$	$MJ\,m^{-3}$
Cement	7.8	15 210
Cement mortar	2.0	3 200
Fibre cement board	9.5	13 550
Soil-cement	0.42	819
Ceramic		–
Brick	2.5	5 170
Brick, glazed	7.2	14 760
Pipe	6.3	–
Tile	2.5	5 250
Concrete		–
Block	0.94	–
Brick	0.97	–
GRC	7.6	14 820
Paver	1.2	–
Pre-cast	2.0	–
Ready mix, 17.5 MPa	1.0	2 350
30 MPa	1.3	3 180
40 MPa	1.6	3 890
Roofing tile	0.81	–
Copper	70.6	631 160
Earth, raw		–
Adobe block, straw, stabilized	0.47	750
Adobe, bitumen stabilized	0.29	–
Adobe, cement stabilized	0.42	–
Rammed soil cement	0.80	–
Pressed block	0.42	–
Fabric		–
Cotton	143	–
Polyester	53.7	–
Glass		–
Float	15.9	40 060
Toughened	26.2	66 020
Laminated	16.3	41 080
Tinted	14.9	375 450
Insulation		–
Cellulose	3.3	112
Fibreglass	30.3	970
Polyester	53.7	430
Polystyrene	117	2 340
Wool (recycled)	14.6	139
Lead	35.1	398 030
Linoleum	116	150 930
Paint	90.4	118 per litre

(*Continued*).

Material	$MJ\,kg^{-1}$	$MJ\,m^{-3}$
Solvent based	98.1	128 per litre
Water based	88.5	115 per litre
Paper	36.4	33 670
Building	25.5	–
Kraft	12.6	–
Recycled	23.4	–
Wall	36.4	–
Plaster, gypsum	4.5	6 460
Plaster board	6.1	5 890
Plastics		–
ABS	111	–
High-density polyethelene (HDPE)	103	97 340
Low-density polyethelene (LDPE)	103	91 800
Polyester	53.7	7 710
Polypropylene	64.0	57 600
Polystyrene, expanded	117	2 340
Polyurethane	74.0	44 400
PVC	70.0	93 620
Rubber		–
Natural latex	67.5	62 100
Synthetic	110	–
Sand	0.10	232
Sealants and adhesives		–
Phenol formaldehyde	87.0	–
Urea formaldehyde	78.2	–
Steel, recycled	10.1	37 210
Reinforcing, sections	8.9	–
Wire rod	12.5	–
Steel, virgin, general	32.0	251 200
Galvanized	34.8	273 180
Imported, structural	35.0	274 570
Stone, dimension		–
Local	0.79	1 890
Imported	6.8	1 890
Straw, baled	0.24	30.5
Timber, softwood		–
Air dried, roughsawn	0.3	165
Kiln dried, roughsawn	1.6	880
Air dried, dressed	1.16	638
Kiln dried, dressed	2.5	1 380
Mouldings, etc.	3.1	1 710
Hardboard	24.2	13 310
MDF	11.9	8 330

(*Continued*).

Material	$MJ\,kg^{-1}$	$MJ\,m^{-3}$
Glulam	4.6	2 530
Particle bd	8.0	–
Plywood	10.4	–
Shingles	9.0	–
Timber, hardwood		–
Air dried, roughsawn	0.50	388
Kiln dried, roughsawn	2.0	1 550
Vinyl flooring	79.1	105 990
Zinc	51.0	364 140
Galvanizing, per kg steel	2.8	–

APPENDIX V

Absorptivity of Various Surfaces for the Sun's Rays

Surface	*Absorptivity*
White paint	0.12–0.26
Whitewash/glossy white	0.21
Bright aluminium	0.30
Flat white	0.25
Yellow	0.48
Bronze	0.50
Silver	0.52
Dark aluminium	0.63
Bright red	0.65
Brown	0.70
Light green	0.73
Medium red	0.74
Medium green	0.85
Dark green	0.95
Blue/black	0.97
Roofs	
Asphalt	0.89
White asbestos cement	0.59
Copper sheeting	0.64
Uncoloured roofing tile	0.67
Red roofing tiles	0.72
Galvanized iron, clean	0.77
Brown roofing tile	0.87
Galvanized iron, dirty	0.89
Black roofing tile	0.92

RSC Energy Series No. 2
Fundamentals of Photovoltaic Modules and Their Applications
By G. N. Tiwari and Swapnil Dubey

Published by the Royal Society of Chemistry, www.rsc.org

Walls	
White/yellow brick tiles	0.30
White stone	0.40
Cream brick tile	0.50
Burl brick tile	0.60
Concrete/red brick tile	0.70
Red sand line brick	0.72
White sand stone	0.76
Stone rubble	0.80
Blue brick tile	0.88
Surroundings	
Sea/lake water	0.29
Snow	0.30
Grass	0.80
Light-coloured grass	0.55
Light-green shiny leaves	0.75
Grey sand	0.82
Rock	0.84
Green leaves	0.85
Earth (black ploughed field)	0.92
White leaves	0.20
Yellow leaves	0.58
Aluminium foil	0.39
Unpainted wood	0.60
Metals	
Polished aluminium/copper	0.26
New galvanized iron	0.66
Old galvanized iron	0.89
Polished iron	0.45
Oxidized rusty iron	0.38

APPENDIX VI

Heating Values of Various Combustibles and Their Conversion Efficiencies

Fuel	*Heating value (kJ kg^{-1})*	*Efficiency of device*
Coal coke	29000	70
Wood	15000	60
Straw	14000–16000	60
Gasoline	43000	80
Kerosene	42000	80
Methane (natural gas)	50000	80
Biogas (60% methane)	20000	80
Electricity	–	95

RSC Energy Series No. 2
Fundamentals of Photovoltaic Modules and Their Applications
By G. N. Tiwari and Swapnil Dubey

Published by the Royal Society of Chemistry, www.rsc.org

Glossary

Absorber plate: A component of the solar flat-plate collector that absorbs solar radiation and converts it into heat.

Absorptance: The ratio between the radiation absorbed by a surface (absorber) and the total amount of solar radiation striking the surface.

Albedo: The ratio of the amount of light reflected by a surface to the light falling onto it.

Alternating current (AC): An electric current that alternates direction between positive and negative cycles, usually 50 or 60 times per second. Alternating current is the current typically available from power outlets in a household.

Altitude: The Sun's angle above the horizon, as measured in a vertical plane.

Amorphous silicon: A thin-film PV silicon cell having no crystalline structure. Manufactured by depositing layers of doped silicon on a substrate.

Ampere: A unit of electric current; a measure of flowing electrons.

Anemometer: Instrument used for measuring wind speed.

Antifreeze: Substance added to water to lower its freezing point. Solar water heaters usually use a mixture of water and propylene glycol instead of just water to prevent freezing.

Anti-reflection coating: A thin coating of a material, which reduces the light reflection and increases light transmission, applied to a photovoltaic cell surface.

Array: Any number of photovoltaic modules connected together electrically to provide a single electrical output. An array is a mechanically integrated assembly of modules or panels together with support structure (including foundation and other components, as required) to form a free-standing field-installed unit that produces DC power.

Azimuth: Horizontal angle between the Sun and due south in the northern hemisphere, or between the Sun and due north in the southern hemisphere.

Balance of system: Term used in photovoltaics, which represents all components and costs other than the PV modules.

Battery: A collection of cells that store electrical energy; each cell converts chemical energy into electricity or *vice versa*, and is interconnected with other cells to form a battery for storing useful quantities of electricity.

Battery capacity: The maximum total electrical charge, expressed in ampere-hours (AH), that a battery can deliver to a load under a specific set of conditions.

Battery cell: The simplest operating unit in a storage battery. It consists of one or more positive electrodes or plates, an electrolyte that permits ionic conduction, one or more negative electrodes or plates, separators between plates of opposite polarity and a container for all the above.

Battery available capacity: Total maximum charge, expressed in ampere-hours, that can be withdrawn from a cell or battery under a specific set of operating conditions, including discharge rate, temperature, initial state of charge, age and cutoff voltage.

Battery energy capacity: Total energy available, expressed in watt-hours (or kilowatt-hours), that can be withdrawn from a fully charged cell or battery. The energy capacity of a given cell varies with temperature, rate, age and cutoff voltage.

Battery cycle life: Number of cycles, to a specified depth of discharge, that a cell or battery can undergo before failing to meet its specified capacity or efficiency performance criteria.

Black body: A perfect absorber and emitter of radiation. A cavity is a perfect black body. Lampblack is close to a black body, while aluminium (polished) is a poor absorber and emitter of radiation.

Brightness: The subjective human perception of luminance.

Cadmium telluride (CdTe): A polycrystalline thin-film photovoltaic material.

Calorific value: Energy content per unit mass (or volume) of a fuel, which will be released in combustion ($kWh\,kg^{-1}$, $MJ\,kg^{-1}$, $kWh\,m^{-3}$, $MJ\,m^{-3}$).

Candela (cd): An SI unit of luminous intensity. An ordinary candle has a luminous intensity of one candela.

Carbon dioxide (CO_2): The colourless, odourless gas that is produced during normal human breathing. It is also emitted by combustion activities used to produce electricity. CO_2 is a major cause of the greenhouse effect that traps radiant energy near the Earth's surface.

Cell: A device that generates electricity, traditionally consisting of two plates or conducting surfaces placed in an electrolytic fluid.

Celsius: The international temperature scale in which water freezes at 0 [degrees] and boils at 100 [degrees]; named after Anders Celsius.

Charge rate: The current applied to a cell or battery to restore its available capacity. This rate is commonly normalized by a charge control device with respect to the rated capacity of the cell or battery.

Charge controller: A component of a photovoltaic system that controls the flow of current to and from the battery to protect the batteries from over-charge and over-discharge. The charge controller may also indicate the system operational status.

Circuit: A system of conductors (*i.e.* wires and appliances) capable of providing a closed path for electric current.

Clear sky: A sky condition with few or no clouds, usually taken as 0–2 tenths covered with clouds. Clear skies have high luminance and high radiation, and create strong shadows relative to more cloudy conditions. The sky is brightest nearest the Sun, whereas away from the Sun it is about three times brighter at the horizon than at the zenith.

Collector: The name given to the device that converts incoming solar radiation to heat.

Collector efficiency: Ratio of the useful (heat) energy converted by the solar collector to the radiation incident on the device.

Collector plate: A component of the solar flat-plate collector that absorbs solar radiation and converts it into heat.

Condensation: Process of vapour changing into the liquid state. Heat is released in the process.

Conductance (C): A measure of the ease with which heat flows through a specified thickness of a material by conduction. Units are $W\,m^{-2}\,°C$.

Conduction: Process by which heat energy is transferred through materials (solids, liquids or gases) by molecular excitation of adjacent molecules.

Conductivity: Quantity of heat that will flow through one square metre of material, one metre thick, in one second, when there is a temperature difference of 1 °C between its surfaces.

Convection: The transfer of heat between a moving fluid medium (liquid or gas) and a surface, or the transfer of heat within a fluid by movements within the fluid.

Concentrating collector: A solar collector that reflects the solar radiation (direct) to an absorber plate to produce high temperatures.

Crystalline silicon: A type of PV cell made from a single crystal or polycrystalline slices of silicon.

Declination: The angle of the Sun north or south of the equatorial plane.

Depth of discharge (DOD): The ampere-hours removed from a fully charged cell or battery, expressed as a percentage of rated capacity. For example, the removal of 25 ampere-hours from a fully charged 100-ampere-hours rated cell results in a 25% depth of discharge.

Diffuse radiation: The solar radiation reaching the surface due to reflection and scattering effect.

Direct current (DC): The complement of AC, or alternating current, DC presents one unvarying voltage to a load. This is standard in motor vehicles.

Direct radiation: Radiation coming in a beam from the Sun, which can be focused.

Dry bulb temperature: The temperature of a gas or mixture of gases indicated by an accurate thermometer after correction for radiation.

Efficiency: The ratio of output power (or energy) to input power (or energy) expressed as a percentage.

Electromagnetic spectrum: The entire range of wavelengths or frequencies of electromagnetic radiation extending from gamma rays to the longest radio waves including visible light.

Embodied energy: Literally the amount of energy required to produce an object in its present form; an inflated balloon's embodied energy includes the energy required to manufacture it and inflate it.

Energy density: Energy per unit area.

Energy intensity: The ratio of energy use in a sector to activity in that sector, for example, the ratio of energy use to constant dollar production in manufacturing.

Energy: A measure of a system's ability to do work.

EVA (Ethylene Vinyl Acetate) Foil: Used by module production for covering the cells.

Fill factor (FF): For an *I–V* curve, the ratio of the maximum power to the product of the open-circuit voltage and the short-circuit current. Fill factor is a measure of the 'squareness' of the *I–V* curve.

Flat-plate collector: A solar collection device for gathering the Sun's heat, consisting of a shallow metal container covered with one or more layers of transparent glass or plastic; either air or a liquid is circulated through the cavity of the container, whose interior is painted 'black' and exterior is well insulated.

Gallium arsenide (GaAs): A crystalline, high-efficiency semi-conductor/photovoltaic material.

Glare: The perception caused by a very bright light or a high contrast of light, making it uncomfortable or difficult to see.

Glazing: Transparent or translucent materials, usually glass or plastic, used to cover an opening without impeding (relative to opaque materials) the admission of solar radiation and light.

Global radiation: The sum of direct, diffuse and reflected radiation.

Greenhouse effect: The global warming resulting from the absorption of infrared solar radiation by carbon dioxide and other traces of gases present in the atmosphere. (The term is a misnomer in that in actual greenhouses the warming comes primarily from restriction on air flow.)

Greenhouse gases: Gases which contribute to the greenhouse effect by absorbing infrared radiation in the atmosphere. These gases include carbon dioxide, nitrous oxide, methane, water vapour and a variety of chlorofluorocarbons (CFCs).

Heat capacity: The quantity of heat required to raise one kilogram of a substance by one degree Celsius.

Heat exchanger: Device that passes heat from one substance to another; in a solar hot water heater, for example, the heat exchanger takes heat harvested by a fluid circulating through the solar panel and transfers it to domestic hot water.

Heat loss: An unwanted decrease in the amount of heat contained in a space.

***I–V* curve:** The plot of current *versus* voltage characteristics of a solar cell, module or array. *I–V* curves are used to compare various solar cell modules, and to determine their performance at various levels of insolation and temperatures.

Incident radiation: The quantity of radiant energy striking a surface per unit time and unit area.

Insolation (or incident solar radiation): Measure of the amount of solar radiation falling on a given surface area in a given time.

Insulation: A material that keeps energy from crossing from one place to another: on electrical wire, it is the plastic or rubber that covers the conductor; in a building, insulation makes the walls, floors and roof more resistant to the outside (ambient) temperature.

Inverter: Electrical device that changes direct current (DC) into alternating current (AC).

Joule: Unit of energy or work. One joule is equal to one watt-second.

Kilowatt (kW): 1000 watts, energy consumption at a rate of 1000 joules per second.

Kilowatt-hour (kWh): One kilowatt of power used for one hour. A typical house uses 750 kWh per month.

Latitude: the angular position of a location north or south of the equator.

Life cycle costing: A method for estimating the comparative costs of alternative energy or other systems. Life cycle costing takes into consideration such long-term costs as energy consumption, maintenance and repair.

Life cycle costs: The entire cost of an energy device, including the capital cost in present dollars and the cost and the benefits discounted to the present.

Longitude: The angular position east or west of the Greenwich meridian.

Maximum power point (MPP): The voltage at which a PV array produces maximum power.

Maximum power point tracker (MPPT): A power conditioning unit that increases the power of a PV system by ensuring operation of the PV generator at its maximum power point (MPP). The ability to do this can depend on climate and the battery's state of charge.

Module: The smallest self-contained, environmentally protected structure housing interconnected photovoltaic cells and providing a single direct current (DC) electrical output.

Natural convection: The natural convection of heat through the fluid in a body that occurs when warm, less-dense fluid rises and cold, dense fluid sinks under the influence of gravity.

Night sky radiation: A reversal of the daytime insolation principle. Just as the Sun radiates energy during the day through the void of space, so also heat energy can travel unhindered at night from the Earth's surface back into space. On a clear night, any warm object can cool itself by radiating longwave heat energy to the cooler sky. On a cloudy night, the cloud cover acts as an insulator and prevents the heat from travelling to the cooler sky.

NOCT (Nominal Operating Cell Temperature): Estimated temperature of a PV module when operating under 800 W m^{-2} irradiance, 20 °C ambient temperature and wind speed of 1 m s^{-1}. NOCT is used to estimate the nominal operating temperature of a module in its working environment.

Open circuit voltage (V_{oc}): The maximum possible voltage across a solar module or array. Open circuit voltage occurs in sunlight when no current is flowing.

Orientation: The arrangement of solar devices along a given axis to face in a direction best suited to absorb solar radiation.

Photon: The elementary particle of electromagnetic energy; light. (Greek *photos*, light.)

Photovoltaic conversion efficiency: The ratio of the electric power produced by a photovoltaic device to the power of the sunlight incident on the device.

Photovoltaic device: A device that converts light directly into DC electricity.

Photovoltaic module: The basic building block of a photovoltaic device, which consists of a number of interconnected solar cells.

Photovoltaics (PV): A technology for using semi-conductors to convert light directly into electricity.

Polycrystalline silicon: A material used to make PV cells, which consists of many crystals as contrasted with single-crystal silicon.

Power: The rate at which energy is consumed or produced. The unit is the watt.

Radiation: Electromagnetic waves that directly transport energy through space. Sunlight is a form of radiation.

Reflectivity: The ratio of radiant energy reflected by a body to that falling upon it.

Renewable energy: An energy source that renews itself without effort; fossil fuels, once consumed, are gone forever, while solar energy is renewable in that the light from the Sun that we harvest today has no effect on the light we can harvest tomorrow.

Resistor: Any electronic component that restricts the flow of electrical current in circuits.

Semi-conductor: A material such as silicon, which has a crystalline structure that will allow current to flow under certain conditions. Semi-conductors are usually less conductive than metals, but not an insulator like rubber.

Short circuit current (I_{sc}): Current across the terminals when a solar cell or module in strong sunlight is not connected to a load (measured with an ammeter).

Silicon: A semi-conductor material commonly used to make PV cells.

Single-crystal structure: A material having a crystalline structure such that a repeatable or periodic molecular pattern exists in all three dimensions.

Solar altitude: The Sun's angle above the horizon, as measured in a vertical plane.

Solar azimuth: The horizontal angle between the Sun and due south in the northern hemisphere, or between the Sun and due north in the southern hemisphere.

Solar cell: A device that converts light energy or solar radiation (photons) directly into DC electricity.

Solar cell module: Groups of encapsulated solar cells framed in a glass or plastic unit, usually the smallest unit of solar electric equipment available to the consumer.

Solar collector: A device that gathers and accumulates solar radiation to produce heat.

Solar concentrator: A device that uses reflective surfaces in a planar, parabolic trough or parabolic bowl configuration to concentrate solar radiation onto a smaller surface.

Solar constant: The amount of radiation arriving from the Sun at the edge of the Earth's atmosphere. The accepted value is about 1367 watts per square metre.

Solar declination: The angle of the Sun north or south of the equatorial plane.

Solar energy: The electromagnetic radiation generated by the Sun.

Solar incident angle: The angle at which an incoming solar beam strikes a surface.

Solar radiation: The radiant energy received from the Sun, from both direct and diffuse or reflected sunlight.

Solar spectrum: The total distribution of electromagnetic radiation emitted from the Sun.

Solar still: A device consisting of one or several stages in which brackish water is converted to potable water by successive evaporation and condensation with the aid of solar heat.

Solar water heater: A water heater that depends on solar radiation as its source of power.

Temperature: Degree of hotness or coldness measured on one of several arbitrary scales based on some observable phenomenon (such as expansion).

Thermal conductivity: The ability of a material to conduct heat.

Thermal mass: A material used to store heat, thereby slowing the temperature variation within a space.

Thermosyphon: A closed-loop system in which water automatically circulates between a solar collector and a water storage tank above it due to the natural difference in density between the warmer and cooler portions of a liquid.

Thin-film silicon: Usually amorphous (non-crystalline) material used to make photovoltaic (PV) cells.

Tilt angle: The angle at which a solar collector is tilted upwards from the horizontal surface for maximum heat collection.

Transmittance: The ratio of the solar radiation transmitted through glass to the total radiant energy falling on its surface.

U-value: The amount of heat that flows in or out of a system at steady state, in one hour, when there is a one degree difference in temperature between fluid inside and outside.

Ultraviolet radiation: A portion of the electromagnetic radiation in the wavelength range of 4 to 400 nanometres.

Wafer: Raw material for a solar cell; a thin sheet of crystalline semi-conductor material is made by mechanically sawing it from a single-crystal boule or by casting it.

Water heating: The process of generating domestic hot water by employing a flat-plate collector and utilizing solar radiation.

Watt hour (Wh): A common energy measure arrived at by multiplying the power by the hours of use. Grid power is ordinarily sold and measured in kilowatt hours.

Watt: Measure of power (or work) equivalent to 1/746 of a horsepower.

Wavelength: The distance between two similar points of a given wave.

Zenith: The top of the sky dome. A point directly overhead, 90° in altitude angle above the horizon.

Subject Index